Lecture Notes in Computer Sc

Edited by G. Goos, J. Hartmanis, and J. va

Springer

Berlin
Heidelberg
New York
Barcelona
Hong Kong
London
Milan
Paris
Tokyo

Marco Dorigo   Gianni Di Caro
Michael Sampels (Eds.)

Ant Algorithms

Third International Workshop, ANTS 2002
Brussels, Belgium, September 12-14, 2002
Proceedings

 Springer

Series Editors

Gerhard Goos, Karlsruhe University, Germany
Juris Hartmanis, Cornell University, NY, USA
Jan van Leeuwen, Utrecht University, The Netherlands

Volume Editors

Marco Dorigo
Gianni Di Caro
Michael Sampels
IRIDIA, Université de Bruxelles
CP 194/6
Avenue Franklin D. Roosevelt 50
1050 Brussels, Belgium
E-mail: {mdorigo, gdicaro, msampels}@ulb.ac.be

Cataloging-in-Publication Data applied for

Die Deutsche Bibliothek - CIP-Einheitsaufnahme

Ant algorithms : third international workshop ; proceedings / ANTS 2002,
Brussels, Belgium, September 12 - 14, 2002. Marco Dorigo ... (ed.). - Berlin ;
Heidelberg ; New York ; Barcelona ; Hong Kong ; London ; Milan ; Paris ;
Tokyo : Springer, 2002
  (Lecture notes in computer science ; Vol. 2463)
  ISBN 3-540-44146-8

CR Subject Classification (1998): F.2.2, F.1.1, G.1, G.2, I.2, C.2.4, J.1

ISSN 0302-9743
ISBN 3-540-44146-8 Springer-Verlag Berlin Heidelberg New York

Springer-Verlag Berlin Heidelberg New York,
a member of BertelsmannSpringer Science+Business Media GmbH

http://www.springer.de

© Springer-Verlag Berlin Heidelberg 2002

Typesetting: Camera-ready by author, data conversion by PTP-Berlin, Stefan Sossna e.K.
Printed on acid-free paper   SPIN: 10871233   06/3142   5 4 3 2 1 0

Preface

Social insects – ants, termites, wasps, and bees – live in almost every land habitat on Earth. Over the last one hundred million years of evolution they have conquered an enormous variety of ecological niches in the soil and vegetation. Undoubtedly, their social organization, in particular the genetically evolved commitment of each individual to the survival of the colony, is a key factor underpinning their success. Moreover, these insect societies exhibit the fascinating property that the activities of the individuals, as well as of the society as a whole, are not regulated by any explicit form of centralized control. Evolutionary forces have generated individuals that combine a total commitment to the society together with specific communication and action skills that give rise to the generation of complex patterns and behaviors at the global level.

Among the social insects, ants may be considered the most successful family. There are about 9,000 different species, each with a different set of specialized characteristics that enable them to live in vast numbers, and virtually everywhere. The observation and study of ants and ant societies have long since attracted the attention of the professional entomologist and the layman alike, but in recent years, the ant model of organization and interaction has also captured the interest of the computer scientist and engineer. Ant societies feature, among other things, autonomy of the individual, fully distributed control, fault-tolerance, direct and environment-mediated communication, emergence of complex behaviors with respect to the repertoire of the single ant, collective and cooperative strategies, and self-organization. The simultaneous presence of these unique characteristics have made ant societies an attractive and inspiring model for building new algorithms and new multi-agent systems.

In the last 10–15 years ant societies have provided the impetus for a growing body of scientific work, mostly in the fields of robotics, operations research, and telecommunications. The different simulations and implementations described in this research go under the general name of *ant algorithms*. Researchers from all over the world and possessing different scientific backgrounds have made significant progress concerning both implementation and theoretical aspects, within this novel research framework. Their contributions have given the field a solid basis and have shown how the "ant way", when carefully engineered, can result in successful applications to many real-world problems. Much as in the fields of genetic algorithms and neural networks, to take two examples, nature seems to offer us a valuable source of ideas for the design of new systems and algorithms.

A particularly successful research direction in ant algorithms, known as *ant colony optimization (ACO)*, is dedicated to their application to discrete optimization problems. Ant colony optimization has been applied successfully to a large number of difficult combinatorial problems such as the traveling salesman problem, the quadratic assignment problem and scheduling problems, as well as to routing in telecommunication networks. Ant colony optimization algorithms have all been inspired by a specific foraging behavior of colonies of Argentine ants (*Iridomyrmex humilis*) that, as also observed in the laboratory, are able to find if not the shortest, at least a very good path connecting the colony's nest with a source of food. The elements playing an essential role in this ant foraging behavior were thoroughly reverse-engineered and put to work to solve problems of combinatorial optimization in *Ant System*, the algorithm described in the doctoral thesis of Marco Dorigo, at the beginning of the 1990s. This first application stimulated the interest of other researchers around the world to design algorithms for several important optimization and network problems getting inspiration from the same ant foraging behavior, and from Ant System in particular. Recently, in 1998, the ant colony optimization metaheuristic was defined, providing a common framework for describing this important class of ant algorithms.

The growing interest in ant colony optimization algorithms and, more generally, in ant algorithms, led, in 1998, to the organization of *ANTS'98 – From Ant Colonies to Artificial Ants*, the first international workshop on ant algorithms and ant colony optimization, held in Brussels, Belgium, which saw the participation of more than 50 researchers from around the world. On that occasion, a selection of the best papers presented at the workshop were published as a special issue of the *Future Generation Computer Systems* (Vol. 16, No. 8, 2000). The success of the workshop incited us to repeat the experience two years later: ANTS 2000 saw the participation of more than 70 participants and the 41 extended abstracts presented as talks or posters at the workshop were collected in a booklet distributed to participants. Also on that occasion, a selection of the best papers were published as a journal special issue (*IEEE Transactions on Evolutionary Computation*, Vol. 6, No. 4, 2002). Today the "ant algorithms community" continues to grow and we can see the field beginning to show encouraging signs of maturity, even if there is still a long way to go before reaching a deep and solid understanding concerning theoretical foundations and the design of effective implementations.

This volume contains the proceedings of *ANTS 2002 – From Ant Colonies to Artificial Ants: Third International Workshop on Ant Algorithms*, held in Brussels, Belgium, on September 12–14, 2002. These proceedings contain 36 contributions: 17 full papers and 11 short papers presented at the workshop as talks, and 8 extended abstracts presented as posters. These papers were selected out of a total of 52 submissions after a careful review process involving at least two referees for each paper.

We are very grateful to the members of the international program committee for their detailed reviews and for being available for additional comments and opinions, when needed. We hope that readers will agree that the quality of the papers collected in this volume reflects a new maturity in the field of ant algorithms, as well as a strong commitment to high standards of review.

The papers contributing to these proceedings are from authors coming from more than 20 different countries. We thank them, as well as all those contributing to the organization of the workshop, in particular, IRIDIA and the ULB for providing rooms and logistic support.

Finally, we would like to thank our sponsors, the company *AntOptima* (`www.antoptima.com`), and the EC funded Research and Training Network *Metaheuristics Network* (`www.metaheuristics.org`), who financially supported the workshop.

We hope that these proceedings will provide an insightful and comprehensive starting point for the scientist entering the field of ant algorithms, as well as a valuable reference to the latest developments for the experienced practitioner in the field.

July 2002

Marco Dorigo
Gianni Di Caro
Michael Sampels

Organization

ANTS 2002 was organized by IRIDIA, Université Libre de Bruxelles, Belgium.

Workshop Chair

Marco Dorigo, IRIDIA, Université Libre de Bruxelles, Belgium

Program Co-chairs

Gianni Di Caro, IRIDIA, Université Libre de Bruxelles, Belgium
Michael Sampels, IRIDIA, Université Libre de Bruxelles, Belgium

Program Committee

Payman Arabshahi, JPL, NASA, Pasadena, CA, USA
Nigel Bean, University of Adelaide, Australia
Freddy Bruckstein, Technion, Haifa, Israel
Oscar Cordón, University of Granada, Spain
David Corne, University of Reading, UK
Jean-Louis Deneubourg, Université Libre de Bruxelles, Belgium
Luca M. Gambardella, IDSIA, Lugano, Switzerland
Michel Gendreau, Université de Montreal, Canada
Walter Gutjahr, University of Vienna, Austria
Owen Holland, University of West England, Bristol, UK
Vittorio Maniezzo, Università di Bologna, Italy
Alcherio Martinoli, Caltech, Pasadena, CA, USA
Daniel Merkle, University of Karlsruhe, Germany
Peter Merz, University of Tübingen, Germany
Nicolas Meuleau, NASA, CA, USA
Martin Middendorf, Catholic Univ. of Eichstätt-Ingolstadt, Germany
Ben Paechter, Napier University, Edinburgh, UK
Rainer Palm, Siemens, Munich, Germany
Van Parunak, Altarum Institute, Ann Arbor, MI, USA
Carles Sierra, CSIC, Bellaterra, Spain
Thomas Stützle, Technische Universität Darmstadt, Germany
Tatsuya Suda, University of California, Irvine, CA, USA
Guy Theraulaz, Université Paul Sabatier, Toulouse, France
Thanos Vasilakos, ICS-FORTH, Crete, Greece

Local Organizations

Mauro Birattari, IRIDIA, Université Libre de Bruxelles, Belgium
Christian Blum, IRIDIA, Université Libre de Bruxelles, Belgium

Additional Referees

Eleonora Bianchi	Keita Fujii	Erol Şahin
Mauro Birattari	Joshua Knowles	Christine Strauss
Christian Blum	Roberto Montemanni	Jun Suzuki
Thomas Bousonville	Tadashi Nakano	Justin Viiret
Andre Costa	Marc Reimann	Mark Zlochin
Karl Doerner	Andrea Rizzoli	
Alberto V. Donati	Andrea Roli	

Sponsoring Institutions

AntOptima (www.antoptima.com), Lugano, Switzerland.
Metaheuristics Network (www.metaheuristics.org), a Research Network of the
Improving Human Research Potential programme funded by the Commission of
the European Communities.

Table of Contents

Short Papers

Posters

A $\mathcal{MAX}$-$\mathcal{MIN}$ Ant System for the University Course Timetabling Problem

Krzysztof Socha, Joshua Knowles, and Michael Sampels

IRIDIA, Université Libre de Bruxelles
CP 194/6, Av. Franklin D. Roosevelt 50, 1050 Brussels, Belgium
{ksocha,jknowles,msampels}@ulb.ac.be
http://iridia.ulb.ac.be

Abstract. We consider a simplification of a typical university course timetabling problem involving three types of hard and three types of soft constraints. A $\mathcal{MAX}$-$\mathcal{MIN}$ Ant System, which makes use of a separate local search routine, is proposed for tackling this problem. We devise an appropriate construction graph and pheromone matrix representation after considering alternatives. The resulting algorithm is tested over a set of eleven instances from three classes of the problem. The results demonstrate that the ant system is able to construct significantly better timetables than an algorithm that iterates the local search procedure from random starting solutions.

1 Introduction

Course timetabling problems are periodically faced by virtually every school, college and university in the world. In a basic problem, a set of times must be assigned to a set of events (e.g., classes, lectures, tutorials, etc.) in such a way that all of the students can attend all of their respective events. Some pairs of events are edge-constrained (e.g., some students must attend both events), so that they must be scheduled at different times, and this yields what is essentially a form of vertex colouring problem. In addition, real timetables must usually satisfy a large and diverse array of supplementary constraints which are difficult to describe in a generic manner. However, the general university course timetabling problem (UCTP) is known to be NP-hard, as are many of the subproblems associated with additional constraints [5,9,22]. Of course, the difficulty of any particular instance of the UCTP depends on many factors and, while little is known about how to estimate difficulty, it seems that the assignment of rooms makes the problem significantly harder than vertex colouring, in general.

Current methods for tackling timetabling problems include evolutionary algorithms [6], simulated annealing [13] and tabu search [14]. Many problem-specific heuristics also exist for timetabling and its associated sub-problems. These have been used within evolutionary methods and other generic search methods, either as 'hyper-heuristics' [1,7], or to repair or decode indirect solution representations [16]. Local search has also been used successfully within a memetic algorithm to do real-world exam timetabling [3]. Although several ant

M. Dorigo et al. (Eds.): ANTS 2002, LNCS 2463, pp. 1–13, 2002.
© Springer-Verlag Berlin Heidelberg 2002

colony optimization (ACO) algorithms [2,11,21] have been previously proposed for other constraint satisfaction problems [18], including vertex-coloring [8], a full timetabling problem has not been tackled before using ACO.

The work presented here arises out of the Metaheuristics Network[1] (MN) – a European Commission project undertaken jointly by five European institutes – which seeks to compare metaheuristics on different combinatorial optimization problems. In the current phase of the four-year project, a university course timetabling problem is being considered. In this phase, five metaheuristics, including ACO, will be evaluated and compared on instances from three UCTP classes. As a potential entry for evaluation by the MN, we developed a $\mathcal{MAX}$-$\mathcal{MIN}$ Ant System ($\mathcal{MMAS}$) [21] algorithm for the UCTP. In this paper we describe this algorithm, discussing the selection of an appropriate construction graph and pheromone representation, the local search, the heuristic information, and other factors. We then report on experiments in which the performance of the $\mathcal{MMAS}$ algorithm is evaluated with respect to using the local search alone, in a random-restart algorithm.

The remainder of this paper is organized as follows: Section 2 defines the particular timetabling problem considered and specifies how instances of different classes were generated. A brief description of the local search is also included. Section 3 describes the design of the $\mathcal{MMAS}$ algorithm, focusing on the key aspects of representation and heuristics. Section 4 reports the experimental method and computational results of the comparison with the random restart local search. In Sect. 5, conclusions are drawn.

2   The University Course Timetabling Problem

The timetabling problem considered by the MN is similar to one initially presented by Paechter in [15]. It is a reduction of a typical university course timetabling problem [6,16]. It consists of a set of n events E to be scheduled in a set of timeslots $T = \{t_1, \ldots, t_k\}$ ($k = 45$, 5 days of 9 hours each), a set of rooms R in which events can take place, a set of students S who attend the events, and a set of features F satisfied by rooms and required by events. Each student attends a number of events and each room has a maximum capacity. A feasible timetable is one in which all events have been assigned a timeslot and a room so that the following hard constraints are satisfied:

- no student attends more than one event at the same time;
- the room is big enough for all the attending students and satisfies all the features required by the event;
- only one event is in each room at any timeslot.

In addition, a feasible candidate timetable is penalized equally for each occurrence of the following soft constraint violations:

- a student has a class in the last slot of the day;

---
[1] http://www.metaheuristics.org .

- a student has more than two classes in a row;
- a student has exactly one class on a day.

Feasible solutions are always considered to be superior to infeasible solutions, independently of the numbers of soft constraint violations. In fact, in any comparison, all infeasible solutions are to be considered equally worthless. The objective is to minimize the number of soft constraint violations (#scv) in a feasible solution.

2.1   Problem Instances

Instances of the UCTP are constructed using a generator written by Paechter[2]. The generator makes instances for which a perfect solution exists, that is, a timetable having no hard or soft constraint violations. The generator is called with eight command line parameters that allow various aspects of the instance to be specified, plus a random seed. For the comparison being carried out by the MN, three classes of instance have been chosen, reflecting realistic timetabling problems of varying sizes. These classes are defined by the values of the input parameters to the generator, and different instances of the class can be generated by changing the random seed value. The parameter values defining the classes are given in Tab. 1.

Table 1. Parameter values for the three UCTP classes.

Class	small	medium	large
Num_events	100	400	400
Num_rooms	5	10	10
Num_features	5	5	10
Approx_features_per_room	3	3	5
Percent_feature_use	70	80	90
Num_students	80	200	400
Max_events_per_student	20	20	20
Max_students_per_event	20	50	100

For each class of problem, a time limit for producing a timetable has been determined. The time limits for the problem classes small, medium, and large are respectively 90, 900, and 9000 seconds. These limits were derived experimentally.

2.2   Solution Representation and Local Search

To make comparison and evaluation meaningful, all metaheuristics developed for the MN project, including the algorithm described here, employ the same solution representation, neighborhood structure, and local search routine (where applicable), as described fully in [19].

---

[2] http://www.dcs.napier.ac.uk/~benp .

A solution vector is an element of T^E and represents an assignment of events to timeslots. The assignment of each event-timeslot pair to a room is not under the direct influence of the metaheuristics (or local search routine). Instead, the room assignment is carried out using a deterministic network flow algorithm. The local search routine (LS) is a first-improvement search based on two move operators. The first operator moves a single event to a different timeslot. The second swaps the timeslots of two events. The LS is deterministic and ends when a true local optimum is reached. The LS makes extensive use of delta evaluation of solutions so that many neighboring timetables can be considered in a short time.

3   Design of an $\mathcal{MMAS}$ for Timetabling

Given the constraints on the representation discussed in the last section, we can now consider the choices open to us, in designing an effective $\mathcal{MMAS}$ for the UCTP. We must first decide how to transform the assignment problem (assigning events to timeslots) into an optimal path problem which the ants can tackle. To do this we must select an appropriate *construction graph* for the ants to follow. We must then decide on an appropriate pheromone matrix and heuristic information to influence the paths the ants will take through the graph.

3.1   Construction Graph

One of the cardinal elements of the ACO metaheuristic is the mapping of the problem onto a *construction graph* [10,12], so that a path through the graph represents a solution to the problem. In our formulation of the UCTP we are required to assign each of $|E|$ events to one of $|T|$ timeslots. In the most direct representation the construction graph is given by $E \times T$; given this graph we can then decide whether the ants move along a list of the timeslots, and choose events to be placed in them, or move along a list of the events and place them in the timeslots. Fig. 1 and Fig. 2 depict, respectively, these construction graphs.

As shown in Fig. 1, the first construction graph must use a set of virtual timeslots $T' = \{t'_1, \ldots, t'_{|E|}\}$, because exactly $|E|$ assignments must be made in the construction of a timetable, but in general $|T| \ll |E|$. Each of the virtual timeslots maps to one of the actual timeslots. To use this representation then, requires us to define an injection $\iota : T' \to T$, designating how the virtual timeslots relate to the actual ones. One could use for example the injection $\iota : t'_g \mapsto t_h$ with $h = \left\lceil \frac{g \cdot |T|}{|E|} \right\rceil$. In this way, the timetable would be constructed sequentially through the week. However, for certain problems, giving equal numbers of events to each timeslot may be a long way from optimal. Other injection functions are also possible but may contain similar implicit biases.

The simpler representation (Fig. 2), where ants walk along a list of events, choosing a timeslot for each, does not require the additional complication of using virtual timeslots and does not seem to have any obvious disadvantages. In fact, it allows us the opportunity of using a heuristically ordered list of events.

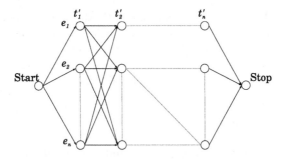

Fig. 1. Each ant follows a list of *virtual* timeslots, and for each such timeslot $t' \in T'$, it chooses an event $e \in E$ to be placed in this timeslot. At each step an ant can choose any possible transition

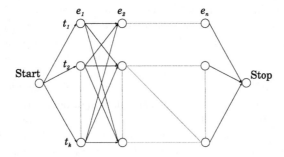

Fig. 2. Each ant follows a list of events, and for each event $e \in E$, an ant chooses a timeslot $t \in T$. Each event has to be put exactly once into a timeslot, and there may be more than one event in a timeslot, so at each step an ant can choose any possible transition

By carrying out some pre-calculation we should be able to order the events so that the most 'difficult' events are placed into the timetable first, when there are still many timeslots with few or no occupied rooms. For these reasons we choose to use this representation.

As the ants traverse our chosen construction graph, they construct partial assignments $A_i : E_i \rightarrow T$ for $i = 0, \ldots, |E|$, where $E_i = \{e_1, \ldots, e_i\}$. An ant starts with the empty assignment $A_0 = \emptyset$. After the construction of A_{i-1}, the assignment A_i is built probabilistically as $A_i = A_{i-1} \cup \{(e_i, t)\}$. The timeslot t is chosen randomly out of T according to probabilities $p_{e_i,t}$ that depend on the pheromone matrix $\tau(A_{i-1}) \in [\tau_{min}, \tau_{max}]^{E \times T}$ ($\tau_{min}, \tau_{max} \in \mathbf{R}$) and any heuristic information $\eta(A_{i-1})$ given by:

$$p_{e_i,t}(\tau(A_{i-1}), \eta(A_{i-1})) = \frac{(\tau_{(e_i,t)}(A_{i-1}))^\alpha \cdot (\eta_{(e_i,t)}(A_{i-1}))^\beta}{\sum_{\theta \in T}(\tau_{(e_i,\theta)}(A_{i-1}))^\alpha \cdot (\eta_{(e_i,\theta)}(A_{i-1}))^\beta} . \quad (1)$$

In this general form of the equation, both the pheromone information τ and the heuristic information η take as argument the partial assignment A_{i-1}.[3] The impact of the pheromone and the heuristic information can be weighted by parameters α and β. In the following sections, we consider different ways of implementing the pheromone and heuristic information.

3.2 Pheromone Matrix

In a first representation, we let pheromones indicate the absolute position where events should be placed. With this representation the pheromone matrix is given by $\tau(A_i) = \tau, i = 1, \ldots, |E|$, i.e., the pheromone does not depend on the partial assignments A_i. Note that in this case the pheromone will be associated with *nodes* in the construction graph rather than *edges* between the nodes.

A disadvantage of this direct pheromone representation is that the absolute position of events in the timeslots does not matter very much in producing a good timetable. It is the relative placement of events which is important. For example, given a perfect timetable, it is usually possible to permute many groups of timeslots without affecting the quality of the timetable. As a result, this choice of representation can cause slower learning because during construction of solutions, an early assignment of an event to an 'undesirable' timeslot may cause conflicts with many supposedly desirable assignments downstream, leading to a poor timetable. This leads to a very noisy positive feedback signal.

In a second representation the pheromone values are indirectly defined. To do this we use an auxiliary matrix $\mu \in \mathbf{R}_+^{E \times E}$ to indicate which events should (or should not) be put together with other events in the same timeslot. Now, the values $\tau_{(e,t)}(A_i)$ can be expressed in terms of μ and A_i by

$$\tau_{(e,t)}(A_i) = \begin{cases} \tau_{max} & \text{if } A_i^{-1}(t) = \emptyset, \\ \min_{e' \in A_i^{-1}(t)} \mu(e, e') & \text{otherwise.} \end{cases}$$

Giving feedback to these values μ, the algorithm is able to learn which events should *not* go together in the same timeslot. This information can be learned without relation to the particular timeslot numbers. This representation looks promising because it allows the ants to learn something more directly useful to the construction of feasible timetables. However, it also has some disadvantages. For solving the soft constraints defined in Sect. 2, certain inter-timeslot relations between events matter, in addition to the intra-timeslot relations. This pheromone representation does not encode this extra information at all.

Some experimentation with the two different pheromone matrices indicated that the first one performed significantly better when the local search procedure was also used. Even though it is not ideal for the reasons stated above, it is capable of guiding the ants to construct timetables which meet the soft constraints as well as the hard ones. The problem of noisy feedback from this representation is also somewhat reduced when using the local search.

---

[3] For τ this is done to allow an indirect pheromone representation to be specified.

Clearly, other pheromone representations are possible, but with the variety of constraints which must be satisfied in the UCTP, it is difficult to design one that encodes all the relevant information in a simple manner. For the moment, the direct coding is the best compromise we have found.

3.3 Heuristic Information

We now consider possible methods for computing the heuristic information $\eta_{(e,t)}(A_{i-1})$. A simple method is the following:

$$\eta_{(e,t)}(A_{i-1}) = \frac{1.0}{1.0 + V_{(e,t)}(A_{i-1})}$$

where $V_{(e,t)}(A_{i-1})$ counts the additional number of violations caused by adding (e, t) to the partial assignment A_{i-1}. The function V may be a weighted sum of several or all of the soft and hard constraints. However, due to the nature of the UCTP, the computational cost of calculating some types of constraint violations can be rather high. We can choose to take advantage of significant heuristic information to guide the construction but only at the cost of being able to make fewer iterations of the algorithm in the given time limit. We conducted some investigations to assess the balance of this tradeoff and found that the use of heuristic information did not improve the quality of timetables constructed by the $\mathcal{MMAS}$ with local search. Without the use of LS, heuristic information does improve solution quality, but not to the same degree as LS.

3.4 Algorithm Description

Our $\mathcal{MAX}$-$\mathcal{MIN}$ Ant System for the UCTP is shown in Alg. 1. A colony of m ants is used and at each iteration, each ant constructs a complete event-timeslot assignment by placing events, one by one, into the timeslots. The events are taken in a prescribed order which is used by all ants. The order is calculated before the run based on edge constraints between the events. The choice of which timeslot to assign to each event is a biased random choice influenced by the pheromone level $\tau_{(e,t)}(A_i)$ as described in (1). The pheromone values are initialized to a parameter τ_{max}, and then updated by a global pheromone update rule. At the end of the iterative construction, an event-timeslot assignment is converted into a candidate solution (timetable) using the matching algorithm. After all m ants have generated their candidate solution, one solution is chosen based on a fitness function. This candidate solution is further improved by the local search routine. If the solution found is better than the previous global best solution, it is replaced by the new solution. Then the global update on the pheromone values is performed using the global best solution. The values of the pheromone corresponding to the global best solution are increased and then all the pheromone levels in the matrix are reduced according to the evaporation coefficient. Finally, some pheromone values are adjusted so that they all lie within the bounds defined by τ_{max} and τ_{min}. The whole process is repeated, until the time limit is reached.

Algorithm 1 $\mathcal{MAX}$-$\mathcal{MIN}$ Ant System for the UCTP

---

input: A problem instance I
$\tau_{max} \leftarrow \frac{1}{\rho}$
$\tau(e,t) \leftarrow \tau_{max} \; \forall \; (e,t) \in E \times T$
calculate $c(e,e') \; \forall \; (e,e') \in E^2$
calculate $d(e)$
sort E according to $\prec$, resulting in $e_1 \prec e_2 \prec \cdots \prec e_n$
while time limit not reached **do**
    for $a = 1$ **to** m **do**
        {construction process of ant a}
        $A_0 \leftarrow \emptyset$
        for $i = 1$ **to** $|E|$ **do**
            choose timeslot t randomly according to probabilities $p_{e_i,t}$ for event e_i
            $A_i \leftarrow A_{i-1} \cup \{(e_i,t)\}$
        end for
        $C \leftarrow$ solution after applying matching algorithm to A_n
        $C_{iteration\ best} \leftarrow$ best of C and $C_{iteration\ best}$
    end for
    $C_{iteration\ best} \leftarrow$ solution after applying local search to $C_{iteration\ best}$
    $C_{global\ best} \leftarrow$ best of $C_{iteration\ best}$ and $C_{global\ best}$
    global pheromone update for τ using $C_{global\ best}$, τ_{min}, and τ_{max}
end while
output: An optimized candidate solution $C_{global\ best}$ for I

Some parts of Alg. 1 are now described in more detail. In a pre-calculation for events $e, e' \in E$ the following parameters are determined:

$c(e, e') := 1$ if there are students following both e and e', 0 otherwise, and
$$d(e) := |\{e' \in E \setminus \{e\} \mid c(e, e') \neq 0\}| \; .$$

We define a total order $\prec$ on the events by

$$e \prec e' :\Leftrightarrow d(e) > d(e') \vee$$
$$d(e) = d(e') \wedge l(e) < l(e') \; .$$

Here, $l : E \to \mathbf{N}$ is an injective function that is only used to handle ties. We define $E_i := \{e_1, \ldots, e_i\}$ for the totally ordered events denoted as $e_1 \prec e_2 \prec \ldots \prec e_n$.

Only the solution that causes the fewest number of hard constraint violations is selected for improvement by the LS. Ties are broken randomly. The pheromone matrix is updated only once per iteration, and the global best solution is used for update. Let $A_{global\ best}$ be the assignment of the best candidate solution $C_{global\ best}$ found since the beginning. The following update rule is used:

$$\tau_{(e,t)} = \begin{cases} (1-\rho) \cdot \tau_{(e,t)} + 1 & \text{if } A_{global\ best}(e) = t, \\ (1-\rho) \cdot \tau_{(e,t)} & \text{otherwise,} \end{cases}$$

where $\rho \in [0, 1]$ is the evaporation rate. Pheromone update is completed using the following:

$$\tau_{(e,t)} \leftarrow \begin{cases} \tau_{min} & \text{if } \tau_{(e,t)} < \tau_{min}, \\ \tau_{max} & \text{if } \tau_{(e,t)} > \tau_{max}, \\ \tau_{(e,t)} & \text{otherwise.} \end{cases}$$

3.5 Parameters

The development of an effective $\mathcal{MMAS}$ for an optimization problem also requires that appropriate parameters be chosen for typical problem instances. In our case, we consider as typical those problem instances made by the generator described in Sect. 2.1. We tested several configurations of our $\mathcal{MMAS}$ on problem instances from the classes listed in Tab. 1. The best results were obtained using the parameters listed in Tab. 2.

Table 2. Parameter configurations used in the comparison.

Parameter	small	medium	large
ρ	0.30	0.30	0.30
$\tau_{max} = \frac{1}{\rho}$	3.3	3.3	3.3
τ_{min}	0.0078	0.0019	0.0019
α	1.0	1.0	1.0
β	0.0	0.0	0.0
m	10	10	10

The values of τ_{min} were calculated so that at convergence (when one 'best' path exists with a pheromone value of τ_{max} on each of its constituent elements, and all other elements in the pheromone matrix have the value τ_{min}) a path constructed by an ant will be expected to differ from the best path in 20 % of its elements. The value 20 % was chosen to reflect the fact that a fairly large 'mutation' is needed to push the solution into a different basin of attraction for the local search.

4   Assessment of the Developed $\mathcal{MMAS}$

To assess the developed $\mathcal{MMAS}$, we consider whether the ants genuinely learn to build better timetables, as compared to a random restart local search (RRLS). This RRLS iterates the same LS as used by $\mathcal{MMAS}$ from random starting solutions and stores the best solution found.

We tested both the developed $\mathcal{MMAS}$ and the RRLS on previously unseen problem instances made by the generator mentioned in Sect. 2.1. For this test study, we generated eleven test instances: five small, five medium, and one large. For each of them, we ran our algorithms for 50, 40, and 10 independent trials, giving each trial a time limit of 90, 900, and 9000 seconds, respectively. All

Table 3. Median of the number of soft constraint violations observed in independent trials of $\mathcal{MMAS}$ and RRLS on different problem instances, together with the p-value for the null hypothesis that the distributions are equal. In the cases where greater than 50 % of runs resulted in no feasible solution the median cannot be calculated. Here, the fraction of unsuccessful runs is given. (In all other cases 100 % of the runs resulted in feasible solutions.) All infeasible solutions are given the symbolic value ∞. This is correctly handled by the Mann-Whitney test.

Instance	Median of #scv		p-value
	$\mathcal{MMAS}$	RRLS	
small1	1	8	$< 2 \cdot 10^{-16}$
small2	3	11	$< 2 \cdot 10^{-16}$
small3	1	8	$< 2 \cdot 10^{-16}$
small4	1	7	$< 2 \cdot 10^{-16}$
small5	0	5	$< 2 \cdot 10^{-16}$
medium1	195	199	0.017
medium2	184	202.5	$4.3 \cdot 10^{-6}$
medium3	248	(77.5 %)	$8.1 \cdot 10^{-12}$
medium4	164.5	177.5	0.017
medium5	219.5	(100 %)	$2.2 \cdot 10^{-16}$
large	851.5	(100 %)	$6.4 \cdot 10^{-5}$

the tests were run on a PC with an AMD Athlon 1100 Mhz CPU under Linux using the GNU C++ compiler gcc version 2.95.3.[4] As random number generator we used ran0 from the Numerical Recipes [17]. For the reproducibility of the results on another architecture, we observed that on our architecture one step of the local search has an average running time of 0.45, 1.4, and 1.1 milliseconds, respectively.

Boxplots showing the distributions of the ranks of the obtained results are shown in Fig. 3. The Mann-Whitney test (see [4]) was used to test the hypothesis H_0 that the distribution functions of the solutions found by $\mathcal{MMAS}$ and RRLS were the same. The p-values for this test are given in Tab. 3, along with the median number of soft constraint violations obtained.

For each of the tested problem instances we got with very high statistical significance the result that $\mathcal{MMAS}$ performs better than RRLS. For some test instances of medium and large size some runs of RRLS resulted in infeasible solutions. In particular, the RRLS was unable to produce any feasible solution for the large problem instance.

5   Conclusions

We devised a construction graph and a pheromone model appropriate for university course timetabling. Using these we were able to specify the first ACO

---

[4] Both algorithms, the test instances and a detailed evaluation of the generated results can be found on http://iridia.ulb.ac.be/~msampels/tt.data .

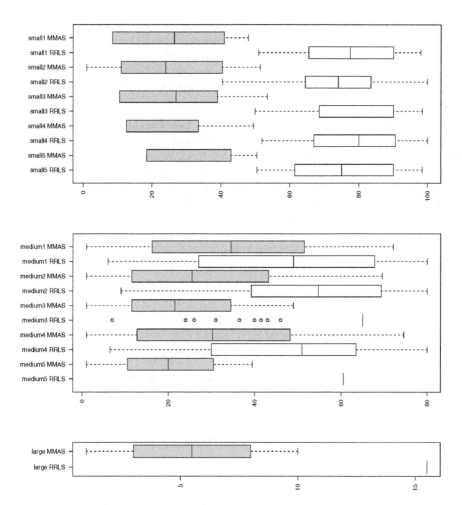

Fig. 3. Boxplots showing the relative distribution of the number of soft constraint violations for $\mathcal{MMAS}$ (shadowed) and RRLS (white) on all test instances. This is the distribution of the ranks of the absolute values in an ordered list, where equal values are assigned to the mean of the covered ranks. A box shows the range between the 25 % and the 75 % quantile of the data. The median of the data is indicated by a bar. The whiskers extend to the most extreme data point which is no more than 1.5 times the interquantile range from the box. Outliers are indicated as circles

algorithm for this problem. Compared to a random restart local search, it showed significantly better performance on a set of typical problem instances, indicating that it can guide the local search effectively. Our algorithm underlines the fact that ant systems are able to handle problems with multiple heterogeneous constraints. Even without using problem-specific heuristic information it is possible to generate good solutions. With the use of a basic first-improvement local search, we found that $\mathcal{MMAS}$ permits a quite simple handling of timetabling problems. With an improved local search, exploiting more problem specific operators, we would expect a further improvement in performance.

Preliminary comparisons indicate that our $\mathcal{MMAS}$ is competitive with the other metaheuristics developed in the Metaheuristics Network for the UCTP. Further results on the comparison of the metaheuristics will appear in [20].

Acknowledgments. We would like to thank Ben Paechter and Olivia Rossi-Doria for the implementation of data structures and routines for the local search. Our work was supported by the *Metaheuristics Network,* a Research Training Network funded by the Improving Human Potential Programme of the CEC, grant HPRN-CT-1999-00106. Joshua Knowles is additionally funded by a CEC Marie Curie Research Fellowship, contract number: HPMF-CT-2000-00992. The information provided is the sole responsibility of the authors and does not reflect the Community's opinion. The Community is not responsible for any use that might be made of data appearing in this publication.

References

1. C. Blum, M. Dorigo, S. Correia, O. Rossi-Doria, B. Paechter, and M. Snoek. A GA evolving instructions for a timetable builder. In *Proceedings of the 4th International Conference on Practice and Theory of Automated Timetabling (PATAT 2002) (to appear)*, 2002.
2. E. Bonabeau, M. Dorigo, and G. Theraulaz. *From Natural to Artificial Swarm Intelligence.* Oxford University Press, 1999.
3. E. K. Burke, J. P. Newall, and R. F. Weare. A memetic algorithm for university exam timetabling. In *Proceedings of the 1st International Conference on Practice and Theory of Automated Timetabling (PATAT 1995)*, LNCS 1153, pages 241–251. Springer-Verlag, 1996.
4. W. J. Conover. *Practical Nonparametric Statistics.* John Wiley & Sons, 3rd edition, 1999.
5. T. B. Cooper and J. H. Kingston. The complexity of timetable construction problems. In *Proceedings of the 1st International Conference on Practice and Theory of Automated Timetabling (PATAT 1995)*, LNCS 1153, pages 283–295. Springer-Verlag, 1996.
6. D. Corne, P. Ross, and H.-L. Fang. Evolutionary timetabling: Practice, prospects and work in progress. In *Proceedings of the UK Planning and Scheduling SIG Workshop, Strathclyde*, 1994.
7. P. Cowling, G. Kendall, and E. Soubeiga. A hyperheuristic approach to scheduling a sales summit. In *Proceedings of the 3rd International Conference on Practice and Theory of Automated Timetabling (PATAT 2000)*, LNCS 2079, pages 176–190. Springer-Verlag, 2001.
8. D. Costa and A. Hertz. Ants can colour graphs. *Journal of the Operational Research Society*, 48:295–305, 1997.
9. D. de Werra. The combinatorics of timetabling. *European Journal of Operational Research*, 96:504–513, 1997.
10. M. Dorigo and G. Di Caro. The Ant Colony Optimization meta-heuristic. In D. Corne, M. Dorigo, and F. Glover, editors, *New Ideas in Optimization*. McGraw-Hill, 1999.

11. M. Dorigo, G. Di Caro, and L. M. Gambardella. Ant algorithms for discrete optimization. *Artificial Life*, 5:137–172, 1999.
12. M. Dorigo, V. Maniezzo, and A. Colorni. The Ant System: Optimization by a colony of cooperating agents. *IEEE Transactions on Systems, Man, and Cybernetics*, 26:29–41, 1996.
13. M. A. S. Elmohamed, P. Coddington, and G. Fox. A comparison of annealing techniques for academic course scheduling. In *Proceedings of the 2nd International Conference on Practice and Theory of Automated Timetabling (PATAT 1997)*, pages 92–115, 1998.
14. L. D. Gaspero and A. Schaerf. Tabu search techniques for examination timetabling. In *Proceedings of the 3rd International Conference on Practice and Theory of Automated Timetabling (PATAT 2000)*, LNCS 2079, pages 104–117. Springer-Verlag, 2001.
15. B. Paechter. Course timetabling. Evonet Summer School, 2001. http://evonet.dcs.napier.ac.uk/summerschool2001/problems.html .
16. B. Paechter, R. C. Rankin, A. Cumming, and T. C. Fogarty. Timetabling the classes of an entire university with an evolutionary algorithm. In *Proceedings of the 5th International Conference on Parallel Problem Solving from Nature (PPSN V)*, LNCS 1498, pages 865–874. Springer-Verlag, 1998.
17. W. H. Press, S. A. Teukolsky, W. T. Vetterling, and B. P. Flannery. *Numerical Recipes in C.* Cambridge University Press, 2nd edition, 1993.
18. A. Roli, C. Blum, and M. Dorigo. ACO for maximal constraint satisfaction problems. In *Proceedings of the 4th Metaheuristics International Conference (MIC 2001)*, volume 1, pages 187–191, Porto, Portugal, 2001.
19. O. Rossi-Doria, C. Blum, J. Knowles, M. Sampels, K. Socha, and B. Paechter. A local search for the timetabling problem. In *Proceedings of the 4th International Conference on Practice and Theory of Automated Timetabling (PATAT 2002) (to appear)*, 2002.
20. O. Rossi-Doria, M. Sampels, M. Chiarandini, J. Knowles, M. Manfrin, M. Mastrolilli, L. Paquete, and B. Paechter. A comparison of the performance of different metaheuristics on the timetabling problem. In *Proceedings of the 4th International Conference on Practice and Theory of Automated Timetabling (PATAT 2002) (to appear)*, 2002.
21. T. Stützle and H. H. Hoos. $\mathcal{MAX}$-$\mathcal{MIN}$ Ant System. *Future Generation Computer Systems*, 16(8):889–914, 2000.
22. H. M. M. ten Eikelder and R. J. Willemen. Some complexity aspects of secondary school timetabling problems. In *Proceedings of the 3rd International Conference on Practice and Theory of Automated Timetabling (PATAT 2000)*, LNCS 2079, pages 18–29. Springer-Verlag, 2001.

ACO Applied to Group Shop Scheduling: A Case Study on Intensification and Diversification

Christian Blum

IRIDIA, Université Libre de Bruxelles
CP 194/6, Av. Franklin D. Roosevelt 50, 1050 Brussels, Belgium
cblum@ulb.ac.be

Abstract. We present a $\mathcal{MAX}$-$\mathcal{MIN}$ Ant System for the Group Shop Scheduling problem and propose several extensions aiming for a more effective intensification and diversification of the search process. Group Shop Scheduling is a general Shop Scheduling problem covering Job Shop Scheduling and Open Shop Scheduling. In general, in Shop Scheduling problems good solutions are scattered all over the search space. It is widely recognized that for such kind of problems, effective intensification and diversification mechanisms are needed to achieve good results. Our main result shows that a basic $\mathcal{MAX}$-$\mathcal{MIN}$ Ant System – and potentially any other Ant Colony Optimization algorithm – can be improved by keeping a number of elite solutions found during the search, and using them to guide the search process.

1 Introduction

Ant Colony Optimization (ACO) [6] is a model–based metaheuristic approach for solving hard combinatorial optimization problems. The inspiring source of ACO is the foraging behavior of real ants which enables them to find shortest paths between a food source and their nest. Model-based search algorithms [24,9] are becoming increasingly popular in the last decade. Among others, this class of algorithms also includes methods such as Estimation of Distribution Algorithms (EDAs)[1] [14,16]. In model-based search algorithms candidate solutions are generated using a parametrized probabilistic model that is updated using the previously seen solutions.

The problem of any search algorithm in general is the guidance of the search process such that *intensification* and *diversification* are achieved. Intensification is an expression commonly used for the concentration of the search process on areas in the search space with good quality solutions, whereas diversification denotes the action of leaving already explored areas and moving the search process to unexplored areas. As several authors point out (see for example [1,11,23]), a fine–tuned balance between intensification and diversification is needed for a metaheuristic to achieve good results. In this paper, we propose a $\mathcal{MAX}$-$\mathcal{MIN}$

---

[1] EDAs are covering several algorithms emerging from the field of Evolutionary Computation.

M. Dorigo et al. (Eds.): ANTS 2002, LNCS 2463, pp. 14–27, 2002.

Ant System based on the Hyper-Cube Framework[2] to tackle the Group Shop Scheduling problem. Furthermore we propose several intensification and diversification mechanisms to be applied by the basic algorithm.

The outline of the paper is as follows. In Section 2 we introduce the Group Shop Scheduling problem. In Section 3 we outline a $\mathcal{MAX}$-$\mathcal{MIN}$ Ant System for the Group Shop Scheduling. In Section 4 we outline additional intensification and diversification mechanisms to be applied by the ACO algorithm. In Section 5 we present results before we conclude in Section 6 with giving a summary and an outlook to the future.

2   Group Shop Scheduling

A general scheduling problem can be formalized as follows: We consider a finite set of operations O which is partitioned into subsets $\mathcal{M} = \{M_1, \ldots, M_m\}$ (machines) and into subsets $\mathcal{J} = \{J_1, \ldots, J_n\}$ (jobs), together with a partial order $\preceq \subseteq O \times O$ such that $\preceq \cap J_i \times J_j = \emptyset$ for $i \neq j$, and a function $p : O \to \mathbf{N}$ assigning processing times to operations. A feasible solution is a refined partial order $\preceq^* \supseteq \preceq$ for which the restrictions $\preceq^* \cap J_i \times J_i$ and $\preceq^* \cap M_k \times M_k$ are total $\forall i, k$. The cost of a feasible solution is defined by

$$C_{\max}(\preceq^*) = \max\{\sum_{o \in C} p(o) \mid C \text{ is a chain in } (O, \preceq^*)\}$$

where $C_{\max}$ is called the makespan of a solution. We aim at a feasible solution which minimizes $C_{\max}$.

M_k is the set of operations which have to be processed on machine k. Further, J_i is the set of operations which belong to job i. Each machine can process at most one operation at a time. Operations must be processed without preemption. Operations belonging to the same job must be processed sequentially. This brief problem formulation covers well known scheduling problems: The restriction $\preceq \cap J_i \times J_i$ is total in the Job Shop Scheduling problem (JSP), trivial ($= \{(o, o) \mid o \in J_i\}$) in the Open Shop Scheduling problem (OSP), and either total or trivial for each i in the Mixed Shop Scheduling problem (MSP) [17,5].

For the Group Shop Scheduling problem (GSP), we consider a weaker restriction on $\preceq$ which includes the above scheduling problems by looking at a refinement of the partition $\mathcal{J}$ to a partition into *groups* $\mathcal{G} = \{G_1, \ldots, G_g\}$. We demand that $\preceq \cap G_i \times G_i$ has to be trivial and that for $o, o' \in J$ ($J \in \mathcal{J}$) with $o \in G_i$ and $o' \in G_j$ ($i \neq j$) either $o \preceq o'$ or $o \succeq o'$ holds. Note that the coarsest refinement $\mathcal{G} = \mathcal{J}$ (groups sizes are equal to job sizes) is equivalent to the OSP and the finest refinement $\mathcal{G} = \{\{o\} \mid o \in O\}$ (group sizes of 1) is equivalent to the JSP. See Fig. 1 for an example of the GSP.

3   A $\mathcal{MAX}$-$\mathcal{MIN}$ Ant System for the GSP

$\mathcal{MAX}$-$\mathcal{MIN}$ Ant System ($\mathcal{MMAS}$) was proposed by Stützle and Hoos in [20] as an improvement of the original Ant System (AS) proposed by Dorigo et al.

---

[2] The Hyper-Cube Framework is a certain way of implementing ACO algorithms.

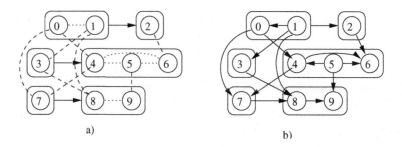

Fig. 1. a) The disjunctive graph representation [10] of a simple instance of the GSP consisting of 10 operations partitioned into 3 jobs, 6 groups and 4 machines (processing times are omitted in this example). The nodes of the graph correspond to the operations, and we have directed arcs symbolized as inter-group connections and undirected arcs between every pair of operations being in the same group or having to be processed on the same machine. In order to obtain a solution, the undirected arcs have to be directed without creating any cycles. b) A solution to the problem (the arcs undirected in a) are directed such that the resulting graph does not contain any cycles).

[8]. $\mathcal{MMAS}$ differs from AS by applying a lower and an upper bound, $\tau_{\min}$ and $\tau_{\max}$, on the pheromone values. The lower bound (a small positive constant) is preventing the algorithm from converging[3] to a solution. Another important feature of $\mathcal{MMAS}$ is that – due to a quite aggressive pheromone update – the algorithm concentrates quickly on an area in the search space. When the system is stuck in an area of the search space[4] the best solution found since the start of the algorithm is used to update the pheromone values in every iteration until the algorithm gets stuck again. After that, a restart is performed. The reason behind that is the hope to find a better solution between the restart best solution and the best solution found since the start of the algorithm. This mechanism is clearly an intensification mechanism. Additional diversification is reached by the original $\mathcal{MMAS}$ in a randomized way by restarting the algorithm.

The framework of our $\mathcal{MMAS}$ algorithm for the GSP is shown in Algorithm 1. The most important features of this algorithm are explained in the following.

In Algorithm 1, $\tau = \{\tau_1, ..., \tau_l\}$ is a set of pheromone values, n_a is the number of ants, and s_j is a solution to the problem constructed by ant j, where $j = 1, ..., n_a$. s_{ib} denotes the iteration best solution, s_{rb} denotes the restart best solution, and s_{gb} denotes the best solution found since the start of the algorithm.

InitializePheromoneValues(τ): In the basic version of our algorithm we initialize all the pheromone values to the same positive constant value 0.5.

---

[3] In the course of this work we refer to convergence of an algorithm as a state where only one solution has a probability greater than 0 to be generated.

[4] This is usually determined by some convergence measure.

Algorithm 1 $\mathcal{MMAS}$ for the GSP

$s_{gb} \leftarrow$ NULL
$s_{rb} \leftarrow$ NULL
$cf \leftarrow 0$
$global\_convergence \leftarrow$ FALSE
InitializePheromoneValues(τ)
while termination conditions not met **do**
    for $j = 1$ to n_a **do**
        $s_j \leftarrow$ ConstructSolution(τ)
        LocalSearch(s_j)
    end for
    $s_{ib} \leftarrow$ argmin($C_{\max}(s_1), ..., C_{\max}(s_{n_a})$)
    Update(s_{ib}, s_{rb}, s_{gb})
    if $global\_convergence ==$ FALSE **then**
        ApplyOnlineDelayedPheromoneUpdate(τ,s_{rb})
    else
        ApplyOnlineDelayedPheromoneUpdate(τ,s_{gb})
    end if
    $cf \leftarrow$ ComputeConvergenceFactor(τ)
    if $cf \geq 0.99$ AND $global\_convergence ==$ TRUE **then**
        ResetPheromoneValues(τ)
        $s_{rb} \leftarrow$ NULL
        $global\_convergence \leftarrow$ FALSE
    else
        if $cf \geq 0.99$ AND $global\_convergence ==$ FALSE **then**
            $global\_convergence \leftarrow$ TRUE
        end if
    end if
end while

ConstructSolution(τ): An important part of an ACO algorithm is the constructive mechanism used to probabilistically construct solutions. We use the well-known list scheduler algorithm proposed by Giffler and Thompson [10] adapted to the GSP. To construct a schedule, this list scheduler algorithm builds a sequence s containing all operations – starting with an empty sequence – by performing $|O|$ steps as follows:

1. Create a set S_t (where t is the step number) of all operations that can be scheduled next.
2. Use a heuristic to produce a set $S'_t \subseteq S_t$.
3. Use a heuristic to pick an operation $o \in S'_t$ to be scheduled next. Set $s[t] = o$.

A sequence s of all operations is a total order on all operations that induces a total order on the operations of each group and of each machine. This unambiguously defines a solution to an instance of the problem. E.g., for the problem instance depicted in Fig. 1, the sequence $1 - 0 - 3 - 5 - 4 - 7 - 8 - 9 - 2 - 6$ defines group order $1 \preceq 0$ in group G_1, $5 \preceq 4 \preceq 6$ in group G_4 and $8 \preceq 9$ in group G_6. It also defines machine orders $0 \preceq 4 \preceq 7$, $1 \preceq 3 \preceq 8$, $2 \preceq 6$ and $5 \preceq 0$.

The construction mechanism applied by the ants choses $S'_t = S_t$ in step 2 of the list scheduler algorithm and in step 3 it choses among the operations in S'_t probabilistically. The probabilities for the operations in S'_t (called transition probabilities) to be chosen, depend on the pheromone model. For our algorithm we chose a pheromone model called $\mathsf{PH_{rel}}$ which was proposed by Blum et al. in [3]. $\mathsf{PH_{rel}}$ appears to be superior to other pheromone models proposed for Shop Scheduling type problems. In particular – as shown in [4] – unlike other pheromone models, $\mathsf{PH_{rel}}$ does not introduce an overly strong model bias potentially leading the search process to low quality areas in the search space. In $\mathsf{PH_{rel}}$ pheromone values are assigned to pairs of *related* operations. We call two operations $o_i, o_j \in O$ *related* if they belong to the same group, or if they have to be processed on the same machine. Formally, a pheromone τ_{o_i, o_j} on a pair of operations o_i and o_j exists, iff $g(o_i) = g(o_j)$ or $m(o_i) = m(o_j)$.

The meaning of a pheromone value τ_{o_i, o_j} is that if τ_{o_i, o_j} is high then operation o_i should be scheduled before operation o_j. The choice of the next operation to schedule is handled as follows. If there is an operation $o_i \in S'_t$ with no related and unscheduled operations left, it is chosen. Otherwise we choose among the operations of set S'_t according to the following transition probabilities:

$$p(s[t] = o \mid s[0,...,t-1], \tau) = \begin{cases} \dfrac{\left(\min_{o_r \in S_o^{rel}} \tau_{o,o_r}\right) \cdot \eta_o^{\beta}}{\sum_{o_k \in S'_t}\left(\min_{o_r \in S_{o_k}^{rel}} \tau_{o_k,o_r}\right) \cdot \eta_{o_k}^{\beta}} & : \text{ if } \quad o \in S'_t \\ 0 & : \text{ otherwise,} \end{cases}$$

where $S_o^{rel} = \{o' \in O \mid m(o') = m(o) \vee g(o') = g(o), o' \text{ not scheduled yet}\}$. As heuristic information η_o we choose the inverse of the earliest starting time of an operation o with respect to the current partial schedule $s[0,...,t-1]^5$, and β is a parameter to adjust the influence of the heuristic information η_o. The meaning of this rule to compute the transition probabilities is the following: If at least one of the pheromone values between an operation $o_i \in S'_t$ and a related operations o_r which is not scheduled yet is low, then the operation o_i probably should not be scheduled now. By using this pheromone model the algorithm tries to learn relations between operations. The absolute position of an operation in the sequence s is not important. Of importance is the relative position of an operation with respect to related operations.

LocalSearch(s_j): To every solution s_j constructed by the ants, a local search based on the neighborhood structure introduced by Nowicki et al. in [15] is applied. The formalization of this local search procedure for the GSP can be found in [19].

Update(s_{ib}, s_{rb}, s_{gb}): This function updates solutions s_{rb} (the restart best solution) and s_{gb} (the best solution found since the start of the algorithm) with the iteration best solution s_{ib}. s_{rb} is replaced by s_{ib}, if $C_{\max}(s_{ib}) < C_{\max}(s_{rb})$. The same holds for s_{gb}.

---

[5] We add 1.0 to all earliest starting times in order to avoid division by 0.

ApplyOnlineDelayedPheromoneUpdate(τ,s): We implemented our algorithm in the Hyper-Cube Framework [2]. The Hyper-Cube Framework is characterized by a normalization of the contribution of every solution used for updating the pheromone values. This leads to a scaling of the objective function values and the pheromone values are implicitly limited to the interval $[0,1]$ (see [2] for a more detailed description). $\mathcal{MMAS}$ algorithms usually apply a pheromone update rule which (depending on some convergence measure) uses the iteration best solution s_{ib}, the restart best solution s_{rb} and s_{gb}, the best solution found since the start of the algorithm. In our algorithm we chose to use only the restart best solution and the global best solution. The reason for that is the different structure of the scheduling problems covered by the GSP. Preliminary experiments showed that for Open Shop Scheduling instances a much higher selection pressure is needed to make the algorithm converge than for Job Shop Scheduling instances. The implications in practice are that the use of the iteration best solution s_{ib} for updating the pheromone values would have to be fine-tuned depending on the problem instance structure. To avoid this we decided against using the iteration best solution.

For a number of solutions $s_1, ..., s_k$, the pheromone updating rule in the Hyper-Cube Framework is the following.

$$\tau_{o_i,o_j} \leftarrow (1-\rho) \cdot \tau_{o_i,o_j} + \rho \cdot \sum_{l=1}^{k} \Delta^{s_l} \tau_{o_i,o_j} \tag{1}$$

where

$$\Delta^{s_l} \tau_{o_i,o_j} = \begin{cases} \dfrac{\frac{1}{C_{\max}(s_l)}}{\sum_{r=1}^{k} \frac{1}{C_{\max}(s_r)}} & \text{if } o_i \text{ before } o_j \text{ in } s_l, \\ 0 & \text{otherwise.} \end{cases} \tag{2}$$

As our algorithm – at any time – only uses one solution (s_{rb} or s_{gb}) for updating the pheromone values, we can specify the pheromone updating rule for our algorithm as follows.

$$\tau_{o_i,o_j} \leftarrow f_{mmas} \left(\tau_{o_i,o_j} + \rho \cdot (\delta(o_i, o_j, s) - \tau_{o_i,o_j}) \right) \tag{3}$$

where

$$\delta(o_i, o_j, s) = \begin{cases} 1 & \text{if } o_i \text{ before } o_j \text{ in } s, \\ 0 & \text{otherwise.} \end{cases} \tag{4}$$

and

$$f_{mmas}(x) = \begin{cases} \tau_{\min} & \text{if } x < \tau_{\min}, \\ x & \text{if } \tau_{\min} \le x \le \tau_{\max}, \\ \tau_{\max} & \text{if } x > \tau_{\max}. \end{cases} \tag{5}$$

We set the lower bound $\tau_{\min}$ for the pheromone values to 0.001 and the upper bound[6] $\tau_{\max}$ to 0.999. Therefore, after applying the pheromone update rule

---

[6] Note that in contrast to the usual way of implementing ACO algorithms, in the Hyper–Cube Framework also the upper bound for pheromone values is necessary to avoid convergence to a solution.

above, we check which pheromone values exceed the upper bound, or are below the lower bound. If this is the case, the respective pheromone values are set back to the respective bound.

ComputeConvergenceFactor(τ): To measure the "extend of being stuck" in an area in the search space we compute after every iteration a so–called *convergence factor cf*. We compute this factor in the following way.

$$cf \leftarrow \left(\left(\frac{\sum_{o_i \neq o_j, \text{ related}} \max\{\tau_{\max} - \tau_{o_i,o_j}, \tau_{o_i,o_j} - \tau_{\min}\}}{\sum_{o_i \neq o_j, \text{ related}} \tau_{\max} - \tau_{\min}} \right) - 0.5 \right) \cdot 2.0 \quad (6)$$

From the formula above it becomes clear that when the algorithm is initialized (or restarted) with pheromone values all 0.5, cf is 0.0 and when all pheromone values are either equal to $\tau_{\min}$ or equal to $\tau_{\max}$, cf is 1.0.

ResetPheromoneValues(τ): In the basic version of our algorithm we reset all the pheromone values to the same positive constant value 0.5.

This concludes the description of the basic $\mathcal{MMAS}$ for the GSP (henceforth identified by U for uniform initialization and resetting of pheromone values). As already mentioned in Section 1, scheduling problems are in general multimodal problems in the sense that good solutions are scattered all over the search space. Therefore we expect to be able to improve the basic algorithm presented in this section with additional intensification and diversification mechanisms.

4   Intensification and Diversification Strategies

Intensification and diversification of the search process are quite unexplored topics in ACO research. There are just a few papers explicitly dealing with the topic. The mechanisms already existing can be divided into two different categories. The first one consists of mechanisms changing in some way the pheromone values, either on-line (e.g., [7,18]) or by resetting the pheromone values (e.g., [20,22]). The second category consists of algorithms applying multiple colonies and exchanging information between them in some way (e.g., [13,12]). In contrast to that, most of the intensification and diversification mechanisms to be outlined in the following are based on a set of elite solutions found by the algorithm in the history of the search process.
As a first strategy to introduce more intensification and diversification into the the search process performed by Algorithm 1 we changed the functions to initialize and reset the pheromone values. In an algorithm henceforth identified by R (for random pheromone setting) we use a pheromone value initialization and resetting function that generates for every pheromone a value uniformly random between 0.25 and 0.75[7]. This introduces more intensification, because

---

[7] Note that the Hyper–Cube framework facilitates a strategy like that, as we explicitly know the space in which the pheromone values move.

right from the start of a restarting phase the algorithm is more focused on an area of the search space. On the other side more diversification is introduced, because with every restart the algorithm focuses probably on a different area of the search space.

Examining the behavior of the $\mathcal{MMAS}$ algorithm presented in Section 3 we got the impression that the algorithm wastes a lot of time by always – at the end of a restart phase – moving toward the best solution found since the start of the algorithm. After some while there are no improvements to be found around this solution. The strategy of always moving toward the best found solution might work well for problems like the Traveling Salesman Problem with a fitness–distance relation, for problems lacking a fitness–distance relation other strategies seem to be required. To change this scheme we keep a list L_{elite} of elite solutions found during the search to be handled and used as described in the following. L_{elite} works as a FIFO list and has a maximum length l. There are several actions to be specified.

Adding solutions to L_{elite}: Every solution s_{rb} at the end of a restart phase (when $cf \geq 0.99$ and $global\_convergence == FALSE$ in Algorithm 1) is added to the list. At the beginning of the algorithm the list is empty. In a situation where the length of L_{elite} is smaller than l, the new solution is just added. In case the length of L_{elite} is equal to l we also remove the first element of L_{elite}.

Usage of solutions in L_{elite}: In Algorithm1 we replace the convergence to the best solution found so far with the following scheme. At the end of a restart phase (when $cf \geq 0.99$ and $global\_convergence == FALSE$) among the solutions in L_{elite} we choose the one, denoted by $s_{closest}$, which is closest to s_{rb}. The distance between two solutions s_1 and s_2 is measured by a kind of Hamming distance d_h which is defined as follows.

$$d_h(s_1, s_2) = \sum_{o_i \neq o_j, \text{ related}} \delta(o_i, o_j, s_1, s_2), \tag{7}$$

where $\delta(o_i, o_j, s_1, s_2) = 1$, if the order between o_i and o_j is the same in s_1 and s_2, and $\delta(o_i, o_j, s_1, s_2) = 0$ if otherwise. In case $s_{closest}$ is different to s_{rb}, s_{rb} is replaced by $s_{closest}$ and the algorithm continues until it gets stuck again. If during this phase $s_{closest}$ is improved, it is removed from the list L_{elite} and the improved solution is added to L_{elite} (as last added element). This phase of the algorithm is henceforth called *phase of local convergence*. When the algorithm gets stuck in the phase of local convergence we choose the best solution s_{best} among the solutions in L_{elite}. If it is different from s_{rb} (the best solution found during the phase of local convergence), we replace s_{rb} with s_{best} and proceed with the algorithm. If during this phase s_{best} is improved, it will be removed from the list L_{elite} and the improved solution is added to L_{elite} (as last added element). This phase of the algorithm is henceforth called *phase of convergence to the best*.

The mechanism above implies that a solution once added to L_{elite} has $|L_{elite}|$ (the length of L_{elite}) restart phases of the algorithm to be improved. If it is improved it has again the same time to be improved again. If a solution in L_{elite} is not improved within $|L_{elite}|$ restart phases of the algorithm, it is dropped from L_{elite} and its neighborhood is regarded as an explored region of the search space.

The intuition behind the usage of a list of elite solutions as described above is the following. Instead of only on one area – as done in the usual $\mathcal{MMAS}$ – our algorithm works on several areas in the search space trying to improve the best solutions found in these areas. If it can't improve them in a certain amount of time they are discarded even if they contain the best solution found since the start of the algorithm. This prevents the algorithm from wasting time and has a diversifying component. This mechanism also incorporates a strong intensifying component by applying the phase of local convergence and the phase of convergence to the best as described above.

An algorithm using the list of elite solutions and the random pheromone initialization and resetting is henceforth identified by ER. Based on the usage of a list of elite solutions we also developed two more ways of resetting the pheromone values. The first one is aiming for a diversification of the search depending on the elite solutions in L_{elite}. In this scheme the pheromone values are reset as follows.

$$\tau_{o_i,o_j} \leftarrow f\left(\frac{\sum_{s \in L_{elite}} \delta(o_i,o_j,s)}{|L_{elite}|}\right) \text{ where } f(x) = \begin{cases} 0.5 + x & \text{if } x < 0.25, \\ 1.0 - x & \text{if } 0.25 \leq x \leq 0.75, \\ x - 0.5 & \text{if } x > 0.75 \end{cases} \text{, (8)}$$

where the delta-function is the same as defined in (4). The intuition of this setting is that we want to reset the pheromone values in order to concentrate the search in areas of the search space different from the current set of elite solutions. However, the more agreement is to be found among the elite solutions about an order between two related operations o_i and o_j, the more we permit ourselves to trust these solutions and the resetting of the corresponding pheromone value is approximating equal chance for both directions (cases $x < 0.25$ and $x > 0.75$ in (8)). An algorithm using the list of elite solutions and this diversification scheme (and random setting for the first three restart phases) is henceforth identified by ED.

Another scheme for resetting pheromone values aims for an intensification of the search process. First the best solution s_{best} among the elite solutions is chosen, then another one s_{second} different to s_{best} is chosen from L_{elite} uniformly random. Using these two solutions the resetting of pheromone values is done as follows.

$$\tau_{o_i,o_j} \leftarrow f_{mmas}\left(\frac{\delta(o_i,o_j,s_{best}) + \delta(o_i,o_j,s_{second})}{2.0}\right) \quad (9)$$

where the delta-function is as defined in (4) and the function f_{mmas} which keeps the pheromone values in their bounds is as defined in (5). An algorithm using the list of elite solutions, for the first three restart phases the random setting of pheromone values and flipping a coin for all consecutive restart phases to choose among the diversification setting and the intensification setting of pheromone values, is henceforth identified by EDI.

Table 1. Overview on the different $\mathcal{MMAS}$ versions.

Identifier	Characteristics
U	$\mathcal{MMAS}$ using uniform setting of pheromone values
R	$\mathcal{MMAS}$ using random setting of pheromone values
ER	$\mathcal{MMAS}$ using elite solutions and random setting of pheromone values
ED	$\mathcal{MMAS}$ using elite solutions, random setting for the first three restart phases, and after that the diversification setting of pheromone values
EDI	$\mathcal{MMAS}$ using elite solutions, random setting for the first three restart phases, and after that flipping a coin in every restart phase to choose among diversification setting and intensification setting
EDI_{TS}	The same as EDI with an additional short Tabu Search run for the best ant in every iteration

5   Comparison

To test the usefulness of the intensification and diversification approaches outlined in the previous section we run experiments on the five different versions of the Algorithm 1 outlined in the previous sections. These are: U, R, ER, ED and EDI. Additionally we testet a further enhancement of EDI where we apply a short Tabu Search run – based on the neighborhood the local search uses – to the best solution per iteration. We fixed the length of this Tabu Search run to $|O|/2$. This version of the algorithm is henceforth identified by EDI_{TS}. For an overview of the different algorithm version see Table 1. We tested these six versions on three problem instances. We chose the first problem tai_15_15_1_jsp with 15 jobs and on 15 machines introduced by Taillard in [21] for the JSP, and the first problem tai_15_15_1_osp with 15 jobs and on 15 machines also introduced in [21] for the OSP. The optimal solutions for these problems are known to be 1231, 937 respectively. As a third problem we chose whizzkids97, a difficult GSP instance on 197 operations which was subject of a competition held at the TU Eindhoven in 1997. The optimal solution for this problem is 469. We run each of the six versions of our algorithm 10 times for 18000 seconds on each problem instance. In these 18000 seconds all the algorithms except EDI_{TS} do about 350 restarts. EDI_{TS} does about 300 restarts. The parameter setting for all the algorithms was as follow: $\rho = 0.1$, $n_a = 10$ (number of ants) and $\beta = 10.0$. For EDI and EDI_{TS} we set $|L_{elite}| = |O|/20$. The results are shown in Figures 2,3 and 4. There are several observations to be mentioned. The short Tabu Search run on the iteration best ant improves the algorithm considerably. Version EDI_{TS} finds the optimal solutions for the JSP and the OSP instance by Taillard and produces a solution which is only 2.77% above the optimal solution for the instance whizzkids97. Furthermore, the results on the whizzkids97 instance show that the versions R, ER, ED and EDI clearly improve on the basic version U of our algorithm. Among these four improved versions, version EDI has a clear advantage over the other three versions. These observations are supported by the results of the pairwise Wilcoxon rank sum test. This test produced probabilties between 89% and 98% for version EDI to be different from the other versions. Another indication of the advantage of EDI is the average number of improvements over the run-time (see Table 2). The results on the Taillard instances are similar. However, the statistical significance is lower. This might be caused by the fact that these two instances are easier to solve than the whizzkids97 instance. The diversification setting alone (version ED) and the random settings (versions R and ER) don't seem to improve the basic version U on the Taillard instances.

6   Summary and Outlook

We have presented a $\mathcal{MMAS}$ algorithm for the application to the Group Shop Scheduling problem. Additionally we have presented a method to introduce more diversification and more intensification into the search process performed by our $\mathcal{MMAS}$ algorithm by means of a list of elite solutions found during

Table 2. Average number of improvements over the run-time.

	U	EDI
whizzkids97	34.6	44.3
tai_15_15_1_jsp	27.6	30.2
tai_15_15_1_osp	27.3	28.3

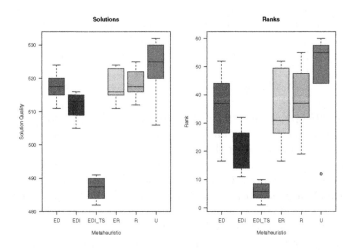

Fig. 2. Comparison of the different ACO versions on the whizzkids97 instance. The absolute values of the solutions generated by each ACO version (left) and their relative rank in the comparison among each other (right) are depicted in two box-plots. A box shows the range between the 25% and the 75% quantile of the data. The median of the data is indicated by a bar. The whiskers extend to the most extreme data point which is no more than 1.5 times the interquartile range from the box. Extreme points are indicated as circles.

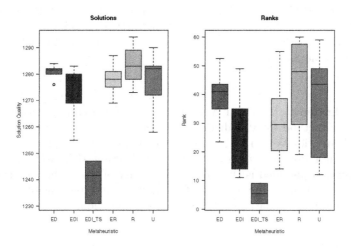

Fig. 3. Comparison of the different ACO versions on the tai_15_15_1_jsp instance. The absolute values of the solutions generated by each ACO version (left) and their relative rank in the comparison among each other (right) are depicted in two box-plots. A box shows the distribution of the results, respectively the ranks of the experiments.

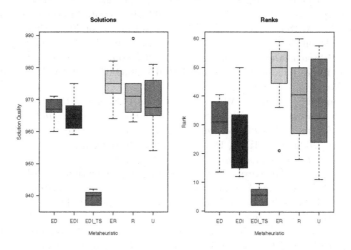

Fig. 4. Comparison of the different ACO versions on the tai_15_15_1_osp instance. The absolute values of the solutions generated by each ACO version (left) and their relative rank in the comparison among each other (right) are depicted in two box-plots. A box shows the distribution of the results, respectively the ranks of the experiments.

the search. The results presented show the usefulness of our diversification and intensification approach. The results also show that our algorithm can be considerably improved by applying a very short run of Tabu Search to the iteration best ant. Our algorithm finds for OSP as well as for JSP instances the optimal solution. In the future we intend to further examine the possibilities of guiding the search of an ACO algorithm by additional diversification and intensification mechanisms.

Acknowledgments. We thank Michael Sampels for useful discussions and for providing his method for visualizing test results. This work was supported by the "Metaheuristics Network", a Research Training Network funded by the Improving Human Potential program of the CEC, grant HPRN-CT-1999-00106. The information provided is the sole responsibility of the authors and does not reflect the Community's opinion. The Community is not responsible for any use that might be made of data appearing in this publication.

References

1. R. Battiti. *Modern Heuristic Search Methods*, chapter Reactive Search: Toward Self–tuning Heuristics. Jossey–Bass, 1996.
2. C. Blum, A. Roli, and M. Dorigo. HC-ACO: The hyper-cube framework for Ant Colony Optimization. In *Proceedings of MIC'2001 – Meta-heuristics International Conference*, volume 2, pages 399–403, Porto, Portugal, 2001. Also available as technical report TR/IRIDIA/2001-16, IRIDIA, Université Libre de Bruxelles.
3. C. Blum and M. Sampels. Ant Colony Optimization for FOP Shop scheduling: A case study on different pheromone representations. In *Proceedings of the 2002 Congress on Evolutionary Computation, CEC'02*, volume 2, pages 1558–1563, 2002. Also available as technical report TR/IRIDIA/2002-03, IRIDIA, Université Libre de Bruxelles.
4. C. Blum and M. Sampels. When Model Bias is Stronger than Selection Pressure. In *Proceedings of the 7th International Conference on Parallel Problem Solving from Nature, PPSN'02 (to appear)*, 2002. Also available as technical report TR/IRIDIA/2002-06, IRIDIA, Université Libre de Bruxelles.
5. P. Brucker. *Scheduling algorithms*. Springer, Berlin, 1998.
6. M. Dorigo and G. Di Caro. The Ant Colony Optimization meta-heuristic. In D. Corne, M. Dorigo, and F. Glover, editors, *New Ideas in Optimization*, pages 11–32. McGraw-Hill, 1999. Also available as Technical Report IRIDIA/99-1, Université Libre de Bruxelles, Belgium.
7. M. Dorigo and L.M. Gambardella. Ant colony system: A cooperative learning approach to the travelling salesman problem. *IEEE Transactions on Evolutionary Computation*, 1(1):53–66, April 1997.
8. M. Dorigo, V. Maniezzo, and A. Colorni. Ant System: Optimization by a colony of cooperating agents. *IEEE Transactions on Systems, Man and Cybernetics - Part B*, 26(1):29–41, 1996.
9. M. Dorigo, M. Zlochin, N. Meuleau, and M. Birattari. Updating ACO pheromones using Stochastic Gradient Ascent and Cross-Entropy methods. In *Proceedings of the EvoWorkshops 2002*, LNCS 2279, pages 21–30. Springer, 2002.
10. B. Giffler and G.L. Thompson. Algorithms for solving production scheduling problems. *Operations Research*, 8:487–503, 1960.
11. F. Glover and M. Laguna. *Tabu Search*. Kluwer Academic Publishers, 1997.

12. H. Kawamura, M. Yamamoto, K. Suzuki, and A. Ohuchi. Multiple Ant Colonies Algorithm Based on Colony Level Interactions. *IEICE Transactions on Fundamentals*, E83–A(2):371–379, February 2000.
13. M. Middendorf, F. Reischle, and H. Schmeck. Information Exchange in Multi Colony Ant Algorithms. In *Proceedings of the Workshop on Bio-Inspired Solutions to Parallel Processing Problems*, LNCS 1800, pages 645–652. Springer Verlag, 2000.
14. H. Mühlenbein and G. Paaß. From recombination of genes to the estimation of distributions. In H.-M. Voigt, W. Ebeling, I. Rechenberg, and H.-P. Schwefel, editors, *Proceedings of the 4th Conference on Parallel Problem Solving from Nature – PPSN IV*, volume 1411 of *Lecture Notes in Computer Science*, pages 178–187, Berlin, 1996. Springer.
15. E. Nowicki and C. Smutnicki. A fast taboo search algorithm for the job-shop problem. *Management Science*, 42(2):797–813, 1996.
16. M. Pelikan, D.E. Goldberg, and F. Lobo. A survey of optimization by building and using probabilistic models. Technical Report No. 99018, IlliGAL, University of Illinois, 1999.
17. M. Pinedo. *Scheduling: Theory, Algorithms, and Systems*. Prentice-Hall, Englewood Cliffs, 1995.
18. M. Randall and E. Tonkes. Intensification and Diversification Strategies in Ant Colony System. Technical Report TR00-02, School of Information Technolony, Bond University, 2000.
19. M. Sampels, C. Blum, M. Mastrolilli, and O. Doria-Rossi. Metaheuristics for Goup Shop Scheduling. In *Proceedings of the 7th International Conference on Parallel Problem Solving from Nature, PPSN'02 (to appear)*, 2002. Also available as technical report TR/IRIDIA/2002-07, IRIDIA, Université Libre de Bruxelles.
20. T. Stützle and H. H. Hoos. $\mathcal{MAX}$-$\mathcal{MIN}$ Ant System. *Future Generation Computer Systems*, 16(8):889–914, 2000.
21. E. Taillard. Benchmarks for basic scheduling problems. *European Journal of Operations Research*, 64:278–285, 1993.
22. E.-G. Talbi, O. Roux, C. Fonlupt, and D. Robillard. Parallel Ant Colonies for the quadratic assignment problem. *Future Generation Computer Systems*, 17:441–449, 2001.
23. M. Yagiura and T. Ibaraki. On metaheuristic algorithms for combinatorial optimization problems. *Systems and Computers in Japan*, 32(3):33–55, 2001.
24. M. Zlochin, M. Birattari, N. Meuleau, and M. Dorigo. Model-based search for combinatorial optimization. Technical Report TR/IRIDIA/2001-15, IRIDIA, Université Libre de Bruxelles, Belgium, 2001.

Agent-Based Approach to Dynamic Task Allocation

Shervin Nouyan

IRIDIA, Université Libre de Bruxelles
CP 194/6, Avenue Franklin Roosevelt 50, 1050 Brussels, Belgium
Lehrstuhl für Nachrichtentechnik, Technische Universität München
Arcisstr. 21, 80290 München, Germany
SNouyan@iridia.ulb.ac.be

Abstract. We investigate a multi-agent algorithm inspired by the task allocation behavior of social insects for the solution of dynamic task allocation problems in stochastic environments. The problems consist of a certain number of machines and different kinds of tasks. The machines are identical and able to carry out each task. A setup, which is linked to a fixed cost, is required to switch from one task to another. Agents, which are inspired by the model of division of labour in social insects, are in charge of the machines. Our work is based on previously introduced models described by Cicirello *et al.* [7] and by Campos *et al.* [6]. Improvements and their effect on the results are highlighted.

1 Introduction

Social insect behaviour has come under increasing attention in the research community in the last years. The number of successful applications in this new area of swarm intelligence has grown exponentially during this period [1]. In combinatorial optimization several implementations of ant based algorithms perform as well as or even better than state of the art techniques, for instance in the traveling salesman problem [11], the quadratic assignment problem [12], the job shop scheduling problem [8], the graph colouring problem [9] or the vehicle routing problem [5]. In the field of telecommunications networks AntNet [10] is an approach to dynamic routing. A set of ants collectively solves the problem by indirectly exchanging information through the network nodes. They deposit information used to build probabilistic routing tables. At the time it was published it outperformed state-of-the-art algorithms for an extensive set of simulated situations. In robotics the recruitment of nestmates in some ant species inspired scientists to program robots in order to achieve an efficient teamwork behaviour [2].

Our particular interest in this paper is the dynamic task allocation in a factory-like stochastic environment. The considered problem consists of machines that perform jobs. These jobs appear dynamically with a certain probability each time step. Jobs are assigned to machines through a bidding mechanism. Machines have a different probability to bid for each kind of job depending on

M. Dorigo et al. (Eds.): ANTS 2002, LNCS 2463, pp. 28–39, 2002.

a dynamic threshold value (see section 2.1). Each time a machine changes the kind of job it performs, a setup is required, which is related to a fixed cost. Machines automatically adapt to the appearances of jobs in the environment in order to minimize the number of setups and to maximize the number of finished jobs. To achieve this we take inspiration from social insects like ants or wasps. In the complex system of their colony an important characteristic is division of labour, i.e. different activities are often performed simultaneously by specialized individuals. This methodology is believed to be more efficient than sequential task performance by unspecialized workers as it avoids unnecessary task switching, which costs time and energy. Of course a rigid specialization of the individual workers would lead to a stiff behaviour at the colony level. So this indivual's specialization is flexible to several internal and external factors such as food availability, climatic conditions or phase of colony development.

A model based on response thresholds was developed by Bonabeau *et al.* [3] in order to explain this flexible division of labour. This model was applied by Cicirello *et al.* [7] and by Campos *et al.* [6] to dynamic task allocation problems and represents the starting point for this work.

In section 2 we give a general description of the threshold model used in [7, 6]. Section 3 introduces to the improvements we propose. We will explain the hypotheses behind the improvements. A detailed description of the problem is delivered in section 4. In section 5 the experimental results are discussed. We conclude in section 6.

2   Model and Application

As mentioned before the characteristic of division of labour in social insects does not lead to rigidity in the colony with respect to tasks performed by individuals. On the contrary the individuals are able to adapt very well to changing demands as shown by Wilson [14].

This flexibility is exhibited for example, by ant species from the *Pheidole* genus. In most species of this genus workers are physically divided into two fractions: The small minors, who fulfill most of the quotidian tasks, and the larger majors, who are responsible for seed milling, abdominal food storage, defense or a combination of these. Wilson [14] experimentally changed the proportion of majors to minors. By diminishing the fraction of minors he observed that within one hour of the ratio change majors get engaged in tasks usually performed by minors.

2.1   Threshold Model

Bonabeau *et al.* [3] have developed a model of response thresholds in order to explain the behaviour observed by Wilson [14]. In this model a set of thresholds is given to each individual performing a task. Each of these thresholds is related to a certain kind of task and the threshold's value represents the level of specialization in that task. In [4] these thresholds remain fixed over time. In [13] the thresholds are dynamically updated with respect to the task being currently performed. For

instance, while an ant is foraging for food the corresponding threshold would decrease whereas the thresholds for all other tasks would increase.
A task emits a stimulus to attract the individuals' attention. Based on this stimulus and on the corresponding threshold, an individual will or will not accept the task. The lower a threshold, the higher the probability to accept a task. Thus lower threshold represents a higher grade of specialization in the task.

Applying this model to the ant species of the *Pheidole* genus with their minor and major workers, the minor workers can be considered to have a basically lower threshold for quotidian tasks than major workers. Decreasing the fraction of the minors will leave their work undone and cause the emitted stimulus of the quotidian tasks to increase. The stimulus will grow until it is high enough to overcome the thresholds of the major workers. By performing the minors' tasks the corresponding thresholds of the major workers will decrease.

2.2   Approach to Dynamic Task Allocation

The dynamic task allocation problem contains a set of machines capable of processing multiple types of jobs[1]. Each machine has an associated insect agent, which is in charge of it. That is, the insect agent decides whether its machine wants to get engaged in a certain job or not. The jobs are created dynamically: each task has a certain probability to appear at each time step. The main values to measure the performance are the number of setups and the number of finished jobs. The number of setups is crucial as it is linked to a high cost in time. Additionally, in a factory environment a setup can cause a monetary cost, that is not considered in this paper. We assume a setup time to be twice the processing time of a job.

Thresholds and Stimuli. Each agent a has a set of response thresholds:

$$\Theta_a = \{\Theta_{a,0}, \ldots \Theta_{a,n}\} \tag{1}$$
$$\Theta_{min} \leq \Theta_{a,j} \leq \Theta_{max} \qquad \forall j \in \{0, \ldots n\}, \tag{2}$$

where $\Theta_{a,j}$ is the response threshold of agent a for task j. Θ_{min} and Θ_{max} are the values for the lower and the upper limit of the thresholds. The threshold value $\Theta_{a,j}$ represents the level of specialization of machine a in task j.

The tasks in the system send to all the machines a stimulus S_j which is equal to the length of time the job is already waiting to be assigned to a machine. Given the stimulus S_j and the corresponding threshold $\Theta_{a,j}$ the agent a will bid for job j with probability:

$$P(\Theta_{a,j}, S_j) = \frac{S_j^2}{S_j^2 + \Theta_{a,j}^2}. \tag{3}$$

---

[1] The model and the equations of this section are taken from, and described in more detail by, Cicirello *et al.* [7].

From (3), we gather that in case the stimulus S_j and the threshold $\Theta_{a,j}$ are equal, the probability for agent a's machine to bid for a job is 50%. Squaring of the stimulus S_j and the threshold $\Theta_{a,j}$ lead to a nonlinear behaviour of $P(\Theta_{a,j}, S_j)$ so that small changes of S_j around $\Theta_{a,j}$ cause high changes in $P(\Theta_{a,j}, S_j)$.

Threshold Update Rules. The threshold values are updated each time step via three rules depending on the the job that is currently being processed. The first rule is given by:

$$\Theta_{a,j} = \Theta_{a,j} - \delta_1, \tag{4}$$

meaning that $\Theta_{a,j}$ is decreased by the constant δ_1 if the machine is currently processing or setting up for a job of type j, in order to give a tendency to accept a job of type j again.

The second update rule is utilized for all thresholds of the jobs that are different from the one currently being processed:

$$\Theta_{a,k} = \Theta_{a,k} + \delta_2, \tag{5}$$

which increases the threshold values by the constant δ_2 in order to diminish the probability to bid for jobs of a type that is different from the current one.

If the machine is currently not processing any job, a third update is used for all thresholds:

$$\Theta_{a,l} = \Theta_{a,l} - \delta_3^t. \tag{6}$$

For idle machines, the thresholds of all types of jobs are decreased by δ_3^t. This update rule encourages the machine to take jobs of any kind with a probability increasing with time. By exponentiating the constant parameter δ_3 with the time t the threshold value is decreased by higher steps the longer the machine stays idle. The thresholds should not immediately be set to a minimum value as this would cause additional setups. It is advantageous that a machine stays idle for a short amount of time instead of taking any job, as a setup determines a very high cost.

Dominance Contest. The dominance contest is used as a rule to decide between several machines in case more than one machines tries to take the same job. In this case a dominance contest is held and the winner is assigned the job. For this, each participant is given a force value F:

$$F_a = 1 + T_{p,a} + T_{s,a}, \tag{7}$$

where $T_{p,a}$ is the sum of the processing times of the jobs in the queue of machine a and $T_{s,a}$ is the sum of the setup times. A lower force value F corresponds to a shorter queue and leads to a higher probability to win in a dominance contest.

The rule to determine the probability for agent 1 to win against agent 2 is:

$$P(F_1, F_2) = \frac{F_2^2}{F_1^2 + F_2^2}, \tag{8}$$

and if the dominance contest consists of more than two competitors, agent 1 will win with probability:

$$P(F_1, \ldots F_n) = \frac{\displaystyle\sum_{i=2}^{n} F_i^2}{(n-1) \displaystyle\sum_{i=1}^{n} F_i^2} \tag{9}$$

This equation was not given in [7], but we deduced it from equation (8).

3   Improvements

In this section we describe the improvements we introduced in order to obtain better results. In section 3.1 we show our modifications on rules already utilized in the original model.

The new rules presented in section 3.2 are designed for situations that still cause problems for the original system. The new rules affect jobs that are not assigned to machines and idle machines, that do not bid for jobs.

3.1   Modifications of Existing Rules

Update Rules (UR). The update rules use the type of the job that is being processed in order to determine which thresholds have to be manipulated. The machines can have a queue of several jobs behind themselves. These jobs in the queue don't necessarily need to be of a same type. For example, a machine might be processing a job of type A and have only one job in its queue that is of type B. That means that, as long as the job of type A is not finished, the corresponding threshold values are decreased and the threshold values that are related to jobs of type B are increased. If the machine is offered two jobs, one of type A and one of type B, the probability to take the job of type A would be higher than the probability to pick the other job. This behaviour is not desirable as it causes an additional setup. If the machine would choose the job of type B this setup could be avoided. We have chosen to modify the update rules by letting the last job in the machines's queue determine which threshold values are updated in order to reduce the number of necessary setups.

Calculation of the Force Variable (CFV). The dominance contest tries to find a good solution to choose between several machines competing for a same job. The solution is found by comparing the force values that represent the queue length of each competitor. Whether a competitor needs to perform a setup or not, in case it is assigned the job, is not taken into consideration. If two machines with the same queue length bid for the same job, the probability for each to win the dominance contest is the same. We maintain that if one of the two machines doesn't need a setup while the other does. We try to take this into account by adding to the force variable the value T_{setup}, which is the setup time of the corresponding job, in case a setup will be required, and zero otherwise:

$$F_a = 1 + T_{a,p} + T_{a,s} + T_{job,s} \qquad (10)$$

$$T_{job,s} = \begin{cases} T_{setup} & \text{if job requires a setup} \\ 0 & \text{if job does not require a setup.} \end{cases} \qquad (11)$$

Dominance Contest (DC). Another problem of the dominance contest is that the more machines compete with each other in a dominance contest, the smaller are the differences between the probabilities to win. For example, if two machines compete and one has a force value of $F_1 = 1$ – what refers to an empty queue – and the other has a force value of $F_2 = 10$ the corresponding probabilities to win following equation (8) are $P_1 = 0.99$ for the machine one and $P_2 = 0.01$ for the other. In case ten machines are competing and the force values are $F_1 = 1$ for machine one and $F_j = 10$ for the others, the probabilities to win according to equation (9) are now $P_1 = 0.111$ and $P_j = 0.099$. In general the probability for one competitor to win in a dominance contest of n competitors is never higher than $\frac{1}{n-1}$. To overcome this problem we use the following rule instead of equation (9):

$$P(F_1, \ldots F_n)_j = \frac{\dfrac{1}{F_j^2}}{\displaystyle\sum_{i=1}^{n} \dfrac{1}{F_i^2}}, \qquad (12)$$

where $P(F_1, \ldots F_n)_j$ is the probability for machine j to win. In the above example with ten machines and the forces $F_1 = 1$ and $F_j = 10$ for the others, the respective probabilities become $P_1 = 0.917$ and $P_j = 0.009$.

3.2 Additional Rules to Optimize the Threshold Values

One major difficulty arrives when the probabilities for the creation of the jobs change during a problem instance. The machines need a long time to adapt to the new situation. Some additional rules have been defined in order to better handle this problem.

No Bid for a Created Job (BCJ). Jobs that were not assigned yet, or that just appeared, are offered to the machines in an allocation process so that every machine is given the possibility to bid for the jobs. It is possible that no machine bids for a job and that it stays unallocated. For instance this situation can appear if the type j of the job that is offered has a low probability to appear, and therefore no machine specializes for this job-type. By increasing the stimulus emitted by the job each time step, the probability that a machine will bid for it also increases over time. As this process is very slow, we introduce an additional method in order to reduce the time that jobs stay unallocated:

$$\Theta_{a,j} = \Theta_{a,j} - \gamma_1 \qquad \forall \text{ agents } a. \qquad (13)$$

Table 1. Probability-mixes for the different sets of tasks. The value n represents the number of jobs.

Problem	$P(Job(1))$ ... $P(Job(\frac{n}{2}))$	$P(Job(\frac{n}{2}+1))$ ... $P(Job(n))$	Simulation Time
16 Jobs-same	0.1333	0.1333	3000
16 Jobs-diff	0.2	0.0666	3000
8 Jobs-same	0.2666	0.2666	3000
8 Jobs-diff	0.4	0.1333	3000

For each agent a, the threshold value $\Theta_{a,j}$ is decreased by the constant value γ_1, if a job of type j was not assigned.

Idle Machine Does Not Bid (IMB). Equation (6) offers an update rule for idle machines in order to encourage the machine to bid for jobs of any type. Another update rule is employed in case an idle machine does not bid for a job if it is offered one:

$$\Theta_{a,j} = \Theta_{a,j} - \gamma_2. \tag{14}$$

In such a situation the corresponding threshold value $\Theta_{a,j}$ of the refused job of type j is decreased by the fixed value γ_2.

4   Experimental Setup

In this section the problem and the experimental design are described in more detail. The dynamic task allocation problem considered here consists of a certain number of machines and different kinds of tasks. Each problem instance contains 32 identical machines that are able to perform all tasks, either eight or sixteen different types of jobs. A machine needs 15 time units to perform a task and 30 time units to perform a setup. Tasks are performed in a first-in-first-out order and are dynamically created in the sense that each task is having a certain probability to appear at each time step. Two different mixes of probabilities have been analyzed: Equal probabilities to all jobs and different probabilities for each half of the jobs. The first mix assigns equal probabilities to all jobs, while the secon mix favors one half of them with a higher probability value. The chosen values are shown in table 1. At each time step there is a certain probability to release one.

The probabilities given in table 1 do not change over simulation time. In order to analyze the adaptation capability of the system we also analyzed a probability mix that changes over time (16 Jobs-chang and 8 Jobs-chang). In this case a simulation time of 5000 is given time steps. For the first 2000 time steps the probabilities are the same as in 16 Jobs-diff and 8 Jobs-diff respectively. Afterwards the jobs swap their probabilities so that the jobs that appeared more frequently, appear more seldom, and vice versa.

The probabilities were chosen so that the average number of jobs appearing during the simulation is equal to the theoretical maximum of jobs that can be performed by the machines.

The same parameter setting was used for all problem instances and all improvements:

- $\Theta_{min} = 1$ Minimum Threshold
- $\Theta_{max} = 2000$ Maximum Threshold
- $\Theta_0 = 500$ Initial Threshold Value for all Machines and all Jobs
- $\delta_1 = 14.0$ Update Rule (4)
- $\delta_2 = 14.0$ Update Rule (5)
- $\delta_3 = 1.01$ Update Rule (6) for idle machines
- $\gamma_1 = 50.0$ New rule: No Bid for a Created Job
- $\gamma_2 = 50.0$ New rule: Idle Machine Does Not Bid

5   System Analysis

The results in this section show a comparison between single improvements, combinations of improvements and the original version of the algorithm used in [7]. Furthermore, we compare these to a static system, in which same types of jobs are always assigned to the same machines according to the frequency of occurence of that job. Comparing the static and the dynamic solution is not fair, as the static algorithm has a knowledge about the global frequency of occurence of the jobs. However, it gives an idea of the upper bound of performance that is possible to reach. The static algorithm outperforms the dynamic algorithms in all measures.

We highlight the improvements **BCJ** and **IMB** (see section 3.2) as standalone improvements as well as in combination with the improvements **UR**, **CFV** and **DC** (see section 3.1). Other combinations of improvements, besides those presented here, were also analyzed. We have chosen these combinations due to their overall performance and stability. All results presented here are averages of 100 runs.

Three measures have been selected, idle time, number of setups and throughput, in order to have a wide range of possible evaluations.

5.1   Idle Time

Depending on the set of thresholds and on the job currently offered, idle machines often prefer to stay idle rather than accepting a job that they are not specialized for. This behaviour is desired as it avoids too many setups.

Figure 1 displays the average time each machine stays idle. Particularly in the changing job mixes (16 Jobs-chang and 8 Jobs-chang), the original model leads in average to longer idle periods. In fact, the probabilities of job appearances change abruptly after 2000 time steps, but machines are able to adjust their threshold values only very slowly. In this long adaptation phase the machines are offered jobs that they are not specialized for. The introduced rules

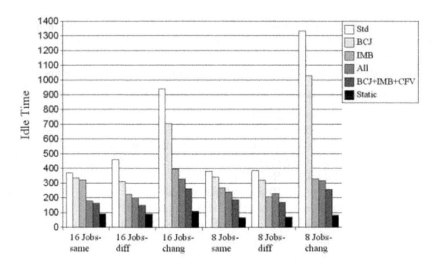

Fig. 1. Average of the time each machine stayed idle during total simulation. Smaller numbers are better. The graph compares the performance obtained by the original algorithm (Std), the static algorithm (static) and the improvements. All indicates, that all improvements were used together. For an explanation of the improvements see section 3. The problems are described in table 1

BCJ and especially **IMB** reduce the time machines stay idle by diminishing the rigidity with that specialization is followed. In fact, **IMB** directly influences idle machines by reducing their thresholds for the jobs they refuse, **BCJ** reduces machines' thresholds when a job is refused by all machines.

5.2   Setups

The number of setups performed by the machines is crucial as a setup always requires a long time. Furthermore, in real world factory environments, a setup can also be connected to a monetary cost. This last factor is not taken into consideration here, as the related monetary costs are not known.

Figure 2 shows the average number of setups for each machine. For the 8 job problems the difference is not appreciable, but for the more complicated 16 job problems, the original model requires far more setups than the modified models. The new introduced rules **BCJ** and **IMB** already diminish the number of required setups when they are applied. However, the system requires even fewer setups if **BCJ** and **IMB** are combined with the modifications on the existing rules. While applying the modifications on existing rules alone does not lead to a better performance, the combination of all improvements makes the whole system more stable and leads to the best results.

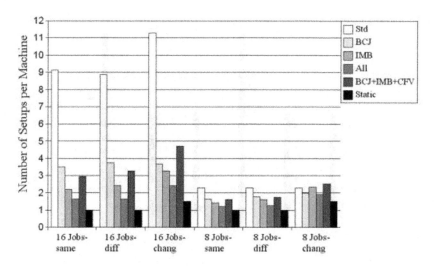

Fig. 2. Average of the number of setups per machine. Smaller numbers are better. The graph compares the performance obtained by the original algorithm (Std), the static algorithm (static) and the improvements. All indicates, that all improvements were used together. For an explanation of the improvements see section 3. The problems are described in table 1

5.3   Throughput

The average number of finished jobs – the throughput – is the most general indicator of the performance. The idle time and the number of setups are more specific measures to evaluate the performance and do not cover all important issues. For instance, if the machines stay idle all the time, they will not need to perform any setup. Obviously the overall performance would be poor in that case, which is well indicated by the throughput. Figure 3 shows a bargraph of the throughput. Again, the modified system performs better than the original one. Amongst the modifications, the combination of the new rules with the modified force variable (**CFV**) performs the best, having a slightly higher performance than the combination of all improvements.

For all problem instances and all improvement combinations we carried out a Student t-test. The null hypothesis was that the distribution of the throughput of the original and of the modified model have the same mean. Except for one, we obtained in all cases p-values smaller than $2.1 \cdot 10^{-5}$. The exception is the **BCJ** improvement in combination with the 16 jobs-chang problem, where we obtain a p-value of $3.35 \cdot 10^{-3}$, which is still small enough to reject the null hypothesis.

6   Conclusion

In this paper we have presented improvements on an adaptive multi-agent system for dynamic factory routing, based on the work presented in [7] and [6]. We have

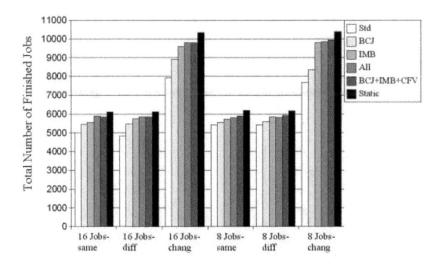

Fig. 3. Average of the total number of jobs performed by all machines. Bigger numbers are better. The graph compares the performance obtained by the original algorithm (Std), the static algorithm (static) and the improvements. All indicates, that all improvements were used together. For an explanation of the improvements see section 3. The problems are described in table 1

introduced new rules and we have modified existing rules in order to speed up the adaptation process and to improve the overall performance, particularly for more complex problems that imply a larger number of different job types and changing probabilities for job appearances. With respect to the analyzed problem configurations, we showed that our modifications perform better than the original system. It is worth noting that we also analyzed problems that imply a lower number of machines (e.g., 2 or 4 machines). For these simple problems the difference between the modified and the original system was much smaller.

Currently, the influence of various parameters on the system performance is under investigation. The values of the parameters of the model were hand fixed. For the parameter analysis we will use a genetic algorithm inspired by the one presented in [6], in order achieve good parameter settings for the different problem configurations. This information can be used to equip several machines with different parameter settings, so that some machines tend to follow a rather stable behaviour, having a higher degree of specialization, and other machines are able to adapt more easily to changing demands.

Furthermore, we intend to investigate real-world problems in order to find constraints that should be taken into account by the system, extending the size of the problems that can be solved by the algorithm. For instance, the system applied in [6] takes the possibility of machine failure and prioritized jobs into account.

Another interesting topic we may consider is online adaptation for the parameter setting of the systen in response to dynamically changing distributions of job arrivals.

References

1. E. Bonabeau, M. Dorigo, and G. Theraulaz. *Swarm Intelligence: From Natural to Artificial Systems*. Oxford University Press, New York, NY, 1999.
2. E. Bonabeau and G. Theraulaz. Swarm Smarts. *Scientific American*, 282(3):54–61, 2000.
3. E. Bonabeau, G. Theraulaz, and J.-L. Deneubourg. Quantitative Study of the Fixed Threshold Model for the Regulation of Division of Labour in Insect Societies. In *Proceedings Roy. Soc. London B*, volume 263, 1996.
4. E. Bonabeau, G. Theraulaz, and J.-L. Deneubourg. Fixed Response Thresholds and the Regulation of Division of Labor in Insect Societies. *Bulletin of Mathematical Biology*, 60:753–807, 1998.
5. B. Bullnheimer, R. F. Hartl, and C. Strauss. An Improved Ant System Algorithm for the Vehicle Routing Problem, 1997.
6. M. Campos, E. Bonabeau, G. Theraulaz, and J. L. Deneubourg. Dynamic Scheduling and Division of Labor in Social Insects. *Adaptive Behavior 2000*, pages 83–96, 2000.
7. V. A. Cicirello and S. F. Smith. Wasp-like Agents for Distributed Factory Coordination. Technical Report CMU-RI-TR-01-39, Robotics Institute, Carnegie Mellon University, Pittsburgh, PA, December 2001.
8. A. Colorni, M. Dorigo, V. Maniezzo, and M. Trubian. Ant System for Job-Shop Scheduling. *JORBEL—Belgian Journal of Operations Research, Statistics and Computer Science*, 34:39–53, 1994.
9. D. Costa and A. Hertz. Ants Can Colour Graphs. *J. Op. Res. Soc.*, 48:295–305, 1997.
10. G. Di Caro and M. Dorigo. AntNet: Distributed stigmergetic control for communications networks. *Journal of Artificial Intelligence Research*, 9:317–365, 1998.
11. M. Dorigo and L. M. Gambardella. Ant Colony System: A Cooperative Learning Approach to the Traveling Salesman Problem. *IEEE Trans. Evol. Comp.*, 1:53–66, 1997.
12. L. M. Gambardella, È. D. Taillard, and M. Dorigo. Ant colonies for the quadratic assignment problem. *Journal of the Operational Research Society*, 50(2):167–176, 1999.
13. G. Theraulaz, E. Bonabeau, and J.-L. Deneubourg. Response Threshold Reinforcement and Division of Labour in Insect Societies. In *Proceedings of the Royal Society of London B*, volume 265, pages 327–335, 1998.
14. E.O. Wilson. The Relation Between Caste Ratios and Division of Labour in the Ant Genus Pheidole (Hymenoptera: Formicidae). *Behav. Ecol. Sociobiol.*, 16:89–98, 1984.

An Ant Colony Optimization Algorithm for the 2D HP Protein Folding Problem

Alena Shmygelska, Rosalía Aguirre-Hernández, and Holger H. Hoos[*]

Department of Computer Science, University of British Columbia
Vancouver, B.C., V6T 1Z4, Canada
{oshmygel,rosalia,hoos}@cs.ubc.ca
http://www.cs.ubc.ca/labs/beta

Abstract. The prediction of a protein's conformation from its amino-acid sequence is one of the most prominent problems in computational biology. Here, we focus on a widely studied abstraction of this problem, the two dimensional hydrophobic-polar (2D HP) protein folding problem. We introduce an ant colony optimisation algorithm for this NP-hard combinatorial problem and demonstrate its ability to solve standard benchmark instances. Furthermore, we empirically study the impact of various algorithmic features and parameter settings on the performance of our algorithm. To our best knowledge, this is the first application of ACO to this highly relevant problem from bioinformatics; yet, the performance of our ACO algorithm closely approaches that of specialised, state-of-the methods for 2D HP protein folding.

1 Introduction

Ant Colony Optimisation (ACO) is a population-based approach to solving combinatorial optimisation problems that is inspired by the foraging behaviour of ant colonies. The fundamental approach underlying ACO is an iterative process in which a population of simple agents ("ants") repeatedly construct candidate solutions. This construction process is probabilistically guided by heuristic information on the given problem instance as well as by a shared memory containing experience gathered by the ants in previous iterations ("pheromone trails"). Following the seminal work by Dorigo *et al.* [3], ACO algorithms have been successfully applied to a broad range of hard combinatorial problems (see, *e.g.*, [4, 5]).

In this paper, we present an ACO algorithm for solving an abstract variant of one of the most challenging problems in computational biology: the prediction of a protein's structure from its amino-acid sequence. Genomic and proteomic sequence information is now readily available for an increasing number of organisms, and genetic engineering methods for producing proteins are well developed. The biological function and properties of proteins, however, are crucially determined by their structure. Hence, the ability to reliably and efficiently predict protein structure from sequence information would greatly simplify the tasks of

---

[*] Corresponding author

M. Dorigo et al. (Eds.): ANTS 2002, LNCS 2463, pp. 40–52, 2002.
© Springer-Verlag Berlin Heidelberg 2002

interpreting the data collected by the Human Genome Project, of understanding the mechanism of hereditary and infectious diseases, of designing drugs with specific therapeutic properties, and of growing biological polymers with the specific material properties.

Currently, protein structures are primarily determined by techniques such as MRI (magnetic resonance imaging) and X-ray crystallography, which are expensive in terms of equipment, computation and time. Additionally, they require isolation, purification and crystallisation of the target protein. Computational approaches to protein structure prediction are therefore very attractive. Many researchers view the protein structure prediction problem as the "Holy Grail" of computational biology; while considerable progress has been made in developing algorithms for this problem, the performance of state-of-the-art techniques is still regarded as unsatisfactory.

The difficulty in solving protein structure prediction problems stems from two major sources: (1) finding good measures for the quality of candidate structures (*e.g.*, energy models), and (2), given such measures, determining optimal or close-to-optimal structures for a given amino-acid sequence. The first of these issues needs to be addressed primarily by biochemists who study and model protein folding processes; the second, however, is a rich source of interesting and challenging computational problems in local and global optimisation. In order to separate these two aspects of protein structure prediction problems, the optimisation problem is often studied for simplified models of protein folding. In this work, we focus on the 2-dimensional hydrophobic-polar (2D HP) model, an extremely simple model of protein structure that has been used extensively to study algorithmic approaches to the protein structure prediction problem. Even in this simplified model, finding optimal folds is computationally hard (NP-hard) and heuristic optimisation methods, such as ACO, appear to be the most promising approach for solving this problem.

The remainder of this paper is structured as follows. In Section 2, we introduce the 2D HP model of protein structure, and give a formal definition of the 2D HP protein folding problem as well as a brief overview of existing approaches for solving this problem. Our new ACO algorithm for the 2D HP protein folding problem is described in Section 3. An empirical study of our algorithm's performance and the role of various algorithmic features is presented in Section 4. In the final Section 5 we draw some conclusions and point out several directions for future research.

2   The 2D HP Protein Folding Problem

The hydrophobic-polar model (HP model) of protein structure was first proposed by Dill [9]. It is motivated by a number of well-known facts about the pivotal role of hydrophobic and polar amino-acids for protein structure [9,13]:

- Hydrophobic interaction is the driving force for protein folding and the hydrophobicity of amino acids is the main force for development of a native conformation of small globular proteins.

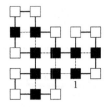

Fig. 1. A sample protein conformation in the 2D HP model. The underlying protein sequence (Sequence 1 from Table 1) is HPHPPHHPHPPHPHHPPHPH; black squares represent hydrophobic amino-acids while white squares symbolise polar amino-acids. The dotted lines represents the H-H contacts underlying the energy calculation. The energy of this conformation is -9, which is optimal for the given sequence.

- Native structures of many proteins are compact and have well-packed cores that are highly enriched in hydrophobic residues as well as minimal solvent-exposed non-polar surface areas.

Each of the twenty commonly found amino-acids that are the building blocks of all natural proteins can be classified as hydrophobic (H) or polar (P). Based on this classification, in the HP model, the primary amino-acid sequence of a protein (which can be represented as a string over a twenty-letter alphabet) is abstracted to a sequence of hydrophobic (H) and polar (P) residues, *i.e.*, amino-acid components. The conformations of this sequence, *i.e.*, the structures into which it can fold, are restricted to self-avoiding paths on a lattice; for the 2D HP model considered in this and many other papers, a 2-dimensional square lattice is used. An example for a protein conformation under the 2D HP model is shown in Figure 1.

One of the most common approaches to protein structure prediction is to model the free energy of the given amino-acid chain depending on its conformation and then to find energy-minimising conformations. In the HP model, based on the biological motivation given above, the energy of a conformation is defined as the number of topological contacts between hydrophobic amino-acids that are not neighbours in the given sequence. More specifically, a conformation c with exactly n such H-H contacts has free energy $E(c) = n \cdot (-1)$; *e.g.*, the conformation shown in Figure 1 has energy -9.

The 2D HP protein folding problem can be formally defined as follows: Given an amino-acid sequence $s = s_1 s_2 \ldots s_n$, find an energy-minimising conformation of s, *i.e.*, find $c^* \in C(s)$ such that $E(c^*) = \min\{E(c) \mid c \in C\}$, where $C(s)$ is the set of all valid conformations for s. It was recently proven that this problem and several variations of it are NP-hard [8].

Existing 2D HP Protein Folding Algorithms

A number of well-known heuristic optimisation methods have been applied to the 2D HP protein folding problem, including Simulated Annealing (SA) [15] and Evolutionary Algorithms (EAs) [8,18,10,17]. The latter have been shown to

Table 1. Benchmark instances for the 2D HP protein folding problem used in this study with known or approximated optimal energy values E^*. (E^* values printed in bold-face are provably optimal.) These instances can also be found at http://www.cs.sandia.gov/tech_reports/compbio/tortilla-hp-benchmarks.html.

Seq. No.	Length	E^*	Protein Sequence
1	20	**-9**	hphpphhphpphphhpphph
2	24	**-9**	hhpphpphpphpphpphpphh
3	25	**-8**	pphpphhppppphhppppphhppppphh
4	36	-14	ppphhpphhppppphhhhhhhpphhppppphhpphpp
5	48	-23	pphpphhpphhppppphhhhhhhhhhpppppphhpphhpphpphhhhhh
6	50	-21	hhphphphphhhhphppphppphppppphppphppphphhhhphphphphh
7	60	-36	pphhhphhhhhhhhhppphhhhhhhhhhhphpppphhhhhhhhhhhhppppphhhhhhphhph
8	64	-42	hhhhhhhhhhhhphphpphhpphhpphpphhpphhpphphphhpphhpphphhhhhhhhhhhhhh
9	20	**-10**	hhhpphphphpphphphpph

be particular robust and effective for finding high-quality solutions to the 2D HP protein folding problem [8].

An early application of EAs to protein structure prediction was presented by Unger and Moult [17,18]. They presented a nonstandard EA incorporating characteristics of Simulated Annealing. Using an algorithm that searches in a space of conformations represented by *absolute* directions and considers only feasible configurations (self-avoiding paths on the lattice), Unger and Moult were able to find high-quality conformations for a set of protein sequences of length up to 64 amino-acids (see Table 1; we use the same benchmark instances for evaluating our ACO algorithm). Unfortunately, it is not clear how long their algorithm ran to achieve these results.

Krasnogor *et al.* [7] implemented another EA in which the conformations are represented using relative folding directions or local structure motifs – the same representation used by our algorithm. Their algorithm found the best known conformations for Sequences 1 through 6 and 9 from Table 1. The best value they achieved for Sequences 7 and 8 were -33 and -39, respectively.

Among the best known algorithms for the 2D HP protein folding problem are various Monte Carlo methods, including the Pruned Enriched Rosenbluth Method of Bastolla *et al.* [1]. Using this method, the best known solution of Sequence 7 ($E^* = -36$) could be found; however, even it failed to obtain the best known conformation for Sequence 8. Other state-of-the-art methods for this problem include the dynamic Monte Carlo algorithm by Ramakrishnan *et al.* [12] and the evolutionary Monte Carlo algorithm by Liang *et al.* [11]. The Core-directed Chain Growth method by Beutler *et al.* was able to find ground states for all benchmark sequences used here, except for Sequence 7 [2]. Currently, none of these algorithm seems to dominate the others.

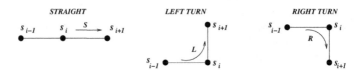

Fig. 2. The local structure motifs which form the solution components underlying the construction and local search phases of our ACO algorithm.

3   Applying ACO to the 2D HP Protein Folding Problem

The ants in our ACO algorithm construct candidate conformations for a given HP protein sequence and apply local search to achieve further improvements. As in [7], candidate conformations are represented using local structure motifs (or relative folding directions) *straight* (S), *left* (L), and *right* (R) which for each amino-acid indicate its position on the 2D lattice relative to its direct predecessors in the given sequence (see Figure 2). Since conformations are invariant *w.r.t.* rotations, the position of the first two amino-acids can be fixed without loss of generality. Hence, we represent candidate conformations for a protein sequence of length n by a sequence of local structure motifs of length $n - 2$. For example, the conformation of Sequence 1 shown in Figure 1 corresponds to the motif sequence LSLLRRLRLLSLRRLLSL.

Construction Phase, Pheromone, and Heuristic Values

In the construction phase of our ACO algorithm, each ant first randomly determines a starting point within the given protein sequence. This is done by choosing a sequence position between 1 and $n - 1$ according to a uniform random distribution and by assigning the corresponding amino-acid (H or P) and its direct successor in the sequence arbitrarily to neighbouring positions on a 2D lattice. From this starting point, the given protein sequence is folded in both directions, adding one amino-acid symbol at a time. The relative directions in which the conformation is extended in each construction step are determined probabilistically using a heuristic function as well pheromone values (also called trail intensities); these relative directions correspond to local structure motifs between triples of consecutive sequence positions $s_{i-1}s_i s_{i+1}$ that form the solution components used by our ACO algorithm; conceptually, these play the same role as the edges between cities in the classical application of ACO to the Travelling Salesperson Problem.

When extending a conformation from sequence position i to the right by placing amino-acid s_{i+1} on the lattice, our algorithm uses pheromone values $\tau_{i,d}$ and heuristic values $\eta_{i,d}$ where $d \in \{S, L, R\}$ is a relative direction. Likewise, pheromone values $\tau'_{i,d}$ and heuristic values $\eta'_{i,d}$ are used when extending a conformation from position i to the left. In our algorithm, we use $\tau'_{i,L} = \tau_{i,R}$, $\tau'_{i,R} = \tau_{i,L}$, and $\tau'_{i,S} = \tau_{i,S}$. This reflects a fundamental symmetry underlying the folding process: Extending the fold from sequence position i to $i + 1$ by placing

s_{i+1} right of s_i (as seen from s_{i-1}) or extending it from position i to $i-1$ by placing s_{i-1} left of s_i (as seen from s_{i+1}) leads to the same local conformation of $s_{i-1}s_is_{i+1}$.

The heuristic values $\eta_{i,d}$ should guide the construction process towards high-quality candidate solutions, *i.e.*, towards conformations with a maximal number of H-H interactions. In our algorithm, this is achieved by defining $\eta_{i,d}$ based on $h_{i+1,d}$, the number of new H-H contacts achieved by placing s_{i+1} in direction d relative to s_i and s_{i-1} when folding forwards (backwards folding is handled analogously and will not be described in detail here). Note that if $s_{i+1} = P$, this amino-acid cannot contribute any new H-H contacts and hence $h_{i,S} = h_{i,L} = h_{i,R} = 0$. Furthermore, for $1 < i < n-1$, $h_{i,d} \leq 2$ and $h_{n-1,d} \leq 3$; the actual $h_{i,d}$ values can be easily determined by checking the seven neighbours of the possible positions of s_{i+1} on the 2D lattice (obviously, the position of s_i is occupied and hence not included in these checks). The heuristic values are then defined as $\eta_{i,d} = h_{i,d} + 1$; this ensures that $\eta_{i,d} > 0$ for all i and d which is important in order not to exclude *a priori* any placement of s_{i+1} in the construction process.

When extending a partial conformation $s_k \ldots s_i$ to s_{i+1} during the construction phase of our ACO algorithm, the relative direction d of s_{i+1} *w.r.t.* $s_{i-1}s_i$ is determined based on the heuristic and pheromone values according to the following probabilities:

$$p_{i,d} = \frac{[\tau_{i,d}]^\alpha [\eta_{i,d}]^\beta}{\sum_{e\in\{L,R,S\}}[\tau_{i,e}]^\alpha [\eta_{i,e}]^\beta} \qquad (1)$$

Analogously, when extending partial conformation $s_i \ldots s_m$ to s_{i-1}, the probability of placing s_{i-1} in relative direction d w.r.t. $s_{i+1}s_i$ is defined as:

$$p'_{i,d} = \frac{[\tau'_{i,d}]^\alpha [\eta'_{i,d}]^\beta}{\sum_{e\in\{L,R,S\}}[\tau'_{i,e}]^\alpha [\eta'_{i,e}]^\beta} \qquad (2)$$

From its randomly determined starting point l, each ant will first construct the partial conformation $s_l \ldots s_1$ and then the partial conformation $s_l \ldots s_n$. We also implemented variants of our algorithm in which all ants start their construction process at the same point (left end, middle, or right end of the protein sequence). Performance results for these alternative mechanisms are reported in Section 4.

Especially for longer protein sequences, infeasible conformations are frequently encountered during the construction process. This happens if an incomplete conformation cannot be extended beyond a given lattice position because all neighbouring lattice positions are already occupied by other amino-acids. Our algorithm uses two mechanisms to address this problem: Firstly, using a simple look-ahead mechanism we never allow an "internal" amino-acid s_i ($1 < i < n$) to be placed such that all its neighbouring positions on the grid are occupied.[1] Secondly, if during a construction step all placements of s_i are ruled out by the

---

[1] This is extremely cheap computationally, since it can be checked easily during the computation of the heuristic values.

look-ahead mechanism, we backtrack half the distance already folded and restart the construction process from the respective sequence position.[2]

Local Search

Similar to other ACO algorithms known from the literature, our new algorithm for the 2D HP protein folding problem incorporates a local search phase. After the construction phase, each ant applies a hybrid iterative improvement local search to its respective candidate conformation. We use two types of neighbourhoods for this local search process:

- the so-called "macro-mutation neighbourhood" described in Krasnogor et al. [8], in which neighbouring conformations differ in a variable number of up to $n - 2$ consecutive local structure motifs;
- a 1-exchange "point mutation" neighbourhood, in which two conformations are neighbours if they differ by exactly one local structure motif.

Our local search algorithm alternates between these two phases. In each iteration, first a macro-mutation step is applied to the current conformation. This involves randomly changing all local structure motifs between two randomly determined sequence positions. All changes are performed in such a way that the resulting conformation is guaranteed to be feasible, i.e., remains a self-avoiding walk on the 2D lattice. If the macro-mutation step results in an improvement in energy, the local search continues from the respective conformation; otherwise, the macro-mutation step has no effect. Next, a sequence of up to $n - 2$ restricted 1-exchange steps are performed. This is done by visiting all sequence positions in random order; for each position, all 1-exchange neighbours that can be reached by modifying the corresponding local structure motif are considered.

Whenever any of these yields an improvement in energy, the corresponding mutation is applied to the current conformation. These local search iterations are repeated until no improvements in solution quality have been achieved for a given number noImpr of search steps. (Of the various hybrid local search methods we implemented and studied, the one described here seemed to work best.)

Update of the Pheromone Values

After each construction and local search phase, selected ants update the pheromone values in a standard way:

$$\tau_{i,d} \leftarrow (1 - \rho)\tau_{i,d} + \Delta_{i,d,c} \tag{3}$$

where $0 < \rho \leq 1$ is the pheromone persistence (a parameter that determines how fast the information gathered in previous iterations is "forgotten") and $\Delta_{i,d,c}$ is the relative solution quality of the given ant's candidate conformation c, if that

---

[2] Various modifications of this backtracking mechanism were tested; the one presented here proved to be reasonably fast and effective.

conformation contains a local structure motif d at sequence position i and zero otherwise. We use the relative solution quality, $E(c)/E^*$, where E^* is the known minimal energy for the given protein sequence (or an approximation based on the number of H residues in the sequence) in order to prevent premature search stagnation for sequences with large energy values.

As a further mechanism for preventing search stagnation, we use an additional "renormalisation" of the pheromone values that is conceptually similar to the method used in MAX-MIN Ant System [16]. For a given sequence position i, whenever the ratio between the maximal and minimal $\tau_{i,d}$ values, τ_i^{max} and τ_i^{min}, falls below a threshold θ, the minimal $\tau_{i,d}$ value is set to $\tau_i^{max} \cdot \theta$ while the maximal $\tau_{i,d}$ value is decreased by $\tau_i^{max} \cdot \theta$. This guarantees that the probability of selecting an arbitrary local structure motif for the corresponding sequence position does not become arbitrarily small.

We implemented various methods for selecting the ants that are allowed to update the pheromone values, including elitist strategies known from the literature. Performance results obtained for these variants are reported in Section 4.

4   Empirical Results

To assess its performance, we applied our ACO algorithm to the nine standard benchmark instances for the 2D HP protein folding problem shown in Table 1; these are the same instances used by Unger and Moult [17,18]. Experiments were conducted by performing a variable number of runs for each problem instance; each run was terminated when no improvement in solution quality had been observed over 10,000 cycles of our ACO algorithm. We used 10 ants for small sequences ($n \leq 25$) and 10–15 ants for larger sequences. Unless explicitly indicated otherwise, we used the following parameter settings for all experiments: $\alpha = 1$, $\beta = 2$, $\rho = 0.6$, and $\theta = 0.05$. The local search procedure was terminated if no solution improvement had been obtained within 100–300 search steps. We used an elitist pheromone update in which only the best 20% of the conformations obtained after the local search phase were used for updating the pheromone values. Additionally, the globally best conformation was used for updating the pheromone values whenever no improvement in solution quality had been seen within 20–50 cycles. Run-time was measured in terms of CPU time and all experiments were performed on PCs with 1GHz Pentium III CPUs, 256KB cache and 1GB RAM.

As can be seen from the results reported in Table 2, our ACO algorithm found optimal solutions for all but the two longest benchmark protein sequences. For Sequence 7, we achieved the same sub-optimal solution quality as Unger and Moult's evolutionary algorithm. For sequences of length 25 and below, our algorithm found optimal solutions in each of multiple attempts, while for longer protein sequences often many solution attempts were required. Following the methodology of Hoos and Stützle [6], we measured run-time distributions (RTD) of our ACO algorithm; for all sequences in which our algorithm found the best known conformation more than once, the respective RTDs are shown in Figure 3. Evidence of search stagnation behavior can be clearly observed for large

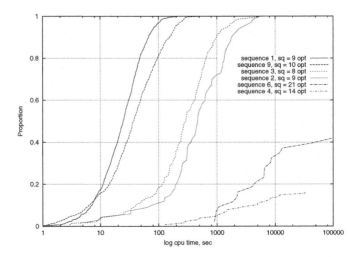

Fig. 3. Run-time distributions of our ACO algorithm applied to several benchmark instances; note stagnation behaviour for large instances.

sequences; in these cases, using a better construction heuristic and/or more aggressive local search may help to improve performance.

To better understand the role of ACO as compared to the local search method in the optimisation process, we also performed experiments in which only the local search method was applied to the same benchmark instances. As seen in Table 2, local search typically takes longer and in many cases fails to find solutions of the same quality as our ACO algorithm. In experiments not reported

Table 2. Comparison of the local search and the ACO, where sq is the best solution quality over all runs, n_{opt} is the number of runs the algorithm finds sq, n_{runs} is the total number of runs, % $suc.$ is the percentage of runs in which solution quality sq was achieved, and t_{avg} is the average CPU time [sec] required by the algorithm to find sq.

Instances			ACO + Local Search				Local Search Only			
Seq. No.	Length	E^*	sq	n_{opt}/n_{runs}	% $suc.$	t_{avg}	sq	n_{opt}/n_{runs}	% $suc.$	t_{avg}
1	20	-9	-9	711/711	100.0	23.90	-9	100/258	38.7	111.43
2	20	-9	-9	596/596	100.0	26.44	-9	8/113	7.0	162.15
3	25	-8	-8	120/120	100.0	35.32	-8	44/129	34.1	125.42
4	36	-14	-14	21/128	16.4	4746.12	-14	5/72	6.9	136.10
5	48	-23	-23	1/151	0.6	1920.93	-21	1/20	5.0	1780.74
6	50	-21	-21	18/43	41.9	3000.28	-20	3/18	16.7	1855.96
7	60	-36	-34	1/119	0.8	4898.77	-33	2/20	10.0	1623.21
8	64	-42	-32	1/22	4.5	4736.98	-33	2/9	22.2	1441.88
9	24	-10	-10	247/247	100.0	43.48	-10	5/202	25.0	134.57

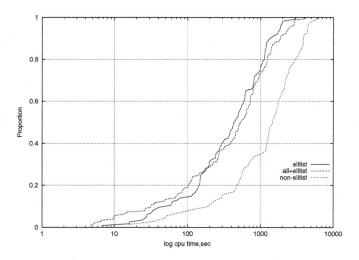

Fig. 4. RTDs for ACO with elitist, all+elitist, and non-elitist pheromone update applied to benchmark instance 2.

here, we also found that ACO without a local search takes substantially more time (cycles) to reach high-quality solutions than ACO with local search.

In order to evaluate the benefit from using a population of ants instead of just a single ant, we studied the impact of varying the number of ants on the performance of our algorithm. While using a single ant only, we still obtained good performance for small problems ($n \leq 25$), the best known solution to Problem 4 ($n = 36$) could not been found within 2 CPU hours.

It has been shown for other applications of ACO that elitist pheromone update strategies can lead to performance improvements. The same appears to be the case here: We tested three different pheromone update strategies: non-elitist – all ants update pheromone values; all+elitist – same, but the best 20% conformations are additionally reinforced; and elitist only – only the best 20% of the conformations are used for updating the pheromone values in each cycle. As can be seen in Figure 4, elitist update results in considerably better performance than non-elitist update. For larger sequences ($n \geq 36$), the best known solution qualities could not been obtained within 2 CPU hours when using non-elitist update. This suggests that the search intensification provided by elitist pheromone update is required for achieving good performance of our ACO algorithm. At the same time, additional experiments (not reported here) indicate that our pheromone renormalisation mechanism is crucial for solving large problem instances, which underlines the importance of search diversification.

In the next experiment, we investigated the influence of pheromone values compared to heuristic information on performance. As illustrated in Figure 5, the results show that both, pheromone values and heuristic information are important for achieving good performance. Both extreme cases, $\alpha = 0$, *i.e.*, pheromone values are ignored, and $\beta = 0$, *i.e.*, heuristic values are ignored,

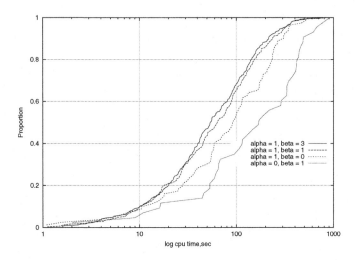

Fig. 5. RTDs for our ACO applied to Sequence 1, using various values of α and β.

lead to performance decreases even for small problem instances. Interestingly, using pheromone information only is less detrimental than solely using heuristic information. This phenomenon becomes more pronounced on larger problem instances; *e.g.*, when ignoring pheromone information ($\alpha = 0$), our ACO algorithm was not able to find the best known solution to Sequence 4 within 3 CPU hours.

Finally, we studied the effect of the starting point for the construction of conformations on the performance of our ACO. It has been shown that real proteins fold by hierarchical condensation starting from folding nuclei; the use of complex and diverse folding pathways helps to avoid the need to extensively search large regions of the conformation space [14]. This suggests that the starting point for the folding process can be an important factor in searching for optimal conformations. We tested four strategies for determining the starting point for the folding process performed in the construction phase of our algorithm: all ants fold forwards, starting at sequence position 1; all ants fold backwards, starting at sequence position n; all ants fold forwards and backwards, starting in the middle of the given sequence; and all ants fold forwards and backwards, starting at randomly determined sequence positions (in which case all ants can fold from different starting points). As can be seen from Figure 6, the best performance is obtained by letting all ants start the folding process from individually selected, random sequence positions. This result is even more prominent for longer sequences and suggests that the added search diversification afforded by multiple and diverse starting points is important for achieving good performance.

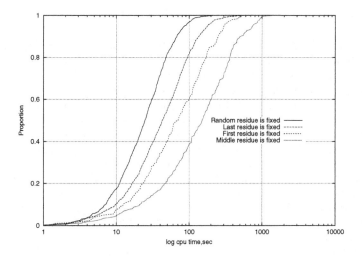

Fig. 6. RTDs for our ACO for Sequence 1, using various strategies for choosing the starting point for constructing candidate conformations.

5   Conclusions and Future Work

In this paper we introduced an ACO algorithm for the 2D HP protein folding problem, an extremely simplified but widely studied and computationally hard protein structure prediction problem, to which to our best knowledge, ACO has not been previously applied. An empirical study of our algorithm demonstrated the effectiveness of the ACO approach for solving this problem and highlighted the impact of various features of our algorithm, including elitist pheromone update and randomly chosen starting points for the folding process.

In this study we presented first evidence that ACO algorithms can be successfully applied to protein folding problems. There are many directions for future research. Clearly, there is substantial room for improvement in the local search procedure. In preliminary experiments with a conceptually simpler local search procedure designed to minimise the occurrence of infeasible configurations we have already observed significant improvements over the results presented here. Furthermore, different heuristic functions should be considered; in this context, techniques that allow the approximation of the size of a protein's hydrophobic core are promising. It might also be fruitful to consider ACO approaches based on more complex solution components than the simple local structure motifs used here. Finally, we intend to develop and study ACO algorithms for other types of protein folding problems, such as the 3-dimensional HP model in the near future [17]. Overall, we strongly believe that ACO algorithms offer considerable potential for solving protein structure prediction problems robustly and efficiently and that further work in this area should be undertaken.

References

1. Bastolla U., H. Fravenkron, E. Gestner, P. Grassberger, and W. Nadler. *Testing New Monte Carlo algorithm for the protein folding problem.* Proteins-Structure Function and Genetics 32 (1): 52-66, 1998.
2. Beutler T., and K. Dill. *A fast conformational search strategy for finding low energy structures of model proteins.* Protein Science (5), pp. 2037–2043, 1996.
3. Dorigo, M., V. Maniezzo, and A. Colorni. *Positive feedback as a search strategy.* Technical Report 91–016, Dip. Elettronica, Politecnico di Milano, Italy, 1991.
4. Dorigo, M. and G. Di Caro. The ant colony optimization meta-heuristic. In *New Ideas in Optimization*, pp. 11–32. McGraw-Hill, 1999.
5. Dorigo, M., G. Di Caro and L.M. Gambardella. *Ant Algorithms for Discrete Optimization.* Artificial Life, 5,2, pp. 137–172, 1999.
6. Hoos, H.H., and T. Stützle. *On the empirical evaluation of Las Vegas algorithms.* Proc. of UAI-98, Morgan Kaufmann Publishers, 1998. IEICE Trans. Fundamentals, Vol. E83-A:2 ,Feb. 2000.
7. Krasnogor, N., D. Pelta, P. M. Lopez, P. Mocciola, and E. de la Canal. *Genetic algorithms for the protein folding problem: a critical view.* In C.F.E. Alpaydin, ed., Proc. Engineering of Intelligent Systems. ICSC Academic Press, 1998.
8. Krasnogor, N., W.E. Hart. J. Smith, and D.A. Pelta. *Protein structure prediction with evolutionary algorithms.* Proceedings of the genetic & evolutionary computation conference, 1999.
9. Lau, K.F., and K.A. Dill. *A lattice statistical mechanics model of the conformation and sequence space of proteins.* Macromolecules 22:3986–3997, 1989.
10. Patton, A.W.P. III, and E. Goldman. *A standard GA approach to native protein conformation prediction.* In Proc. 6th Intl. Conf Genetic Algorithms, pp. 574–581. Morgan Kauffman, 1995.
11. Liang F., and W.H. Wong. *Evolutionary Monte Carlo for protein folding simulations.* J. Chem. Phys. 115 (7), pp. 3374–3380, 2001.
12. Ramakrishnan R., B. Ramachandran, and J.F. Pekny. *A dynamic Monte Carlo algorithm for exploration of dense conformational spaces in heteropolymers.* J. Chem. Phys. 106 (6), 8 February, pp. 2418–2424,1997.
13. Richards, F. M. *Areas, volumes, packing, and protein structures.* Annu. Rev. Biophys. Bioeng. 6:151–176, 1977.
14. Rose, G. D. *Hierarchic organization of domains in globular proteins.* J. Mol. Biol. 134:447–470, 1979.
15. Sali, A., E. Shakhnovich and M. Karplus, *How Does a Protein Fold?* Nature, 369, pp. 248–251, May 1994.
16. Stützle, T., and H.H. Hoos. *Improvements on the ant system: Introducing MAX-MIN ant system.* In Proc. Intel. Conf. on Artificial Neural Networks and Genetic Algorithms, pp. 245–249. Springer Verlag, 1997.
17. Unger, R., and J. Moult. *A genetic algorithm for three dimensional protein folding simulations.* In Proc. 5th Intl. Conf. on Genetic Algorithms, pp. 581–588. Morgan Kaufmann, 1993.
18. Unger, R., and J. Moult. *Genetic algorithms for protein folding simulations.* J. of Molecular Biology 231 (1): 75–81, 1993.

An Experimental Study of a Simple Ant Colony System for the Vehicle Routing Problem with Time Windows

Ismail Ellabib, Otman A. Basir, and Paul Calamai

Department of Systems Design Engineering, University of Waterloo
Waterloo, Canada, N2l 3G1

Abstract. The Vehicle Routing Problem with Time Windows (VRPTW) involves scheduling and routing of a vehicle fleet to serve a given set of geographically distributed requests, subject to capacity and time constraints. This problem is encountered in a variety of industrial and service applications, ranging from logistics and transportation systems, to material handling systems in manufacturing. Due to the intrinsic complexity of the problem, heuristics are needed for analyzing and solving it under practical problem sizes. In this paper, a model of an Ant Colony System (ACS) is proposed to solve the VRPTW. The aim here is to investigate and analyze the performance of the foraging model of a single colony ACS to solve the VRPTW from an experimental point of view, with particular emphasis on different initial solution techniques and different visibility (*desirability*) functions. Finally, experimental analyses are performed to compare the proposed model to other metaheuristic techniques. The results show that the single colony ACS algorithm, despite its simple model, is quite competitive to other well know metaheuristic techniques.

1 Introduction

Transportation systems are the response to the ever-growing needs for contacts between individuals, companies, or societies. The internal structure and the distribution management of the transportation systems, and their importance and effects on economical, social, and environmental aspects are the subject of many specialized studies and publications. However, the vehicle fleet planning is responsible for an important fraction of the economical, social and environmental aspects.

We view the problem as a combined vehicle routing and scheduling problem which often arises in many real-world applications. This combination is often known as the Vehicle Routing Problem with Time Windows (VRPTW). It is focused on the efficient use of a fleet of vehicles that must make a number of stops to serve a set of customers, and to specify which customers should be served by each vehicle and in what order so as to minimize cost, subject to vehicle capacity and service time restrictions. In fact, The VRPTW is a generalization of the well-known problem of vehicle routing (VRP) in which the time dimension incorporated in the form of customer-imposed time-window constraint. The problem involves assignment of vehicles to trips such that the assignment cost and the corresponding routing cost are minimum. We maintain that a hierarchical objective function can be used, where the minimization of vehicles is the primary criterion and the total length of their tours is considered as secondary objectives.

M. Dorigo et al. (Eds.): ANTS 2002, LNCS 2463, pp. 53–64, 2002.

The reader may consult the review surveys on the VRPTW reported in [6,7,8]. The complexity of different vehicle routing and scheduling problems are investigated and reviewed by Lenstra and Rinnooy [14]. It was found that almost all problems of the vehicle routing and scheduling are **NP**-hard. Moreover, it was recently proved that to solve the VRPTW is NP-hard in the strong sense [11]. The first work proposing an exact algorithm for solving the VRPTW was reported in 1987. This algorithm was based on the dynamic programming technique. Subsequently, more successful approaches were introduced based on column generation and Lagrange Relaxation techniques. Only a small set of Solomon's 56 test problems were solved to optimality. Recognizing the complexity of the problem, many researches have recently focused on approximation algorithms and heuristics to solve the problem. Solomon in 1987 was among the first to generalize VRP heuristics for solving the VRPTW. Thangiah et al in 1994 applied different approaches based on Tabu Search, Simulated Annealing, and Genetic algorithms. They were able to achieve improved solution quality by hybridizing these approaches [16]. Rochat and Taillard in 1995 presented a probabilistic technique for the diversification and intensification of the Tabu Search [8]. Chiang and Russell in 1996 applied Simulated Annealing [4]. Taillard et al. in 1997 used adaptive memory with different neighborhood structure [12]. Chiang and Russell in 1997 applied reactive Tabu Search that dynamically adjusts its parameter settings during the search [6]. Gambardella et al. in 1999 applied two ant colonies to solve the problem based on the Ant Colony System (ASC) [13]. More recently, Tan et al. applied the Tabu Search, Simulated Annealing and Genetic algorithms to solve the VRPTW based on their basic models [17].

Recently, a few models of natural swarm-intelligence are presented and transformed into useful artificial swarm-intelligent systems such as ant systems [1]. Those systems are inspired by the collective behaviour of social insect colonies and other animal societies. They have many features that make them a particularly appealing and promising approach to solve hard problems cooperatively. Furthermore, swarm intelligence systems are more appealing for distributed optimization, in which the problem can be explicitly formulated in terms of computational agents. A number of algorithms inspired by the foraging behavior of ant colonies have been recently applied successfully to solve many hard combinatorial optimization problems, and it has been shown that the improvement attained by ACS algorithms can make them a competitive to other meta-heuristic techniques [1], [10].

The aim of this work is to investigate and analyze the performance of the foraging model of a single colony ACS to solve the VRPTW from an experimental point of view, with particular emphasis on different initial solution techniques and different visibility (desirability) functions. The methodology starts by generating an initial solution for the ACS by using a constructive heuristic technique, and allow the ants improve the solution without applying any local search improvement.

The remainder of this paper is organized as follows: a mathematical formulation for the VRPTW is given in Section 2, the concept of the Ant Colony system is described in Section 3, and the methodology of solving the VRPTW based on the Ant Colony System model with different components of the algorithm is described in Section 4. The computational results based on well-known data sets are presented in Section 5 along with a comparative performance analysis involving other metaheuristic Section 6 provides some concluding remarks.

2   Mathematical Formulation for the VRPTW

The VRPTW can be formally stated as follows: given the graph $G=(V, A)$, where V denotes the set of all vertices in the graph; it consists of the subset C, plus the nodes 0 and $n+1$, which represent the depot. $C=\{1,2,.....,n\}$ is the set of customers to be served. Given K the set of available vehicles to be routed and scheduled. The vehicles are identical with the capacity Q. Every customer $i \in C$ has a positive demand d_i, service time s_i, and a time window $\{e_i, l_i\}$ in which the service should start. e_i and l_i are the earliest and the latest service time respectively allowed to serve the customer i. A cost C_{ij} and a travel time t_{ij} are associated with each arc. At each customer location, the start of the service must be within the time window. A vehicle must also leave and return to the depot within the time window $\{e_0, l_0\}$. x^k_{ij} is one if vehicle k drives from node i to node j, and 0 otherwise. b^k_i denotes the time for which a vehicle k starts to service customer i. The service of the customers must be feasible with respect to the capacity of the vehicles, and the time windows of the customers serviced. The objective is to first minimize the number of routes or vehicles, and then the total distance of all routes. However, the problem can be stated mathematically as:

$Minimize$:

$$\sum_{k\in K}\sum_{i\in N}\sum_{j\in N}c_{ij}x^k_{ij} \tag{1}$$

$subject\ to$:

$$\sum_{k\in K}\sum_{j\in N}x^k_{ij} =1 \qquad \forall i\in C \tag{2}$$

$$\sum_{i\in C}d_i\sum_{j\in N}x^k_{ij} \le Q \qquad \forall k\in K \tag{3}$$

$$\sum_{j\in N}x^k_{0j} =1 \qquad \forall k\in K \tag{4}$$

$$\sum_{i\in N}x^k_{i,n+1} =1 \qquad \forall k\in K \tag{5}$$

$$\sum_{j\in N}x^k_{ih} - \sum_{j\in N}x^k_{hj} = 0 \qquad \forall h\in C, \forall k\in K \tag{6}$$

$$b^k_i + s_i + t_{ij} + M(1 - x^k_{ij}) \le b^k_j \ \forall i, j\in N, \forall k\in K \tag{7}$$

$$e_i \le b^k_i \le l_i \qquad \forall i\in N, \forall k\in K \tag{8}$$

$$x^k_{ij} \in \{0,1\} \qquad \forall i,j\in N, \forall k\in K \tag{9}$$

Constraint (2) states that every customer is visited exactly once, and constraint (3) states that no vehicle is loaded with more than the capacity allowed. Constraints (4), (5), and (6) ensure that each vehicle leaves the depot 0; after arriving at customer node the vehicle leaves again, and finally arrives at the depot $n+1$. The inequality constraint (7) states that a vehicle k cannot arrive at node j before time $\{b_i^k + s_i + t_{ij}\}$ elapses, if it is traveling from node i to node j. M in this constraint is a large scalar. Constraint (8) ensures that the service of each customer starts within a customer time window. Finally, constraint (9) is the integrality constraint.

3   The Ant Colony System

The basic idea of the Ant Colony Optimization (ACO) is that a large number of simple artificial agents are able to build good solutions to hard combinatorial optimization problems via low-level based communications; Real ants cooperate in their search for food by depositing chemical traces (*pheromones*) on their path. An artificial ant colony simulates the behavior in which artificial ants cooperate by using a common memory that corresponds to the pheromone deposited by real ants. The artificial pheromone is accumulated at run-time through a learning mechanism. Artificial ants are implemented as parallel processes whose role is to build problem solutions using a constructive procedure driven by a combination of artificial pheromone. ACO is the result of research on computational intelligence approaches to combinatorial optimization originally conducted by Dr. Marco Dorigo, in collaboration with Alberto Colorni and Vittorio Maniezzo [9]. We refer to Dorigo et al. [10] and Bonabeau et al. [1] for more on the concept of Ant System algorithms and their applications. In this paper, an Ant Colony System (ACS) algorithm is proposed to solve the VRPTW. This algorithm was first proposed by Dorigo and Gambardella (1996) to solve TSP. Where, an artificial ant is considered as an agent that moves from city to city on a TSP graph. The agents traveling strategy is based on a probabilistic function that considers two things. Firstly, it counts the edges it has traveled accumulating their length and secondly it senses the trail (pheromone) left behind by other ant agents. An agent selects the next city j among a candidate list based on the following transition rule:

$$j = \begin{cases} \arg\max_{u \in J_i^k} \{ [\tau_{iu}(t)].[\eta_{iu}(t)]^\beta \} & \text{if } q \leq q_0; \\ J & \text{if } q \geq q_0, \end{cases} \quad (10)$$

$$P_{ij}^k(t) = \frac{[\tau_{ij}(t)].[\eta_{ij}(t)]^\beta}{\sum_{l \in J_i^k} [\tau_{il}(t)].[\eta_{il}(t)]^\beta}, \quad (11)$$

where, q is a random variable uniformly distributed over [0,1], q_0 is a tunable parameter in the interval [0,1], and J belongs to the candidate list and is selected based on the above probabilistic rule as in Equation (11). Each agent modifies the environment in two different ways:

Local trail updating: As the agent moves between cities, it updates the amount of pheromone on the edge by the following formula:

$$\tau_{ij}(t) = (1-\rho).\tau_{ij}(t-1) + \rho.\tau_0 \quad (12)$$

The value τ_0 is the initial value of pheromone trails and can be calculated as $\tau_0 = (n.L_{nn})^{-1}$, where n is the number of cities and L_{nn} is the length of the tour produced by one of the construction heuristics.

Global trail updating: When all agents have completed a tour the agent that found

$$\tau_{ij}(t) = (1 - \rho).\tau_{ij}(t-1) + \frac{\rho}{L^+} \tag{13}$$

the shortest route updates the edges in its path using the following formula:
where L^+ is the length of the best tour generated by one of the agents.

4   ACS Model of VRPTW

The first ant system is applied by Bullnheimer in 1997, and 1999 to one of the vehicle routing problems called Capacitated Vehicle Routing Problem (CVRP) [2,3]. Recently, Gambardella et al. [13] applied two ant colony systems for VRPTW to successively optimize the number of vehicles and the total length in a hierarchy manner. They improved some of the best-known solutions by the cooperation between the two colonies.

In order to investigate the performance of the most recent ant system (ACS), a simple model of a single colony system is proposed in this work to solve the VRPTW with hierarchical objective function. The basic idea is to let the ACS perform its search in the space of local minima rather than in the search space of all feasible tours. The model of ACS is applied to the VRPTW by transforming it closely to the traditional Traveling Salesman problem (TSP) as proposed by Gambardella et al. [13]. The transformation can be done by staring from the initial solution and assigning a number of depots equal to the initial number of vehicles, the distances between copies of the depots are set to zero, and with same coordinates. Figure 1 depicts this representation.

The advantage of this representation is that the trails in terms of pheromone update are less attractive than in the case of single depot representation. The approach starts by applying a tour construction heuristic for creating a good initial solution, and then let the ACS operates on the search space of local optima to guide the search toward the global optimum. The ant constructive procedure is similar to the ACS constructive procedure designed for the TSP in [10]. In this procedure, each agent (ant) starts from a randomly chosen depot, and moves to the feasible unvisited node (customer) based on the transition rule until finish all the remaining unvisited nodes.

In each agent step, exploration and exploitation mechanism is applied for the diversification and intensification balance, visibility (desirability) is computed for the transition rule, and the pheromone of the selected edge is updated locally. The global update rule is update at the end of all ant tours in which the pheromone of the best solution edges is update. However, the mount of pheromone updated in Equations 12 and 13 does not only depend on the length of the tour as considered in TSP but on the number of depots (number of vehicles).

4.1   Initial Solution Techniques

Several researchers have pointed out the importance of the quality of initial heuristics on the performance of different metaheuristics such as the construction heuristics proposed by Solomon in 1987. In this paper, two effective heuristics are applied to generate the initial solution for the ACS, namely, the insertion heuristic (I1), and the Nearest Neighbor (NN) [15].

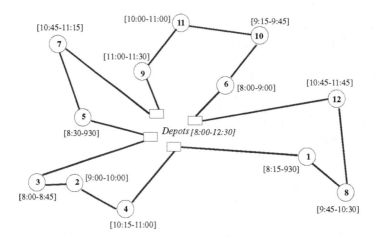

Fig. 1. Transformation of VRPTW instance to the corresponding TSP

Insertion (I1) heuristic. The insertion heuristic considers the insertion of an unvisited customer u between two adjacent customers i_{p-1} and i_p in a partially finished route. We focus on the most effective Solomon the sequential insertion heuristic called (I1) [15]. This heuristic applies two criteria one for selecting best position of the unvisited customer and the other for the customer who has the best cost. The cheapest insertion cost and the associated insertion place for each unvisited customer are calculated using the following formulas:

$$C_0 = l_i + d_{0i} \tag{14}$$
$$C_{11} = d_{iu} + d_{uj} - \mu.d_{ij} \tag{15}$$
$$C_{12} = b_{ju} - b_j \tag{16}$$
$$C_1 = \alpha_1.C_{11} + \alpha_2.C_{12} \tag{17}$$
$$C_2 = \lambda.d_{ou} - C_1 \tag{18}$$
$$where, \quad \alpha_1 + \alpha_2 = 1, \quad \mu \geq 0, and \; \lambda \geq 0$$

Notation C_0 refers to the cost of the first customer inserted in a new a route, C_1 refers to the cost of the best position, and C_2 refers to the cost of best customer. l_i refers to the latest service time of customer i, d_{0i} is the distance from the depot to the customer i. d_{iu}, d_{uj} and d_{ij} refer to the distance between the corresponding pair of customers. b_j and b_{ju} refer to the begin of service before and after the insertion respectively.

Nearest Neighbor heuristic. The most natural heuristic for the routing is the famous Nearest Neighbor algorithm (NN). NN algorithm belongs to class of sequential, and tour building heuristics. It first introduced by Flood in 1956 and first applied to VRPTW by Solomon in 1987 [15]. However, the simplicity and the low computation time are the most advantages of this algorithm. The algorithm starts every route by finding the closest unvisited customer in terms of geographical and temporal aspects.

4.2   Visibility Functions

Visibility (η_{ij}) is based on strictly local information and it measures the attractiveness of the next node to be selected. On the other hand, it represents the heuristic desirability of choosing next customer j after customer i. In the ant system algorithms, visibility is used with the amount of pheromone to direct the search based on the transition rule. It is important to include the visibility term in the transition rule because without the heuristic desirability the algorithm will lead to the rapid selection of tours that may give bad performance [1]. Three types of cost functions are presented and implemented, and the inverse of the cost is used as the visibility measure ($\eta_{ij}=1/C_{ij}$).

Type1: This function is introduced by Solomon in 1987 [15] to evaluate the cost of visiting the next customer in the Nearest Neighbor (NN) heuristic. It measures the direct distance, the time difference between the completion of service at customer i and the beginning of service at customer j, and the urgency of delivery to customer j as in Equation 21.

$$T_{ij} = b_j - (b_i + s_i) \tag{19}$$
$$V_{ij} = l_j - (b_i + s_i + t_{ij}) \tag{20}$$
$$C_{ij} = w_1.d_{ij} + w_2.T_{ij} + w_2.V_{ij} \tag{21}$$
$$where, \quad w_1 + w_2 + w_3 = 1$$

Notation d_{ij} refers to the Euclidean distance between customers i and j. T_{ij} refers to the time difference between the completion of service at customer I, and the beginning of service at customer j. V_{ij} refers the urgency of delivery to customer j.

Type2: This function is introduced in the work of Gambardella et al, [13], and it was used as the visibility measure to optimize the total tour time. It is computed by taking the product of the time difference between the completion of service at i and the beginning of service at j (T_{ij}), and the urgency of delivery to customer j (V_{ij}).

$$V_{ij} = l_j - (b_i + s_i) \tag{22}$$
$$C_{ij} = T_{ij}.V_{ij} \tag{23}$$

where, T_{ij} is calculated according to Equation 19 and V_{ij} is calculated according to Equation 22.

Type3: The above functions are defined in terms of the travel time and the time delay due to the selection of the next customer j. The travel time can be incorporated in the time difference between the completion of service at i and the beginning of service at j, as in Equation 19, and the waiting time can be incorporated in the urgency of delivery to customer j as in Equation 22. However, it has been shown from the preliminary computational experiments, the above functions are not best suited for some problems, especially in random problem sets. In fact, the positions of the customers are scattered in the Euclidean space, and they are identified by x and y coordinates. Therefore, it is useful if the position angle is considered in the attractive measure of the next customer. In order for the *Type2* function to be modified the difference between the position angle of the current customer and the candidate customer is introduced as a second term as seen in Equation 25.

$$P_{ij} = \left| \theta_j - \theta_i \right| \qquad (24)$$

$$C_{ij} = w_1.(T_{ij}.V_{ij}) + w_2.P_{ij} \qquad (25)$$

where, θ_i and θ_j are the polar coordinate angles of the customer i, and j respectively. The first term of the cost function represents the travel time T_{ij} and the time delay V_{ij}, and they are calculated according to Equation 19 and Equation 22 respectively. The second term represents the difference in the position angle between the current customer i and the candidate customer j, and it is calculated according to Equation 24.

5   Computational Results and Comparisons

ACS model of VRPTW is implemented in C++. The Experiments have been performed using 6 well-known data sets. The performance was studied on each of the six data sets by running 10 computer simulations for each of the 56 problem instances. It has been found from the these experiments that the best values of the ACS parameters are: Number of ants=100, β=1, ρ=0.15, q0=0.85 with the number of cycles (Iterations)=300. The following are the values of weights that are used in all problem instances:

Insertion heuristic (I1): μ=1, λ=2, α1=0.9, α2=0.1
Visibility function (Type1): W_1=0.2, W_2=0.7, W_3=0.1
Visibility function (Type3): W_1=0.95, W_2=0.05

5.1   Problem Data Sets

In order to demonstrate and confirm the improvement in the solution of VRPTW based on the proposed algorithm, performance tests on problems with different properties were carried out. Standard benchmark problem instances are available to support empirical investigations and facilitate meaningful comparison in the performance. Solomon in 1987 introduced 56 problem instances [15]. The instances are subdivided into six problem classes (C1, C2, R1, R2, RC1, RC2) with specific properties as shown in Table 1. The problems vary in fleet size, vehicle capacity, travel time, spatial, and temporal customer distribution (position, time window density, time window width, and service time). Each problem instance involves a central depot of vehicles, one hundred customers to be serviced, as well as constraints imposed on vehicle capacity constraint, customer visit or delivery time (time windows), and total route time. C1 and C2 data sets are characterized by a clustered customer distribution whereas *R1* and R2 refer to randomly distributed customers. Data sets RC1, RC2 represent a combination of a random and a clustered customer distribution. Problems in class1 are characterized by narrow time window and small vehicle capacity while problems in class2 are characterized by the wide time window and large vehicle capacity. C1, R1, and RC1 include short scheduling horizon and small vehicle capacity problems. The latter types C2, R2, and RC2 include long scheduling horizon and larger vehicle capacity. The data sets use distances based on the Euclidean metric and assume travel time is equal to travel distance.

Table 1. Summary of the data sets.

Data set	Customers	Instances	Comments
R1	100	12	Random, Short routes.
R2	100	11	Random, Long routes.
C1	100	9	Clustered, short routes.
C2	100	8	Clustered, Long routes.
RC1	100	8	Random, with clusters, Short routes.
RC2	100	8	Random, with clusters, Long routes

5.2 ACS Performance Using Different Initial Feasible Heuristics and Different Visibility Functions

A comparison on the performance of ACS with different methods is summarized in Table (2). The performance here is measured in terms of the average best solution seen so far for each data set. The table obtained the results of running ACS algorithm with different methods. The insertion (I1) and Nearest Neighbor (NN) are used to obtain initial solutions and the different visibility functions are used in the ACS (Type1, Type2, and Type3). The solution quality is based on minimizing the number of routes followed by the total distance. That is, a solution with k number of routes is better than $k+1$ routes, even if the total distance for the k route is greater than $k+1$ routes. The best average solutions in terms of routes number and total distances for all methods with respect to each data set are highlight in bold.

For problems in which the customers are uniformly distributed and semi clustered (data sets R1, R2, RC1, and RC2), the ACS+I1 with the visibility function (Type3) obtains better average solutions in comparison to the ACS with the other methods. While the ACS+NN with the visibility, function (Type3) does well for problems in which the customers are clustered (data sets C1, and C2). In general, it was found that the best solutions, for all the data sets, are found by using visibility function *Type3* in ACS with both initial solution methods.

Table 2. Average number of tours and length obtained for the six data sets using two heuristics with different visibility functions.

Data set	ACS-+						ACS+					
	Type1		Type2		Type3		Type1		Type2		Type3	
	Tours	Length	Tours	Length	Tours	Length	Tours	Length	Tours	Length	Tours	Length
R1	13.25	1474.4	13.25	1448.0	13.08	1441.08	13.42	1452.2	13.17	1434.8	13.42	1436.1
R2	3.18	1361.0	3.18	1309.0	3.18	1294.09	3.18	1413.1	3.18	1333.5	3.18	1329.4
C1	10.33	1150.2	10.44	1126.7	10.44	1118.6	10.00	1004.0	10.00	953.11	10.00	949.67
C2	3.38	830.00	3.13	710.00	3.25	698.88	3.25	799.13	3.00	727.63	3.00	717.88
RC1	13.25	1611.7	13.00	1560.8	12.88	1566.88	13.00	1625.2	13.00	1593.0	13.00	1620.6
RC2	3.63	1617.3	3.38	1570.8	3.38	1521.25	3.50	1643.2	3.38	1560.8	3.50	1536.2

5.3   Synergistic Effects (Cooperative within the Low Level Agents)

In this simple experiment, we tried to illustrate how the set of agents cooperatively improve the solution of the problem within their low level of cooperation. We realized that in the most of the problem instances the pheromone update has a significant contribution in the improvement of the current solution compared with solution obtained without pheromone update as you see in one of the problem instances (Figure 2).

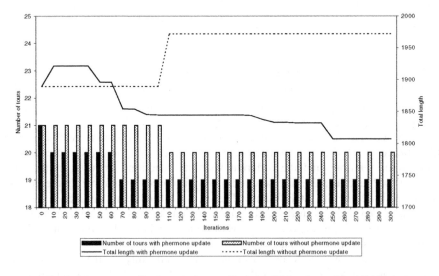

Fig. 2. Improvement in the solution quality by ACS for the problem R101.txt

5.4   Comparison with other Heuristics

Table 3 compares the average solution obtained by five different heuristics and the ACS. As shown, the ACS+I1 with the visibility function (Type3) obtains the best average solution for data sets R1, R2, RC1, and RC2 in comparison to the five competing heuristics. The heuristic TSAS of Thangiah [16] obtains the best average solution for the data sets C1, and C2, and ACS+NN with the visibility function (Type3) obtained the same number of tours with large total distance in these data sets.

Consistently, ACS produces higher performance solutions relative to competing heuristics on problems that are uniformly distributed and semi clustered when the insertion heuristic is applied to generate the initial solution. In terms of computation time, it is very difficult to compare the various approaches, due to different languages and hardware. However, ACS looks a little bit computationally expensive compared with the other heuristics.

Table 3. Average number of tours and length obtained for the six data sets by the ACS algorithm and five other competing heuristics

Data set	SA^1		TS^1		GA^1		SA^2		$TSSA^2$		ACS+I1		ACS+NN	
	Tours	Length	Tours	Length	Tours	Length	Tours	Length	Tours	Length	Tours	Length	Tours	Length
R1	14.50	1420.12	13.83	1266.24	14.42	1314.84	13.70	1252.00	13.30	1242.00	13.08	1441.08	13.42	1436.17
R2	3.64	1278.97	3.82	1080.23	5.64	1093.49	3.20	1169.00	3.20	1113.00	3.18	1294.09	3.18	1329.45
C1	10.11	958.57	10.00	870.93	10.11	860.65	10.00	883.00	10.00	831.00	10.44	1118.67	10.00	949.67
C2	3.25	766.46	3.25	634.82	3.25	623.49	3.00	687.00	3.00	663.00	3.25	698.88	3.00	717.88
RC1	14.75	1648.77	13.63	1458.18	14.63	1512.93	13.40	1454.00	13.00	1413.00	12.88	1566.88	13.00	1620.63
RC2	4.25	1641.89	4.25	1293.38	7.00	1282.47	3.80	1249.00	3.90	1257.00	3.38	1521.25	3.50	1536.25

Legend:
AS^1, TS^1, and GA^1: best solution obtained using Simulated Annealing, Tabu Search, and Genetic Algorithm metaheuristics of Tan et al. [17].
SA^2, and $TSSA^2$: Simulated Annealing, and hybridizing approach between the Tabu Search and the Simulated Annealing metaheuristics of Thangiah et al.[16].

6   Conclusion and Further Work

A simple Ant Colony System (ACS) model is proposed to solve the VRPTW. The model is applied to the well-known data sets, and the performance is evaluated for each data set based on the average number of tours and the total length. In addition, the performance of the ACS to solve the VRPTW is improved by modifying the visibility functions that are introduced before.

In comparison with the other metaheuristics, ACS performs uniformly better than the other metaheuristics with the exception of the clustered data sets. However, ACS obtains solutions that are as close as the best solutions obtained by ASTS for clustered data sets. ACS does not tend to compete the best solution published so far. This is to be expected, as the ACS algorithm requires local search procedure to improve the solutions that are generated from the ants as proposed in the formal algorithm.

References

1. Bonabeau, E., Dorigo, M. and Theraulaz, G.: Swarm Intelligence: From Natural to Artificial Systems. New York: Oxford University Press (1999).
2. Bullnheimer, B., Hartl, R. and Strauss, C: Applying the Ant System to the Vehicle Routing Problem. In: Voss S., Martello S., Osman I.H., Roucairol C. (eds.), Meta-Heuristics: Advances and Trends in Local Search Paradigms for Optimization, Kluwer:Boston (1997).

3. Bullnheimer, B., Hartl, R. and Strauss, C.: An improved ant system algorithm for the vehicle routing problem. Paper presented at the Sixth Viennese workshop on Optimal Control, Dynamic Games, Nonlinear Dynamics and Adaptive Systems, Vienna (Austria), May 21-23, 1997, to appear in: Annals of Operations Research, Dawid, Feichtinger and Hartl (eds.): Nonlinear Economic Dynamics and Control, (1999).
4. Chiang, W. and Russell, R.: Simulated Annealing Metaheuristics for the Vehicle Routing Problem with Time Windows, Annals of Oper. Res. 63, (1996), 1-29.
5. Chiang, W. and Russell, R.: A reactive tabu search metaheuristic for the vehicle routing problem with time windows, INFORMS J. on Computing 9, 4 (1997).
6. Cordeau, J.-F., Desrosiers ,G., Solomon , M. and Soumis, F., "The VRP with Time Windows", in The Vehicle Routing Problem, Chapter 7, Paolo Toth and Daniele Vigo (eds), SIAM Monographs on Discrete Mathematics and Applications, 157-193, 2002.
7. Desrochers, M., Lenstra, J., Savelsbergh, J. and Soumis, F.: Vehicle routing with time windows: Optimization and approximation, in: B.L. Golden and AA. Assad(eds.), Vehicle Routing: Methods and Studies, North-Holland, Amsterdam, (1988), 85-105.
8. Desrosiers, J. Dumas, Y., Solomon, M., Soumis, F., Time constrained routing and scheduling, in: M. Ball, T. Magnanti, M. Monma, G. Nemhauser (Eds.), Handbooks in Operations Research and Management Science, vol. 8: Network Routing, Netherlands, Amsterdam, (1995), 35-139.
9. Dorigo, M., Maniezzo V., and Colorni, A.: The Ant System: Optimization by a Colony of Cooperating Agents, IEEE Trans. Sys. Man Cyb. B 26 (1996), 29-41.
10. Dorigo, M., Gambardella, L.: Ant Colony System: A Cooperative Learning Approach to the Traveling Salesman Problem, IEEE Trans. Evol. Comp. 1, No.1 (1997), 53-66.
11. Drop, M., Note on the complexity of the shortest path models for column generation in VRPTW, Operations Research 42, 5 (1994).
12. Taillard, E. Badeau, P. Gendreau, M., Guertin, F. and Potvin, J.: A Tabu Search Heuristic for the Vehicle Routing Problem with Soft Time Windows, Transportation Science 31, 2, (1997).
13. Gambardella, L., Taillard, E. and Agazzi, G.: MACS-VRPTW: A multiple Ant Colony system for vehicle routing problems with time windows. In D. Corne, M. Dorigo and F. Glover, editors, New Ideas in Optimization. McGraw-Hill (Also available as, Tech. Rep. IDSIA-06-99 , IDSIA, Lugano, Switzerland), (1999).
14. Lenstra, J. and Rinnooy Kan, A.: Complexity of Vehicle Routing and Scheduling Problems, Networks 11, (1981), 221-227.
15. Solomon, M.: Algorithms for the vehicle Routing and Scheduling Problems with Time Window constraints, Operations research 35, (1987), 254-265.
16. Thangiah, S., Osman, I. and Sun, T.,:Hybrid Genetic Algorithm, Simulated Annealing, and Tabu Search Methods for Vehicle Routing Problems with Time Windows. Technical Report 27, Computer Science Department, Slippery Rock University (1994).
17. Tan, K.C., Lee,Q.L., Zhu,K.,Ou: Heuristic methods for vehicle routing problem with time windows, Artificial Intelligence in Engineering 15, Elsevier, (2001), 281-295.

Ant Algorithms for Assembly Line Balancing

Joaquín Bautista and Jordi Pereira

ETSEIB, Universitat Politécnica de Catalunya
Avda. Diagonal 647, Planta 7, 08028 Barcelona, Spain
{Joaquin.bautista,jorge.pereira}@upc.es

Abstract. The present work is focused on the assembly line balancing design problems whose objective is to minimize the number of stations needed to manufacture a product in a line given a fixed cycle time, equivalent to a fixed production rate. The problem is solved using an ACO metaheuristic implementation with different features, obtaining good results. Afterwards, an adaptation of the previous implementation is used to solve a real case problem found in a bike assembly line with a hierarchical multi-objective function and additional constraints between tasks.

1 Introduction

The assembly line balancing problem ALBP consists on assigning the tasks in which a final product can be divided to the work stations in its assembly line. Using the classification given by Baybards [4], the assembly line balancing problems can be divided in two general categories: SALBP (Simple Assembly Line Balancing Problems) and GALBP (Generalized Assembly Line Balancing Problems).

The first set of problems (SALB) can be formulated as: find an assignment of tasks to a set of work stations (each one consisting on a worker, a workgroup or robots) with identical cycle time or production rate, from a set of elementary tasks with previously defined duration. Each task can only be assigned to one work station and a set of precedence relationships must be fulfilled. The problem is a generalization of the Bin Packing problem where precedence constraints are added. Three formulations of the objective are found in the literature:

Minimizing the number of stations given a fixed work rate, cycle time of the line; problem known as SALBP-1 which will be the problem studied in this work.

Minimize the cycle time given to each workstation to do their tasks given a fixed number of workstations, problem known as SALBP-2.

To minimize the total idle time of the assembly line, equivalent to maximize the line efficiency, with a lower and upper bound to the number of stations. This problem is known as SALBP-E

The second category, GALBP is composed by the rest of problems, including those problems showing extra constraints, different hypothesis and different or multiple objective functions. Some models in this category have been studied in the literature including problems with parallel workstations [6], task grouping, [7] and incompatibilities between tasks [1]. As the objective function for these problems are

M. Dorigo et al. (Eds.): ANTS 2002, LNCS 2463, pp. 65–75, 2002.
© Springer-Verlag Berlin Heidelberg 2002

partially or completely the same function from their SALBP counterparts, we will keep the same notation for the objective.

It's important to remark that all SALBP and GALBP instances have the property of reversibility, a solution to an instance where precedence relationships between tasks have been reversed is also a solution to the original problem. Reversing an instance consists on substituting all precedence constraints of the original instance by their reverse precedence constraints; for each pair of tasks i and j where task i is a predecessor of task j, the reverse instance will have a precedence relationship where task j is a predecessor of task i, while maintaining all other constraints untouched. This property has been used in some exact approaches, [17].

Several exact approaches for assembly line balancing problems have been proposed. The most effective approaches are those based on implicit enumeration, [12], [13] and [17]. Even though branch and bound procedures perform well, problems with big dimensions found in industrial settings are usually solved using heuristics, allowing smaller running times to obtain good solutions and facilitating the adaptation to real industrial concerns not found in the academic models. Heuristic approaches have been formulated in the literature for the SALBP-2, [20], and SALBP-1 family of problems, [3], [5] and [19].

A first class of heuristic procedures applied to the problem are greedy heuristics based on priority rules [11]. Priority rules usually refer to or combine characteristics of the instances, as the processing time of tasks, the number of successors of a task, lower and upper bounds on workstations, etc. to decide which tasks should be assigned first. Annex I show a list of different priority rules found in the literature.

The greedy procedure assign tasks to a station using the priority rule to determine the most appropriate task between a set of tasks compatible with tasks already assigned until the station is filled. The set of candidate tasks is composed by those tasks whose predecessors has already been assigned, its processing time is not bigger than the remaining available time in the station and fulfill any problem specific constraints. Once the set of candidate tasks becomes empty, no other task can be assigned to the current station, the procedure creates a new station and begins to assign tasks to it.

These greedy procedures usually use a single priority rule and provide good mean results, usually improving as the number of characteristics from the instance taken into account by the rule increases, but no single rule dominates all other rules for all instances of the problem.

A second class of heuristic procedures is constituted by the GRASP (*Greedy Randomized Adaptive Search Procedure*) [3], and ACO metaheuristics, that allow the generation of several solutions due to the incorporation of randomness in the procedure. On each iteration of the greedy procedure, a task is chosen from a subset of the candidates using a probabilistic rule which may take into account a priority rule and information obtained by previous iterations.

Finally, a third class of heuristics are local search procedures. Between them: the hill-climbing procedures (HC), the simulated annealing (SA), the tabu search (TS), [16], and the genetic algorithms (GA) [3]. This class of heuristics provide alternative ways to find new solutions in a solution space limited by the neighborhood definition.

This work is divided in the following sections. In section 2 a constructive heuristic and a local search procedure for SALBP-1 problems are shown. Section 3 proposes an ACO heuristic approach to the SALBP-1 problem based on the AS heuristic [9], for the traveling salesman problem, analyzing different models of trail information maintenance and the hybridization with a local search heuristic shown in section 2. Section 4 covers a computational experiment with a benchmark instance set to the problem, [16], for the previous proposed heuristics. Finally section 5 details the adaptation of the implemented heuristics to a special real GALBP application found in the bike assembly industry to finish in section 6 with the conclusions of this work.

2 Greedy Heuristics for SALB-1 Problems

Figure 1 shows a schematic model of a constructive, greedy, heuristic to generate solutions for SALB and GALB problems.

The heuristic works as follows: tasks are selected one by one, using the associated priority rule (or rules) which identify the algorithm, from a set of tasks satisfying precedence relationships, time limitations and incompatibilities with the already constructed part of the solution. After a complete solution is found, a local search procedure can be applied to the solution to improve it.

Obviously, to define the heuristic at least a priority rule must be selected: Annex I shows a list of thirteen heuristics rules found in the literature and an example of its usage.

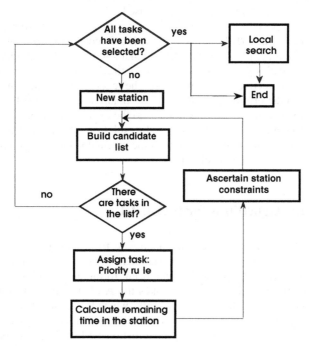

Fig. 1. Procedure scheme for obtaining solutions to SALB-GALB problems.

The final results obtained are the number of stations needed, an assignment of tasks to each station and, if needed by any other part of the algorithm, an ordered list containing when the tasks were selected.

2.1 Hill Climbing Procedure

The proposed local search procedure starts with a solution to the problem and tries to reduce the objective function of the SALBP-1 minimizing the number of stations needed. Unfortunately due to the structure of the solution space of SALBP-1, many solutions exists with the same objective function value, and usually the whole solution neighborhood have the same objective function value, making direct approaches not useful. Instead of this, the algorithm tries to achieve the goal using a different strategy consisting on minimizing the idle time in the first stations of the solution, and thus maximizing the idle time in the later stations. Eventually, and as a side effect, the last station might end up with no assigned tasks and thus it can be eliminated, reducing the objective function value.

The neighborhood is defined by two possible operators: (1) exchange the station of two tasks if the task duration which is moving to the station nearer to the initial station is greater than the task duration moving to the station nearer to the final station, and (2) moving a task from a station to a previous station. Both operators are forced to construct feasible solutions discarding any not feasible solution they might reach to, and not to accept any exchange whose solution is worse than the previous one. The algorithm is applied in such a way that once an improvement is found it is automatically applied on the solution. At the end of the procedure if the final station is completely idle the objective function is decreased.

A different approach is given by [16], consisting on a transformation of the SALBP-1 instance in a SALBP-2 instance with a fixed number of stations equal to one less than the number of stations of the original solution. Tasks assigned to the last station in the original solution are assigned to the last station of the transformed solution, and the objective is to minimize the cycle time under the cycle time constraint given to the original SALBP-1. If we obtain a solution for the transformed problem with a cycle time equal to or inferior to the cycle time given in the SALBP-1 problem, this assignment of tasks is a valid solution for the original SALBP-1 problem with smaller objective function than the original solution. The process can be repeated until no improvement is obtained.

3 Ant Colony Algorithms for SALBP-1

During the development of the ACO metaheuristic from its beginning [9], up to the last implementations [18], different variations, some slight some other more important, of the original approach have been formulated, and several different algorithms have appeared. Our approach takes the AS heuristic as presented by [10] as its starting point. In each iteration m ants, we will call them subcolony, build solutions to the problem. Each ant iteratively chooses tasks using the constructive approach shown in section 2, where a probabilistic rule based on a priority rule and the trail information deposited by previous ants are used to guide the search. The heuristic information, indicated as η_i, where i is a task, and the trail information, indicated as τ, are in advance indicators of how good seems to assign the task i in the current decision. As more than a single priority rule are available in the literature for this problem, we opt for assigning a different rule to each ant of the subcolony, and additionally to use the inversability property, some ants will solve the direct instance while others the reverse instance. As there are thirteen rules available from the literature and two ants of each subcolony use the same priority rule, one for the direct problem and the other for the reverse problem, each iteration, or subcolony, is formed by twenty-six ants.

The different policies to manage in which characteristics of the solution the trail information will be left and two techniques to read the information have been tested. The three trail information usage policies are: (1) The trail information is deposited between correlative tasks (i,i+1) in an ordered assignment list, (τ_{ij} meaning the existing trail information between task j and task i), called trail between pair of tasks, (2) the trail information is deposited between the task and the iteration in which the task was selected (meaning the trail information existing between the task i and the position in an ordered list of decisions j) called trail between task and position, and (3) the trail information is left between the task and its assigned station (where τ_{ij} mean the existing trail information between task j and station i), and called trail between task and station.

To keep information in such a way that it's possible to use it for ants solving the direct and the reverse instance, the trail information for reverse instances is deposited in the following ways: (1) The trail information is left between correlative pairs of tasks (i,i-1), (2) The task is left between a task and the position (number_of_tasks - chosen_position), and (3) the trail information is left between a task and the station (upper_station_bound – chosen_station).

The two ways used to read information are: (1) direct trail information reading, called direct evaluation, and (2) summed trail information reading, as [15], called accumulative evaluation.

Under the previous hypothesis, the probability of assigning task j from a set D of candidate tasks is as follows:

$$p_{ij} = \frac{[\tau_{ij}]^\alpha [\eta_j]^\beta}{\sum_{h \in D} [\tau_{ih}]^\alpha [\eta_h]^\beta} \tag{1}$$

Where τ_{ij} and τ_{ih} will depend on the trail information updating policies and η_j, η_h the priority of each task for the given rule. Due to the differences between the range of values each priority rule can have, the heuristic information η_j is linearly normalized between 1 and the number of candidate tasks for each task i. This approach was firstly proposed in [14].

In case of accumulative evaluation, the probability to choose a task is as follows:

$$p_{ij} = \frac{\left(\sum_{k=1}^{i} [\tau_{kj}]\right)^\alpha \cdot [\eta_j]^\beta}{\sum_{h \in D} \left(\sum_{k=1}^{i} [\tau_{kh}]\right)^\alpha \cdot [\eta_h]^\beta} \tag{2}$$

Where k represent, depending on the trail information updating policy, all previously selected tasks, all previous assignment positions or all previous stations.

In both cases, α and β are two constants determining the relative influence of the heuristic information and the trail information in the ants decision process.

Additionally, the management policy to read the trail information takes into account if the direct or reverse instance of the problem is being solved, eg. if the trail information policy is to keep trail information between tasks and stations in an accumulative fashion, the reading of trail information for the reverse instance to assign the task i to the station j τ_{ij} will be the summation of trails between the station (upper_bound-j) and station (upper_bound).

The trail information is updated after each subcolony have generated and evaluated all their ants, and only the best ants for the direct and reverse problem leave trail information, following the elitist ant scheme from [8] for the ACS algorithm. For each task j, the amount of trail information left equals (3) which take into account the objective function of the solution (noted as solution) and the best solution previously obtained by the procedure (noted as upper bound of the problem).

$$\tau_{ij} = \tau_{ij} + \frac{upper\_bound\_problem}{solution} \cdot \rho \tag{3}$$

Where ρ is a parameter related to the evaporation factor used to harden premature convergence problems.

The trail information is left between task j and i, where i is the previously assigned task in case of trail between tasks policy is used, the position in which the task has been assigned in case the trail information is left between tasks and positions, or the station to which the task has been assigned in case the trail information between tasks and stations is used. To harden convergence problems from appearing, before depositing the trail information an evaporation occurs following the formula (4).

$$\tau_{ij} = (1 - \rho) \cdot \tau_{ij} \tag{4}$$

The algorithm is run until the final condition is reached, which may be a fixed number of iterations or a maximum running time.

3.1 Hill Climbing Optimization Strategy

The procedures have been tested with, and without, an additional local search procedure inserted. The local search procedure is shown in the section 2.1, but as the time required to apply the local search procedure is bigger than the time spent to build solutions, the local search is only applied to the best solutions of each subcolony for the direct and reverse instances and randomly to 1% of all solutions constructed. A similar strategy was also used by [2], [8] and [15].

The number of tested heuristics is, thus, twelve depending on the three trail information management policies, the two trail information reading procedures and the use or not of the local search procedure.

4 Computational Experience

To compare the results given by the proposed procedures, we compare their results with those present in the webpage www.bwl.tu-darmstadt.de/bwl3/forsch/projekte/

alb/index.htm by Scholl for his instance set [16], and composed of a total of 267 instances. The next table shows the differences between the best know solution and the number of optimal solutions found by each procedure, with a time limit of one and five minutes and parameters α=0.75, β=0.25 and ρ=0.1. After the results are shown, the differences between the presence of a local search procedure or not, the trail information management policy and the reading policy and running times are analyzed.

Table 1. Results obtained by each procedure. The number of optimal solutions, the mean variation between the optimum solution and the obtained solution of the optimal or best solution known is reported. The trail information management policy used is referred as task-station (TS), task-position (TP) and task-task (TT). The reading trail information procedure is refered as direct (D) or accumulative (A) and the use of the local search procedure (LS) or not (NLS).

Procedure	Optimal Sol. (1 min.)	Deviation Rate (1 min.)	Optimal Sol. (5 min.)	Deviation Rate (5 min.)
TSD-NLS	169	0.461 %	173	0.404 %
TSD-LS	172	0.436 %	175	0.379 %
TSA-NLS	162	0.493 %	168	0.45 %
TSA-LS	161	0.493 %	164	0.475 %
TTD-NLS	172	0.454 %	175	0.426 %
TTD-LS	160	0.514 %	171	0.454 %
TTA-NLS	178	0.426 %	182	0.394 %
TTA-LS	177	0.436 %	178	0.415 %
TPD-NLS	164	0.46 %	172	0.433 %
TPD-LS	165	0.472 %	171	0.44 %
TPA-NLS	180	0.39 %	182	0.379 %
TPA-LS	177	0.418 %	180	0.401 %

4.1 Presence of a Local Search Procedure

With a limited running time of one minute, four algorithms with local search perform worse than their counterparts without local search. This number even increases when the running time is limited to five minutes.

As shown, the incorporation of this element is not very attractive for the problem. The time spent in the intensive exploration of the neighborhood of a solution doesn't help the ants very much to obtain the same results exploiting the trail information. Only for the case where trail information is left between tasks and stations, the algorithm obtains better results than their counterpart without local search, and only if the reading trail information policy is the direct one, even though this difference is reduced when the algorithm is given more computation time. This is probably a side effect of the proposed local search algorithm who tend to put tasks in the first stations helping to create a fast and good trail information, than the obtained without the local search.

A different approach like the transformation proposed by [16], may lead to an improvement in the number of optimal solutions found.

4.2 Trail Information Management and Reading Policies

Even if all procedures give similar results, the trail information management policies between tasks and positions and between tasks seem to be more fitted than not the task station policy. The accumulative task to task trail information management policy which takes into account trail information between already assigned tasks and the new assigned task is the best procedure for a direct trail information reading, while the accumulative algorithms seem to be better than their direct counterparts.

In case of task station management policy, direct reading seems to be better.

4.3 Running Time

Small improvements are obtained when the running time is increased to five minutes, indicating the high quality of one minute processing. Most solutions are only one station over the best known solution even with one minute runs, showing how robust the heuristic is, even with small computation times.

5 A Real Case Study

The following problem comes from a previous collaboration between our department and a bike assembly industry from Barcelona, who facilitated data of the production line, tasks attributes, precedence relationships and cycle times for each model produced by the factory. The problem consisted on assigning one hundred and three tasks with a cycle time of 327 seconds, one bike produced each five minutes and a half, trying to minimize the number of work stations needed with the given cycle time.

The real problem had several additional differentiated characteristics from the academic model shown before, which included constraints related to incompatibilities between tasks, some tasks couldn't be done in the same work station with other tasks which needed special working conditions. That included a division of tasks between dirty hand jobs, like some jobs related to motor assembly, clean hand jobs, like embellishment tasks, and don't care jobs, those were no special consideration was kept. Additionally, the tasks possessed an additional attribute marking them as right side jobs or left side jobs, depending on which side of the chassis the task had to be performed. This attribute didn't imposed a constraint, as the chassis is moved along the line by a platform allowing this movement, but some nuisances and wasted time was caused to the worker when turning the bike. As the number of turns was hard to calculate exactly, the worker does its assigned tasks in the order he is more comfortable with, we decided to try to maximize the sum of differences between tasks done in one side of the chassis and the other side in each station.

The final objective function was a weighted function between minimizing the number of required workstations and minimizing the number of turns in the solution. As the primary objective was minimizing the number of stations, a multiplying factor was applied to minimize the significance of turns in the objective solution compared to the primary objective.

To solve the problem, the heuristics from section 3 were used, with some special subroutines designed to keep incompatibility constraints between tasks during the construction of solutions, the evaluation of the number of turn existing in the solution and a hierarchical objective function.

The local search improvement method was also modified to take care of the additional constraints and to focus the minimization of turns present in the solution, once we saw the optimal number of stations, nine, was easy to obtain by the procedures without local search.

The following table shows a bound in the number of turns (calculated as a sum of each task with a chassis side associated), the solution found by a previous MILP solution to the problem, stopped after four days of computation, and the proposed heuristics for one and five minutes runs.

Table 2. Obtained results for each procedure in the real case problem. Additionally a trivial bound to the problem and the solution obtained using a mixed integer linear programming formulation is shown.

Procedure	Turns (1 min.)	Turns (5 min.)	Procedure	Turns (1 min.)	Turns (5 min.)
Bound	52	---	MILP	---	41
ESPD-NML	38	40	ESPD-ML	42	42
ESPA-NML	40	40	ESPA-ML	42	44
PRPD-NML	38	38	PRPD-ML	42	44
PRPA-NML	38	38	PRPA-ML	40	44
POPD-NML	46	46	POPD-ML	42	42
POPA-NML	40	40	POPA-ML	42	44

The results show that the result obtained by almost every heuristic is better than the obtained by the exact procedure after a longer computation time, even if no local search procedure was used, and also very near to the known problem bound that doesn't take into account task incompatibilities or precedence relationships between tasks.

6   Conclusions

The present work has proposed several procedures based on the ACO metaheuristic for the assembly line balancing problem. After showing the available greedy procedure to solve the problem and a local search procedure, several trail information management policies and trail information reading techniques are studied. Some new ideas are also described as solving the direct and reverse instance of a problem concurrently and using several priority rules together. Finally the heuristics are modified to handle a real life case found in the bike assembly industry with positive results.

Appendix: Heuristic Rules

Nomenclature

i, j	Tasks index
N	Number of tasks
C	Cycle time
t_i	Duration of task i.
IS_i	Set of immediate successors of task i.
S_i	Set of successors of task i.
TP_i	Set of predecessors of task i.
L_i	Level of task i in the precedence graph.

Assign the task z^* : $v(z^*) = \max_{i \in z}[v(i)]$.

Name	Priority Rule
1. Longest Processing Time	$v(i) = t_i$
2. Number of Immediate Successors	$v(i) = \mid IS_i \mid$
3. Greatest Number of Successors	$v(i) = \mid S_i \mid$
4. Greatest Ranked Positional Weight	$v(i) = t_i + \sum t_j \; (j \in S_i)$
5. Greatest Average Positional Weight	$v(i) = (t_i + \sum t_j \, (j \in S_i)) / (\mid S_i \mid + 1)$
6. Smallest Upper Bound	$v(i) = -UB_i = -N - 1 + [(t_i + \sum t_j \, (j \in S_i))/C]^+$
7. Smallest Upper Bound / Successors	$v(i) = -UB_i / (\mid S_i \mid + 1)$
8. Processing Time / Upper Bound	$v(i) = t_i / UB_i$
9. Smallest Lower Bound	$v(i) = -LB_i = -[(t_i + \sum t_j \, (j \in TP_i)) / C]^+$
10. Minimum Slack	$v(i) = -(UB_i - LB_i)$
11. Maximum Number Successors/Slack	$v(i) = \mid S_i \mid / (UB_i - LB_i)$
12. Bhattcharjee & Sahu	$v(i) = t_i + \mid S_i \mid$
13. Kilbridge & Wester Labels	$v(i) = -L_i$

Example: Let's suppose an instance with five tasks (A,B,C,D and E) with a duration of 3,5,4,1 and 11s. respectively and a cycle time of 12s. The precedence relationships between tasks are: A precedes B, C and D, C precedes E and D precedes E. In the reverse instance B precedes A, C precedes A, and E precedes C and D, while keeping the same task duration and cycle time. The heuristic weights for each task, using rule 2, are 3,0,1,1 and 0 for the direct instance and 0,1,1,1 and 2 for the reverse instance. Using the greedy algorithm and a direct lexicographic order, for the direct instance, or reverse lexicographic order, for the reverse instance to break ties, the solution to the direct instance will have 3 stations composed by (station 1) tasks A, C and D, (station 2) task B and (station 3) task E, while the reverse instance will have 2 workstations composed by (station 1) tasks E and D, and (station 2) tasks C, B and A.

Acknowledgements. This work has been partially funded by CYCIT grant DPI2001-2169. We would also like to thank two anonymous reviewers for their suggestions to improve this work.

References

1. Agnetis, A., A. Ciancimino, M. Lucertini and M. Pizzichella: Balancing Flexible Lines for Car Components Assembly. *International Journal of Production Research* (1995) 33: 333-350.
2. Bauer, A., B. Bullnheimer, R.F. Hartl and C. Strauss: An Ant Colony Optimization Approach for the Single Machine Total Tardiness Problem. In *Proceedings of the 1999 Congress on Evolutionary Computation (CEC'99)*. IEEE Press, Piscataway, NJ. (1999) 1445-1450.
3. Bautista, J., R. Suárez, M. Mateo and R. Companys: Local Search Heuristics for the Assembly Line Balancing Problem with Incompatibilities Between Tasks. In *Proceedings of the 2000 IEEE International Conference on Robotics and Automation, ICRA 2000*, (2000) 2404-2409.
4. Baybars, I.: A Survey of Exact Algorithms for the Simple Assembly Line Balancing Problem. *Management Science* (1986) 32 (8): 909-932.
5. Boctor F.F.: A Multiple-rule Heuristic for Assembly Line Balancing. *Journal of the Operational Research Society*, (1995) 46: 62-69.
6. Daganzo, C.F and D.E. Blumfield: Assembly System Design Principles and Tradeoffs, *International Journal of Production Research* (1994) 32: 669-681
7. Deckro, R.F.: Balancing Cycle Time and Workstations. *IIE Transactions* (1989) 21: 106-111
8. Dorigo M. and L. M. Gambardella: Ant Colony System: A Cooperative Learning Approach to the Traveling Salesman Problem. *IEEE Transactions on Evolutionary Computation* (1997) 1(1): 53-66.
9. Dorigo M., V. Maniezzo and A. Colorni: The Ant System: An Autocatalytic Optimizing Process. *Technical Report 91-016 Revised*, Dipartimento di Electronica, Politecnico di Milano, Italy (1991).
10. Dorigo M., V. Maniezzo and A. Colorni: The Ant System: Optimization by a Colony of Cooperating Agents. *IEEE Transactions on Systems, Man, and Cybernetics – Part B*, (1996) 26(1): 29-41
11. Hackman, S.T. M.J. Magazine and T.S. Wee: Fast, Effective Algorithms for Simple Assembly Line Balancing Problems. *Operations Resarch* (1989) 37, 916-924.
12. Hoffmann, T.R.: Eureka. A Hybrid System for Assembly Line Balancing. *Management Science*, (1992) 38 (1): 39-47.
13. Johnson R.V.: Optimally Balancing Assembly Lines with „FABLE". *Management Science* (1988) 34: 240-253
14. Daniel Merkle, Martin Middendorf: An Ant Algorithm with a New Pheromone Evaluation Rule for Total Tardiness Problems, *Proceeding of the EvoWorkshops* (2000)
15. Merkle, D., M. Middedorf, and H. Schmeck: Ant Colony Optimization for Resource Constrained Project Scheduling. *Proceedings of GECCO-2000*. (2000)
16. Scholl, A.: Balancing and Sequencing of Assembly Lines, *Physica-Verlag*, Heidelberg (1999)
17. Scholl, A. and R. Klein: Balancing Assembly Lines Effectively – A Computational Comparison. *European Journal of Operational Research* (1999) 114: 50-58
18. Taillard É. D.: Ant Systems, in: P. Pardalos & M. G. C. Resende (eds.), *Handbook of Applied Optimization*, Oxford Univ. Press, (2002) 130-137
19. Talbot,F.B., J.H. Patterson and W.V. Gehrlein: A Comparative Evaluation of Heuristic Line Balancing Techniques, *Management Science* (1986) 32: 430-454.
20. Ugurdag, H.F., R. Rachamadugu, and A. Papachristou: Designing Paced Assembly Lines with Fixed Number of Stations. *European Journal of Operational Research*, (1997) 102(3): 488-501.

Ant Colonies as Logistic Processes Optimizers[*]

Carlos A. Silva[1,2], Thomas A. Runkler[1], João M. Sousa[2], and Rainer Palm[1]

[1] Siemens AG, Corporate Technology
Information and Communications, CT IC 4
81730 Munich, Germany
{carlos.silva.external,thomas.runkler,rainer.palm}@mchp.siemens.de
[2] Technical University of Lisbon, Instituto Superior Técnico
Dep. Mechanical Engineering - Control, Automation and Robotics Group
Av. Rovisco Pais, 1049-001 Lisbon, Portugal
jmsousa@ist.utl.pt

Abstract. This paper proposes a new framework for the optimization of logistic processes using ant colonies. The application of the method to real data does not allow to test different parameter settings on a trial and error basis. Therefore, a sensitive analysis of the algorithm parameters is done in a simulation environment, in order to provide a correlation between the different coefficients. The proposed algorithm was applied to a real logistic process at Fujitsu-Siemens Computers, using the set of parameters defined by the analysis. The presented results show that the ant colonies provide a good scheduling methodology to logistic processes.

1 Introduction

In supply chain management, logistics can be defined as the subprocess of the supply chain process that deals with the planning, handling, and control of the storage of goods between the manufacturing point and the consumption point. In the past, goods were produced, stored and then delivered on demand. Nowadays, many companies do not work with stocks, using instead *cross-docking centers* [1]. The goods are transported from the suppliers to these cross-docking centers, stored, and then shipped to the customers. The lack of storage may increase the delivery time, but it considerably reduces the volume of invested capital and increases the flexibility of the supply chain. The key issue is to deliver the goods in time by minimizing the stocks. The goods should be delivered at the correct date (not earlier or later) in order to ensure the customers satisfaction.

The scheduling algorithm has to decide which goods are delivered to which customers. Nowadays, there is an increasing need to replace the centralized static scheduling strategies by dynamic distributed methods [2]. One way is to assign individual agents to the orders and let the population of agents interactively find an optimal scheduling solution [3]. The interaction between the agents is realized by exchanging information about quantity, desired date and arriving date, but in logistic problems, the quantity of information that has to be exchanged is very large. Multi-agent algorithms based on social insects can avoid this complexity. In [4], a new ant algorithm for logistic processes

---

[*] This work is supported by the German Ministry of Education and Research (BMBF) under Contract no.13N7906 (project Nivelli) and by the Portuguese Foundation for Science and Technology (FCT) under Grant no. SFRH/BD/6366/2001.

M. Dorigo et al. (Eds.): ANTS 2002, LNCS 2463, pp. 76–87, 2002.

was proposed. Here, we extend the application of this algorithm, tune the algorithm parameters and apply it to a real logistic process at Fujitsu-Siemens Computers.

The paper is organized as follows. The next section presents a global description of a logistic process and a standard scheduling algorithm is briefly described, giving the motivation for a distributed dynamic scheduling algorithm. Then, we introduce the new framework of the ant colonies algorithm applied to the scheduling of logistic processes. Further a simulation example and a sensitivity analysis of the algorithm parameters is presented. Finally, the algorithm performance is tested in a real world example. The closing section concludes this paper and defines the future research work.

2 The Logistic Process

Figure 1 presents a schematic representation of a logistic process, which can be described in probabilistic terms. In fact, the birth process of the system (arrival of new orders in a certain period of time) and the death process (delivery of orders per unit of time, or the time it took them to be processed), can be described by the classical theory of queuing processes [5]. For the process being studied, this theory asserts the Poisson distribution for the model of the birth process,

$$p(x, \lambda T) = \frac{(\lambda T)^x}{x!} e^{-\lambda T} \tag{1}$$

where x is the random variable *number of orders* and λT is the parameter indicating the probability of this event *occur on a certain time* T. The death process is modeled by the exponential distribution

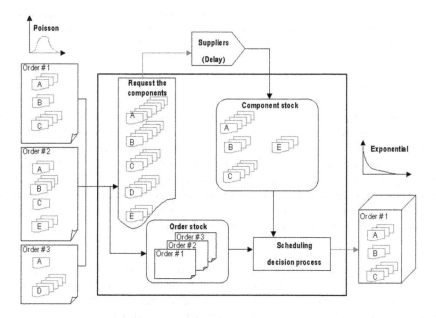

Fig. 1. General representation of the logistic process

$$p(T, \mu) = \mu e^{-\mu T} \tag{2}$$

where the time T is the random variable and μ is the death rate.

2.1   Process Description

The process can be divided into five sequential steps:

- **Order arrival**. The client buys a product, called *order*, which is a set of one or more different items, the *components* c_i. An order must contain a *desired delivery date*, which is the date required by the client for the order to be delivered.
- **Component request**. The different components must be requested from the external suppliers. Each component is characterized by a certain quantity.
- **Component arrival**. Each component takes some time to be delivered to the logistic system. This time is called the *supplier delay*. After this time, the component is delivered to the so-called cross-docking places, e.g. airports [1]. A *component stock* list is built at these places, which contains the available components and their quantity.
- **Component assignment**. Usually the components are not all available at the same time. For this reason, the orders have to wait for all the required components to be available. This waiting list is called the *order stock*. Each order has its own desired delivery date. The decision process has to decide which orders are going to be delivered, taking into account the availability of their components. This is normally performed once per day. The focus of this paper is to optimize this component assignment process.
- **Order delivery** The order is delivered to the client, with a certain delay d, where *delay* stands for the difference between the delivery date and the desired date. This delay should always be close to zero.

2.2   Scheduling Policies

The component assignment is the key issue in logistic processes. The company can not influence the arrival rates of the orders (birth process), or the suppliers delay. The service rates (death process) of the orders are the only control variable by considering the assignment of components to the orders. The control goal is to generate particular death rates for each order by some decision process using external information like desired times or internal information, such as the stock levels [6]. The scheduling assigns components to orders at each day. This paper compares the *pre-assignment* method with a dynamic decentralized approach using the ant colonies.

Pre-assignment (p.a.). When the components arrive at the cross-docking center from the external suppliers, they are already assigned to specific orders. The components are stored there until all the missing components arrive and then the order is completed and can be delivered. This strategy can not deal efficiently with disturbances, e.g. a delay in the component arrival. This assignment method is called a static scheduling method.

Distributed approach. The distributed approach works with a global stock. The agents associated with orders and components exchange information between each other. This information can be the desired delivery dates, the quantity of components in the stock, or the number of orders that will be delivered. After exchanging this information, the agents jointly decide which orders will be delivered. This approach is more flexible than pre-assignment, because it allows the evaluation of the scheduling result, and it can modify this scheduling before delivery.

3   Scheduling Using Ant Colonies

The *ant colonies* algorithm were introduced in [7] to solve different classes of NP-hard problems. The optimization of the scheduling process described in Section 2 is also a NP-hard problem, for which the optimization heuristics need to process a large amount of information, such as the arrival and desired delivery dates, which can be extremely time-consuming. The ant colonies algorithm provides an optimization method where all the problem's information can be translated into an indirect form, the *pheromones*, and used by all the interacting agents in order to achieve a good global solution. In this way, a new framework of the ant colonies algorithm, which was introduced in [4], is extended and analyzed in this paper. The notation will follow as close as possible the one used in [7], in order to make clear the new proposed algorithm.

Two different sets of entities can be identified: the orders and the components. Each type of component stands for a type of food source and each order stands for a nest. The nests may require different types of food in different amounts, as orders require different types of components in different quantities. Different nests may require the same type of food, as different orders may require common types of components. Conceptually, the job of the ants from this colony is very different from the foraging job described in [7]. Here, there are m ants, one per food source. The job of these ants is to distribute the quantity of food in the food sources to the n nests. Once again, in every iteration t of the algorithm, the ants have to choose with some probability p which is the nest to visit first. Finally, they deposit a pheromone τ in the path from the food source to the nest. One tour is complete when all the m ants from each of the food source have visited all the n nests.

Each ant delivers an amount q_j^i from the total amount q^i of component $i \in \{1, \ldots, m\}$ to an order $j \in \{1, \ldots, n\}$. Since there are several nests to visit, the ant k chooses the path to a particular nest with a probability

$$p_{ij}^k(t) = \begin{cases} \dfrac{\tau_{ij}{}^\alpha \cdot \eta_{ij}{}^\beta}{\sum\limits_{r \notin \Gamma^k} \tau_{ir}{}^\alpha \cdot \eta_{ir}{}^\beta} & \text{if } j \notin \Gamma^k \\ 0 & \text{otherwise} \end{cases} \tag{3}$$

where τ_{ij} is the amount of pheromone connecting component type i to order j, η_{ij} is a visibility function and the Γ^k is the tabu list of the kth ant. The tabu list is the list of the orders that the ant k has already visited plus the orders that do not need component type i (thus the visit can be avoided). The visibility function η_{ij} is defined as:

$$\eta_{ij} = e^{d_j} \tag{4}$$

where d_j is the *delay* of order j, i.e. the difference between the actual day and the desired date. This function reflects the objective of providing the ants with information regarding the delay of a certain order, since the objective of the problem is not only to fulfill the highest number of orders per day, but also to deliver them on the desired day (not before and not after). Orders that should be delivered on the actual date have visibility $\eta_{ij} = 1$. Orders that can still wait will have a lower visibility, and orders that are already delayed will have a higher one, so the ants will preferably feed delayed orders. The pheromone trails indicate the prior attempts from other ants. The parameters α and β express the relative importance of trail pheromone (experience) with visibility (knowledge), respectively.

After choosing the order to visit, the ant will deposit a pheromone on the trail connecting the type of component i to the order j. At this point, the ant can find two different situations:

1. When the total amount of components q^i of component type i is not enough to fulfill the amount of components q_j^i of same type needed by order j, the ant should not reinforce this path, so the pheromone value of this connection remains the same;
2. When sufficient components q^i exist to deliver to order j, the ant should reinforce this path, adding a pheromone value to the existing pheromone concentration on this path.

The local update of the pheromone concentration is then given by

$$\delta_{ij}^k(t) = \begin{cases} \tau_c & \text{if } q_j^i \leq q^i \\ 0 & \text{otherwise} \end{cases} \tag{5}$$

where τ_c is some small constant. In this way, the pheromones are locally updated with the information considering the amount of components i to deliver to the order j. After the ant visits order j, we decrement q^i by q_{ij} before the next ant from resource i visits order $j+1$. Note also that the ant always delivers all the q_j^i components that order j needs, to avoid the deadlock situation where all orders have part of the needed components, but none has the complete amount. In this way, the algorithm only finds feasible solutions. At the end of a complete tour the change of pheromones in all paths is given by

$$\Delta\tau_{ij} = \sum_{k=1}^{m} \delta_{ij}^k. \tag{6}$$

At this point, when all ants have delivered all possible components, the solution can be evaluated using a performance measure

$$z = \sum_{j=1}^{n} a_j, \text{ where } \begin{cases} a_j = 1 \text{ if } d_j = 0 \\ a_j = 0 \text{ otherwise} \end{cases} \tag{7}$$

where n is the number of orders and d_j is the delay of order j. This index is used to update globally the pheromones.

At each tour N of the algorithm (where each tour has $n \times m$ iterations t), a z is computed and stored in the set $Z = \{z(1), \cdots, z(N)\}$. If $z(N)$ is higher than the previous $z \in Z$, then the actual solution has improved and the used pheromones should

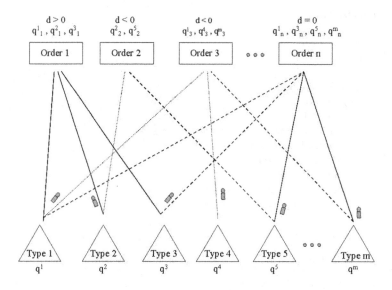

Fig. 2. Example of an ant colony applied to the logistic process with pheromone concentration level on the trails: High (—), Medium (- -) and Low (··)

be increased. If it is worse, they should be reduced. This is done once again by the global pheromone update, where the evaporation coefficient $(1 - \rho)$ avoids the solution to stagnate. In this way, the global pheromone update done after each complete tour N is the following:

$$\tau_{ij}(N \times n \times m) = \begin{cases} \tau_{ij}(n \times m) \times \rho + \Delta\tau_{ij} & \text{if } z(N) \geq \max Z \\ \tau_{ij}(n \times m) \times \rho & \text{otherwise} \end{cases} \quad (8)$$

Figure 2 represents schematically this framework. This particular framework of the ant colonies algorithm can be implemented using the general algorithm described in Fig. 3. Notice that the algorithm is initialized with a τ_0 value of pheromone concentration in every path, and that the algorithm runs a maximum number of tours N_{max}.

4   Simulation Results and Sensitivity Analysis of the Algorithm

In [4], the new algorithm was successfully applied to a simulation example. There, a logistic system was considered with the following parameters: the number of orders arriving at each day is a Poisson distribution (1) with $\lambda T = 10$; each order can have at the most 7 different types of components c_i; the quantity for each component varies randomly between 1 and 20; each type of component has a constant supplier delay, which are $1, 3, 2, 3, 1, 2, 6$ days for components type $c_1, \cdots, c_7$ respectively. For each order a desired date is generated using an exponential distribution (2) with $\mu = 7$. The simulation refers to an interval of 6 months. At each day the system adds the components arriving at that day to the stock. These components are then distributed to the orders following

---

Initialization:
  Set for every pair (i, j): $\tau_{ij} = \tau_0$
  Set $N = 1$ and define a N_{max}
  Place the m ants
  While $N <= N_{max}$
    Build a complete tour
    For $i = 1$ to m
      For $k = 1$ to n
        Choose the next node using $p_{ij}^k(t)$ using (3)
        Update locally using $\delta_{ij}^k(t)$ using (5)
        Update the **tabu list** Γ^k
    Analyze solutions
    Compute performance index $z(N)$ using (7)
    Update globally $\tau_{ij}(N \times n \times m)$ using (8)

Fig. 3. Ant colonies optimization algorithm for logistic processes

the methods described before. The results are presented in Fig. 4 and Table 1, with the parameters $\alpha = 1$, $\beta = 10$, $\rho = 0.9$ and $N_{max} = 20$.

Figure 4 presents the histograms of the order delay for both scheduling methods. They show the number of orders with a specific delay d: if the orders were delivered earlier, $d < 0$; if they were delivered at the correct date, $d = 0$; and if they were delivered with some delay, $d > 0$. Table 1 quantifies this analysis in number of orders and also indicates the spread for each method in number of days. It is clear that the ant colonies algorithm yields a higher number of deliveries on the correct date ($\#d = 0$), and a lower number of delayed orders ($\#d > 0$). It also yields a lower spread between maximum and minimum delays ($\max d - \min d$). This means that, even if the orders are delivered before or after the correct date, it is more likely with this method that the delays, negative or positive, are closer to the correct desired date. This shows clearly, that the ant algorithm is a better scheduling method than the pre-assignment for this application. In [4], these two methods were compared with two other simple dynamic heuristics (First In First Served and First Desired First Served), and the ant algorithm proved to be the best scheduling method.

4.1 Tuning the Parameters

The ant colonies algorithm has a large set of parameters that has to be tuned in order to provide the best possible solutions. In this section, we present a study where the

Table 1. Comparison between the scheduling methods in number of orders

Scheduling	$\#d < 0$	$\#d = 0$	$\#d > 0$	$\min d$	$\max d$
p.a.	275	162	833	-32	7
Ants	536	266	468	-12	6

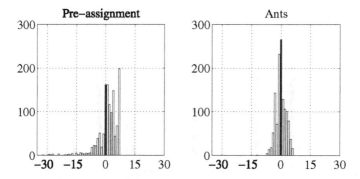

Fig. 4. Histograms of the order delay d for the scheduling methods. On the left hand side, the results for the pre-assignment method and on the right hand side, the results for the ants. The orders delivered at the correct date, are represented with the black bar

parameter variation is done partially: the parameters who are coupled between each other are changed at the same time, while the decoupled parameters remain constant, e.g., α and β are coupled, so the analysis is done with a fixed value of ρ. Thus, we intend to provide some insight on how the different parameters influence the algorithm.

Relative weights α and β. As explained in Section 3, the parameters α and β weight the relative importance of the pheromone concentration on the trails and the visibility, when the ants choose which path to take. In simpler words, they weight the importance of the experience against the knowledge for each ant. The following analysis is done using a fixed value of evaporation coefficient $(1 - \rho)$ with $\rho = 0.9$.

If $\alpha > \beta$, the ants give more importance to their previous experiences than to actual knowledge about the environment; on the other hand if $\alpha < \beta$, the ants tend to decide based on the knowledge. In this case, the ants are not using what they are learning. Figure

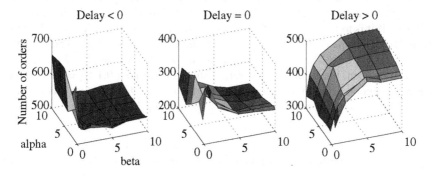

Fig. 5. Number of orders delivered for a fixed ρ and varying α and β. On the left hand side, the orders delivered before the desired date $(d < 0)$, at the center the orders delivered at the correct date $(d = 0)$ and on the right hand side, the orders delivered after the desired date $(d > 0)$

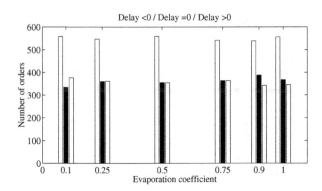

Fig. 6. Number of orders in the right day, for different sets of fixed α and β and varying ρ

5 shows how these two factors interact by presenting the number of orders delivered and the respective delay for different combinations of α and β. The first conclusion that can be taken is that the number of orders delivered on the correct date is high, as long as β is small. In this case, the number of orders delivered with delay ($d > 0$) is also smaller. However, this index β should not be zero, in other words, the ants should have some visibility. If $\beta = 0$, this means that the visibility function is not being used and the results are worse; the number of orders delivered on the correct day decreases and the number of orders delivered earlier ($d < 0$) increases. The variation of the pheromone concentration, given by α, is important in terms of orders delivered on the correct date ($d = 0$). It has an optimal value for $\alpha = 1$.

We can conclude that the α parameter tunes the number of orders in the right day and β controls the spread around that value. Table 2 presents the improvement in the simulation results achieved with the tuning of parameters α and β.

Evaporation coefficient $(1-\rho)$. Another parameter that may influences also the results is the evaporation coefficient. This coefficient is strongly related with the parameter α. It accounts for the learning from the previous colony to the next colony. If it is too strong ($\rho \sim 0$) the increment received by the new ants it will influence greatly the paths of the next ants. If it is to weak ($\rho \sim 1$) the solution can rapidly stagnate. Figure 6 shows how this index affects the results, in terms of total number of orders delivered before, at, and after the correct date. The analysis is done for the fixed values of $\alpha = 1$ and $\beta = 0.5$. As it can be seen, the value of evaporation should be around 0.1 ($\rho = 0.9$), in order to achieve a good solution, i.e. more orders in the correct date and less orders early or delay. With $\rho = 0.9$, the solution is neither forgotten nor does it disable the ants to find better solutions.

Table 2. Compare results for different values of α and β

α	β	$d < 0$	$d = 0$	$d > 0$
1	10	536	266	468
1	0.5	559	356	355

Number of colonies per day N_{max}. Each day is a new optimization problem and the number of ant colonies released each day (i.e. the number of tours performed by the algorithm) will influence the optimization result: if very few colonies are released, the algorithm does not have enough iterations to learn and to find a good solution; if the number is too large, the algorithm may take too much time to find a solution. On the other hand, if the ants are learning and passing the knowledge to other ants by changing the concentration on the pheromone trails, it is expectable that for each day the function being maximized (the number of orders delivered on the correct day) converges. Figure 7 shows the evolution of the solution for different number of ant colonies. We can see that using more and more colonies provides better results, but with great increase in the computational effort. Using 20 colonies or more did not improve significantly the solution, but increased severely the computational cost.

5   Real World Example

The analysis presented in Section 4 showed that in order to obtain an optimized solution, the relation between α and β should be such that $\alpha > \beta$, e.g. $\alpha = 1$ and $\beta = 0.5$. The evaporation coefficient should be close to, but smaller than 1, e.g. $\rho = 0.9$, and the number of colonies should be around 20. These were the parameters that achieved the best results for the simulation environment.

The question that we faced was if the optimal parameters for the simulation example also yield good results for the corresponding real-world problem: a scheduling process at Fujitsu-Siemens Computers. The real problem has a higher number of types of different components (in the simulation environment we have only 7 different types), and the quantities of each component can reach 100 (in the simulation environment we have at the most 20). The fact that there are much more different types of components, thus a much higher number of ants, increases very much the complexity of the algorithm. Therefore, we have tested only a small data set of the data from the year 1999. Figure 8 and Table 3 show the results achieved with pre-assignment and with the ant algorithm.

With the real data, the ants proved once again that they yield a better solution than the pre-assignment: more orders are delivered at the correct day, less orders are early and

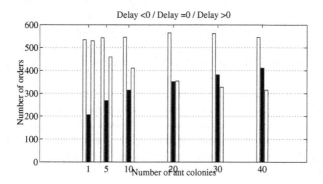

Fig. 7. Evolution of the solution for different number of colonies

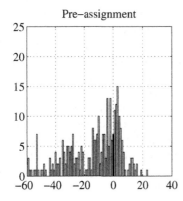
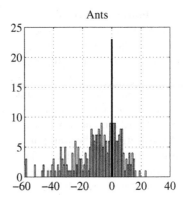

Fig. 8. Histograms of the orders delay d for the scheduling methods

Table 3. Comparison between the strategies for the real data

Scheduling	$\#d < 0$	$\#d = 0$	$\#d > 0$	min d	max d
p.a.	253	7	83	-60	24
Ants	190	23	79	-60	24

less orders are delayed. However, in the real case the difference between both results is smaller than in the simulation. As in the real data the number of different types of components is much larger, there are less possibilities of exchanging components than in the simulation environment. Nevertheless, the ants showed that they are a good alternative to the standard pre-assignment scheduling method.

6   Conclusions

This paper analyzes the performance of a new framework of the ant colonies optimization algorithm to apply in the optimization of logistic processes. The ant colony optimization algorithm uses a large set of parameters and the tuning of these parameters influences greatly the performance of the algorithm. This paper explored the possible correlations between the several parameters and their role in the algorithm. This sensitivity analysis on the parameters aimed also the application of the algorithm to the real data. The large set of real data, does not allow to run the optimization algorithm with different parameterizations, thus a careful set of parameters has to be chosen before the implementation on the real data.

The parameter analysis was done using a simple but illustrative simulation example, where the new algorithm is compared to the standard method used in practice in the scheduling process. The results show how the analysis is able to improve the algorithm performance, and explain the reasons for that improvement. Finally, the algorithm was applied to a real data set, and the ant algorithm proved to be a better scheduling method than the pre-assignment.

The ant algorithm has still some space for improvement. The use of a different cost function z, or the incorporation in the ants with some sort of prediction mechanism can lead to better performances. Moreover, there is still some information in the process, like priority of the orders, that was not used so far and can influence the scheduling of the logistic process.

References

1. Jayashankar M. Swaminathan, S.F.S., Sadeh, N.M.: Modeling supply chain dynamics: A multiagent approach. Decision Sciences Journal **29** (1998) 607–632
2. McKay, K., Pinedo, M., Webster, S.: A practice-focused agenda for production scheduling research. Production and Operations Management **10** (2001)
3. Palm, R., Runkler, T.: Multi-agent control of queuing processes. In: To appear in Proceedings of World Conference on Automatic Controlo - IFAC'2002, Barcelona, Spain. (2002)
4. Silva, C.A., Runkler, T., Sousa, J.M., Palm, R.: Optimization of logistic processes using ant colonies. In: Proceedings of Agent-Based Simulation 3. (2002) 143–148
5. Wolff, R.W.: Stochastic Modeling and the Theory of Queues. Prentice -Hall (1989)
6. Palm, R., Runkler, T.: Decentralized control of hybrid systems. In: Proceedings of ISADS-2001, Fifth International symposium on autonomous decentralized systems. (2001)
7. Dorigo, M., Maniezzo, V., Colorni, A.: Ant system: Optimization by a colony of cooperating agents. IEEE Transactions on Systems, Man, and Cybernetics-Part B **26** (1996) 29–41

Ant Systems for a Dynamic TSP
Ants Caught in a Traffic Jam

Casper Joost Eyckelhof[1] and Marko Snoek[2]

[1] Multigator vof
Enschede, The Netherlands
casper@eyckelhof.nl
[2] Department of Computer Science, University of Twente
P.O. Box 217, 7500 AE, Enschede, The Netherlands
snoek@cs.utwente.nl

Abstract. In this paper we present a new Ants System approach to a dynamic Travelling Salesman Problem. Here the travel times between the cities are subject to change. To handle this dynamism several ways of adapting the pheromone matrix both locally and globally are considered. We show that the strategy of smoothing pheromone values only in the area containing a change leads to improved results.

1 Introduction

In nature, many systems consisting of very simple parts demonstrate a remarkable complexity as a whole. An example of this is ant colonies: every single ant just seems to walk around independently, however the colony itself is organised very well. This effect is also known as *emergent behaviour*. Ants' foraging behaviour shows that travel between nests and sources of food is highly optimised. Based on this phenomenon, Ant Colony Optimisation (ACO) algorithms have been developed [10,2,5], of which Ant System (AS) was the first [4]. These ACO algorithms have been successfully applied to a variety of combinatorial optimisation problems [1].

Like other metaheuristics ACO algorithms have proven their use for static problems, but less is known about their behaviour on dynamic problems. In dynamic problems the goal is not to find desirable solutions, but to track and/or find new desirable solutions.

After the initial emphasis on static problems, some of the focus is now shifting towards dynamic variants of combinatorial optimisation problems. Recently some research is being done on ACO for dynamic problems [8,7,9].

This paper gives an outline of research done as part of a graduation project at the University of Twente [6]. The main goals of that research were to test the hypothesis that ant systems can be applied successfully to dynamic problems and to adapt the original ant system algorithm in order to increase its performance on a particular dynamic problem.

In this paper we will show that both goals were reached: ant systems can be applied successfully to dynamic problems and enhancements to the original

M. Dorigo et al. (Eds.): ANTS 2002, LNCS 2463, pp. 88–99, 2002.
© Springer-Verlag Berlin Heidelberg 2002

algorithm are proposed for a chosen dynamic problem. The problem we study in this research is based on the Travelling Salesman Problem (TSP). TSP is a well-known combinatorial optimisation problem and the original ant system algorithm was designed for TSP.

The paper is structured as follows. In §2 the TSP as well as the ant system for this problem are explained. In §3 we introduce our adaptation to TSP in order to create a dynamic problem and we will explain how ant systems can be modified to perform better on this dynamic problem. Next the test setup for the experiments is explained in §4. In §5 the experimental results are presented. We will finish with our conclusions in §6 and ideas for future work in §7.

2   The Ant System Approach to TSP

The *Travelling Salesman Problem* is a well-known problem among computer scientists and mathematicians. The task basically consists of finding the shortest tour through a number of cities, visiting every city exactly once. Formally, the symmetric TSP is defined as:

> Given a set of n nodes and costs associated with each pair of nodes, find a closed tour of minimal total costs that contains each node exactly once, the cost associated with the node pairs $\{i, j\}$ and $\{j, i\}$ being equal[1].

It is also possible to discard that last condition and allow the distance from city i to city j to be different from the distance between city j and city i. We refer to that case as the asymmetric travelling salesman problem. Our focus will be on the symmetric TSP.

The popularity of TSP probably comes from the fact that it is a very easy problem to understand and visualise, while it is very hard to solve. Many other problems in the $\mathcal{NP}$-Hard class are not only hard to solve, but also hard to understand. With TSP, having n cities, there are (n-1)! possible solutions for asymmetric TSP and (n-1)!/2 possible solutions for symmetric TSP. For small instances it is no problem to generate all solutions and pick the shortest, but because the number of possible solutions 'explodes' when the number of cities increases. Within this domain heuristics that find acceptable solutions using an acceptable amount of resources are necessary.

In their book [2], Bonabeau et al. give a good explanation of *Ant System (AS)*, the algorithm we will use as a basis for our own algorithm. It has become common practice to name the algorithms with their field of application, so this version of AS is called AS-TSP.

In AS-TSP m ants individually construct candidate solutions in an incremental fashion. The choice of the next city is based on two main components: pheromone trails and a heuristic value, called visibility in TSP. At the start all possible roads are initialised with a certain amount of pheromone: τ_0. Then each ant constructs a solution by choosing the next city based on the observed

---

[1] Using nomenclature corresponding to the metaphor of a travelling salesman we will use city for node, distance or travel time for cost, and road for node pair.

pheromone levels (τ) and visibility (η) until it has visited all cities exactly once. The visibility is a proximity measure, defined as the inverse of the distance between two cities i and j: $\eta_{ij} = 1/d_{ij}$, where η_{ij} is the visibility associated with choosing city j when in city i and d_{ij} is the distance between these two cities.

The choice for the next city is probabilistic. The more pheromone there is on a certain road the bigger the chance that this road will be taken. The same goes for visibility: a higher visibility yields a higher chance of being visited next. Cities that have already been visited have a zero chance of being visited again, because of a tabulist mechanism. This assures the construction of valid tours. The parameters α and β control the relative weight of pheromone trail intensity and visibility. If $\alpha = 0$ the algorithm behaves as a standard greedy algorithm, with no influence of the pheromone trails. If $\beta = 0$ only pheromone amplification will occur and the distance between cities has no direct influence on the choice. Usually a trade-off between these two factors is best.

Formally, if during the tth iteration the kth ant is located in city i, the next city j is chosen according to the probability distribution over the set of unvisited cities J^k defined by:

$$p_{ij}^k(t) = \frac{[\tau_{ij}(t)]^\alpha \cdot [\eta_{ij}]^\beta}{\sum_{l \in J_i^k} [\tau_{il}(t)]^\alpha \cdot [\eta_{il}]^\beta}$$

When each ant has completed a tour, the amounts of pheromones are updated. On every road some fraction of the pheromone evaporates, while on roads, that have been visited by at least one ant, new pheromones are deposited. The amount of pheromone that is deposited is based upon the number of ants that visited the road, the length of the tours the road is part of and a parameter Q^2. The more a road is visited and the shorter the tour, the more pheromone is deposited. The fraction of pheromone that evaporates is based on the parameter ρ.

All roads on the best tour so far get an extra amount of pheromone. This is called an *elitist ant* approach: the best ants reinforce their tour even more. The balance between positive reinforcement through depositing pheromone by an (elite) ant and pheromone evaporation, which is a negative reinforcement, is crucial for finding good solutions.

This process goes on until a certain condition, such as a certain number of iterations, amount of CPU time, or solution quality, has been achieved.

3   Ant System Approaches to Dynamic TSP

The Dynamic Travelling Salesman Problem (DTSP) that was used in this project is a variation on the TSP in the sense that the original TSP metaphor is extended to include traffic jams. If we look at the distance between cities as travel times, they no longer need to be fixed. By introducing a traffic jam on a certain road the associated travel time is increased[3]. We mention that several other variants

---

[2] Since the value of Q only weakly influences the result, it will not be discussed here.

[3] Since we are not adapting distances *as if we are moving a city*, the resulting TSP structure is not Euclidian. For instance the Cauchy-Schwarz inequality does not always hold.

of DTSP, such as the DTSP resulting from the insertion or deletion of cities, are also possible [8,7,9].

In our DTSP, ants regularly come across traffic jams. This is guaranteed since traffic jams will be introduced only on roads that are in the best tour of the AS at that moment. Moreover, these traffic jams have a somewhat natural way of increasing for a while and then decreasing until they are resolved. More details will be given in the next paragraph. Let us now discuss ways to adapt the AS to handle this kind of dynamics.

The combination of positive and negative reinforcement, as mentioned in §2, works well for static problems. In the beginning there is relatively much exploration. After a while all roads that are not promising will be slowly cut off from the search because they do not get any positive reinforcement and the associated pheromones have evaporated over time. In dynamic situations however, solutions that are bad before a change in the environment, might be good afterwards. Now, if the ant system has converged to a state where those solutions are ignored, very promising roads will not be sampled and the result will be a suboptimal solution. During our research this was one of the first things we ran into. We have devised several ways to counter this effect, which we will now discuss.

We used a lower boundary on the amount of pheromone on every road in AS-DTSP. This prevents the chances of a road to be chosen by an ant to approach zero beyond a certain point [10]. Initial testing showed that using τ_0 as a lower boundary seemed to do what we wanted.

Another change to AS-TSP we introduced in AS-DTSP was what we called *shaking*. It is a technique in which the environment is 'shaken' to smooth all pheromone levels in a certain way. If the amount of pheromones on a road becomes much higher than on all other roads going out of a city, this road will almost certainly always be chosen. That is a way for the static case to ensure a good road will always be followed, but it prevents ants from taking another road when a traffic jam occurs on it.

Shaking changes the ratio between the amount of pheromone on all roads, while it ensures that the relative ordering is preserved: if $\tau_{ij}(t) > \tau_{i'j'}(t)$ holds before shaking, it also holds after shaking. The formula used for shaking is a logarithmic one:

$$\tau_{ij} = \tau_0 \cdot (1 + log(\tau_{ij}/\tau_0))$$

See figure 1. This formula will cause pheromone values close to τ_0 to move a little towards τ_0 and higher values to move relatively more to τ_0. See figure 1. Note that τ_0 is the minimum value for τ, so the condition $\tau_{ij} \geq \tau_0$ is preserved.

A possible problem with this shake might be that it has a global character, i.e. it affects the pheromone values on all roads. When we have a big problem instance where one traffic jam occurs somewhere near the edge of the map there is a high probability that routes only have to change in the vicinity of the traffic jam [10]. So, too much information might be lost in the global shaking process. For this reason we define *local shaking*. Here, the same formula is used, but it is only applied to roads that are closer than $p \cdot MaxDist$ to one of the two cities where the traffic jam is formed between. MaxDist is the maximum distance between

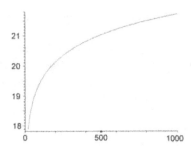

Fig. 1. Shaking function, pheromone level before and after shaking

any two cities in the original problem (without traffic jams) and $0 < p < 1$. In pseudo code:

> **if** the distance between cities (a, b) changed **then**
>> **for** every road (i, j) **do**
>>> **if** $(d_{ai} < p \cdot MaxDist) \vee (d_{bi} < p \cdot MaxDist) \vee$
>>> $(d_{aj} < p \cdot MaxDist) \vee (d_{bj} < p \cdot MaxDist)$ **then**
>>> $\tau_{ij} = \tau_0 \cdot (1 + log(\tau_{ij}/t_0))$
>>> **endif**
>> **endfor**
> **endif**

Note we only introduce traffic jams on roads that are in the current best tour of the salesman. Therefore these two cities are generally close to each other and the effect of local shaking with the mentioned values for p and global shaking differ therefore significantly. Another observation is that for $p = 0$ no pheromone values are affected. Global shaking is equivalent with $p = 1$ since all pheromone values are now subject to shaking.

In this paragraph, we discussed our DTSP and presented the proposed changes to AS-TSP. Summarizing, these are:[4]

- After pheromone evaporation, we make sure that the amount of pheromone does not become less than τ_0: $\forall i, j, i \neq j : \tau_{ij} = max(\tau_{ij}, \tau_0)$.
- We introduce a shaking operator, that through smoothing of a part of the pheromone matrix increases the amount of exploration done after a change has occurred. Depending on the setting of the 'shaking-percentage' p, this operator acts globally with $p = 1$ or locally with $0 < p < 1$. Setting p to zero yields no shaking.

4   Test Setup

A problem with evaluating our results is the complete lack of benchmarks. For static TSP there are benchmarks available: usually just the best scores and

---

[4] A complete high level description of AS-DTSP is given in [6].

sometimes the time it took to reach those scores with a particular algorithm (on particular hardware). For dynamic TSP there are no such benchmarks. As a result we had to choose our own criteria to measure the performance of an algorithm.

There are some obvious things to look at for dynamic problems. Because the problem changes we want our algorithm to respond to the changes rapidly. In our tests this would mean that when a traffic jam is created the ants quickly find a better route that does not contain the congested road. Because all traffic jams in our tests are introduced in a few incremental steps, this is visible in the graphs as low up-peaks: before the traffic jam is at its highest point, the ants have already changed their route.

After the ants changed their route to avoid the traffic jam, usually some more parts of the route have to be adjusted: a single change has impact on a bigger part of the route. This means that we would like to see a high (negative) gradient after a peak or low down-peaks, meaning that some more optimisations are being performed before the next change.

And finally the values for the length of the complete tour should be as low as possible. For dynamic problems this would mean a low average value. When discussing the results of various experiments, we will explain which values were exactly compared to each other.

We tested AS-DTSP with two different case studies, one with 25 cities, the other with 100 cities. For both case studies we compared five approaches:

- *Original* is the original AS-TSP with a lower bound on the pheromone values: here, no special measures are taken when a traffic jam occurs.
- *With Reset* is a copy of *original*, but when a traffic jams occurs, we reset the complete pheromone matrix to τ_0.
- *With Shake* is equal to *original* extended with shaking as explained in the previous paragraph: both global with $p = 1$ and local shaking with $p = 0.1$ and $p = 0.25$ are tested.

4.1   25 City Problem

The problem instance used for this case study has 25 cities placed on a 100x100 grid. See [6] for the problem data. All cities are located on an integer intersection of the gridlines. At certain points in time, traffic jams occur between cities that are very close to each other and are likely to be on the ideal route.

These locations have been chosen empirically by running the initial problem through a classic AS-TSP implementation. This way we found certain pairs of cities that are almost always next to each other in the best solution found. Using this technique we are sure that all traffic jams have influence on the best solution so far. We will introduce three traffic jams. In this case study there are two interesting intervals: from iteration 100 to 150 one traffic jam is introduced on a certain road. From iteration 200 to 250 this traffic jam disappears, while simultaneously two others are introduced on other roads.

All traffic jams appear and disappear in increments and decrements of 10 units, one incremental step every 5 iterations. This way the algorithms get a chance to respond to small changes. If a traffic jam is introduced all at once, an algorithm cannot show that it is able to switch its route even while the traffic jam is still growing.

The AS-parameters were set to: $\alpha = 1, \beta = 6, m = n = 25, Q = 100, \tau_0 = 10^{-6}$.

4.2   100 City Problem

To check how DTSP performs on a similar, but extended problem, we constructed a bigger and more dynamic test case. This case study has *one* static 100 city TSP problem at its foundation. See [6] for the problem data. This instance was constructed by placing 100 cities randomly in a 100x100 grid. After a short start-up period, we create a new traffic jam every 50 iterations by increasing the length of a road by a number of units in 5 equal steps. During the same 25 iterations the last traffic jam created, disappears also in 5 equal steps. The traffic jam occurs on a road in the currently best route. This orderly way of introducing jams is not realistic, but it has the advantage that we can clearly compare approaches.

Assume a solution to the problem without traffic jams, which is reached after some time of optimisation, has length x. Now we start introducing a traffic jam every 50 iterations. For every newly created traffic jam we simultaneously remove the previous one. The result is that without any further optimisation, the length of the original solution is $x + $ *length of one full traffic jam*. Because all jams are created in 5 equal steps of y units in our test setup, there is always a $5 * y$ traffic jam (except for the first 20 iterations). This process of appearing and disappearing traffic jams is shown in figure 2. Using a pre-established arbitrary ordering of the cities, in each simulation the first traffic jam occurs on the outgoing road of city number 1, the second on the outgoing road of city number 2, and so on. The traffic jams disappear in the same order. This way we have problems that are somewhat similar each run. Every experiment runs for 5000 iterations after the start-up period, since exactly 100 traffic jams are created. Each experiment is done 10 times.

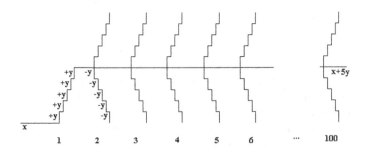

Fig. 2. Traffic jam appearance and disappearance in 100 city problem. The tour length of the fixed solution during the dynamic part of the simulation is $x + 5y$, except for a small time period in the beginning.

For this problem we will judge our strategies on the low and high peaks as well as the average tour length during the dynamic part of the simulation. Good scores will require good reactions to both kinds of change. For a road with an increased road length an alternative route becomes worthwhile during its steady growth. Contrarily, a road with a decreasing road length might become attractive to re-enter into the tour.

The structure of the dynamics of this problem has as an advantage that it is meaningful to compare our strategies with the strategy of ignoring all changes. This strategy simply means using the solution to the static problem and putting up with the traffic jams as they occur. Since the dynamics of the problem involves no more than two traffic jams at all times, the problem can be said to 'wobble around' the static instance. For this reason, the seemingly trivial strategy of ignoring all changes gives us a means of comparison.

The stepsize y is set to 10. The AS-parameters were set to: $\alpha = 1, \beta = 6, m = n = 100, Q = 100, \tau_0 = 10^{-6}$.

5   Experimental Results

5.1   25 City Problem

The average tour length for *shake, p = 1* as well as some points for which the scores will be measured are shown in figure 3. Until point 1 this problem instance is static. Two interesting points are located directly after 100 and 200 iterations: those are the dynamic parts of the problem. At both times, we see a peak in the graph. Low peaks suggest a quick recovery from a growing traffic jam. Two other interesting values that say something about the capability to recover from changes are the result right before the second peak and the result after the last iteration.

When running the problem without any traffic jams, the average solution after 500 iterations is 441.4. It would be nice if we get to that same score when running our algorithms on the problem with traffic jams. After the last change has occurred, there are 250 iterations left to reach that value. The scores of three of the five strategies during the entire simulation are given in figure 4. The results on the four checkpoints are shown in figure 5.

The results are very close to each other, and it is difficult to point out a clear winner. Although *reset* has multiple best values, some others are very close, like *shake*. As described in §4.1, the first change around point 2 is less dynamic than the second change around point 4. This can explain the fact that *reset* scores well at point 4, while it does not score well at point 2. For a major change it is better to reset the pheromone matrix. For point 2 the three shake algorithms perform best. For this change it is better to retain some of the pheromone matrix (compared with *reset*), but not all of it (compared with *original*). For the highly dynamic situation at point 4 the relative scores make sense when we consider the amount of exploration involved. *Reset* has to rely fully on exploration and achieves the best score. It is closely followed by *global shake*, that also uses a lot of exploration. In the middle are the *local shake* strategies. Last is *original* that

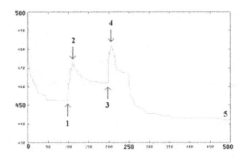

Fig. 3. Example of tour length during 500 iterations, averaged over 10 runs with locations of interesting checkpoints 2-5 marked. Results are from *Shake, (p = 1)* strategy.

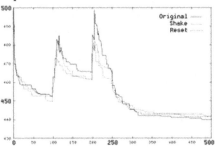

Fig. 4. The results for *Orginal, Shake (p = 1),* and *Reset* averaged over 10 runs.

Strategy	2	3	4	5
Original	485.2	465.4	498.7	**440.1**
With Shake, $p = 0.1$	477.7	463.0	489.0	443.4
With Shake, $p = 0.25$	479.5	465.0	485.0	440.6
With Shake, $p = 1$	**472.0**	461.8	481.8	441.9
With Reset	483.6	**461.2**	**480.0**	441.6

Fig. 5. Results at various checkpoints of the 25 city DTSP.

does not increase exploration because of changes. An interesting observation is that all strategies score comparable at point 5. Their performance is also comparable to an AS-TSP that did not encounter any dynamics. During the traffic jams *original* performs worse than all other algorithms, but in the end it reaches the best score.

Based on this case study we see support for the claim made by [2] and [8] that ant systems are very robust and even able to solve dynamic problems. Next to that, we have also shown that extending AS-TSP does improve performance in certain dynamic situations.

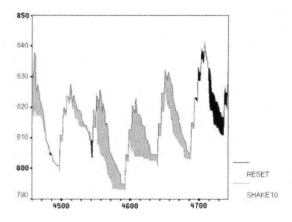

Fig. 6. Comparison of average tour length for *Reset* and *Shake (p = 0.1)* during an average part of the simulation. The grey area indicates the latter outperforming the first and the black vice versa.

5.2   100 City Problem

For the problem described in §4.2 the average results during 300 iterations are given in figure 6 for two strategies. Note, that the first 600 iterations were used as a start-up duration during which the five AS approaches could settle. No scores were recorded during this period.

As can be seen the consecutive growth and decrease of traffic jams lead to alternating up- and down-peaks. In comparison with the previously described test case, the traffic jams are fewer iterations apart. Usually the AS has not converged when the next traffic jam occurs. For each of the strategies the average tour length, average down-peak, and average up-peak are given.

Strategy	Avg peak⁻	Avg length	Extra length	Relative %	Avg peak⁺
Fixed without t.jam	N/A	781.6	0.0	0 %	N/A
Fixed with t.jam	N/A	831.5	49.9	100 %	N/A
Original	805.5*	813.9*	32.3	65 %	830.8
With Shake, $p = 0.1$	**803.0**	**811.8**	**30.2**	**60 %**	**830.2**
With Shake, $p = 0.25$	803.1	812.7	31.1	62 %	831.8*
With Shake, $p = 1$	805.3*	815.6*	34.0	68 %	834.9*
With Reset	804.7*	815.5*	33.9	68 %	835.5*

Fig. 7. Results on the 100 city DTSP. In the down-peak, average length and up-peak columns the scores that differ significantly from the best are marked with "*".

The fixed strategy has an optimised tour length of 781.6. With a traffic jam of length 50 the same solution obviously has a length of 831.6 (resulting in an average of 831.5). It is clear that all AS strategies are able to react to the dynamics and are able to significantly improve upon the fixed strategy. As can be seen, the local shake strategies achieve the best results, although they are closely followed by the other AS strategies. Interpreting these results in the context of our earlier findings, the level of dynamics of this problem seems to be intermediate. While it is beneficial to retain some of your pheromone matrix, exploration near the site of change needs to be encouraged.

Interestingly, *original* comes in third with respect to average performance. A likely explanation of this fact is that in this bigger problem it is very expensive to throw away information as *reset* and *global shake* do. Exploration done far away from the directly affected area will seldom lead to an improvement.

6   Conclusion

In our version of the Dynamic TSP, traffic jams lead to increased travel times on certain roads. By allowing the traffic jams to grow in a stepwise manner we can compare the reaction times of different strategies.

We have looked at various versions of AS applied to two different instances of the dynamic travelling salesman problem. The most striking effect that we have observed is that every strategy we tried performed reasonably well on both problems. This supports the claim we were investigating as stated in §1: Ant Systems are able to perform well on dynamic problems.

But when looking closer at the results, we also were able to achieve the second goal of our research. We created our AS-DTSP by extending AS-TSP through the addition of the shake routine. Especially the strategies with 'local shake' performed well. The higher the frequency of the dynamics gets, the more important it is to preserve some of the information that was collected earlier in the pheromone matrix. This is why resetting the pheromone matrix yields poor results for the 100 city problem: all information is lost and new changes occur before the algorithm is able to find a good solution. The *local shake* algorithm has as a strength that it combines exploitation of the developed pheromone matrix and biased exploration within the area around a change. With the shake parameter p the size of this area can be tweaked.

7   Recommendations for Future Research

A common problem when looking at Ant Systems is the large number of parameters in the algorithm. As we used settings from literature, we did not look at the influence of these regarding dynamic problems.

Several ideas on the introduced shaking routine present interesting leads for further research. First, the shake parameter p was changed a few times ($p = 0.1, 0.25, 1$) in this research, but requires more research in our opinion. We expect the optimal setting of this parameter to be related to problem attributes, such as size and structure, and change attributes, such as change frequency, duration, magnitude, and location.

The ratio behind our shaking routine is to increase the exploration after the occurrence of a change, while at the same time limiting this exploration to the affected area. Alternative ways to implement this thought, such as a more smoothed manner of shaking (e.g. the η- and τ-strategy presented in [9]) or the application of a local updating rule ([3]) near the site of change, might be worth considering for our DTSP.

A third option to deal with changes does not involve smart adaptation of the pheromone matrix but smart utilisation of it. An important characteristic of an Ant Colony System is its ability to directly balance the ratio of exploration versus exploitation [3]. Also for our DTSP it seems logical to increase this ratio around the region most affected by a change.

References

1. Christian Blum and Andrea Roli. Metaheuristics in combinatorial optimization: Overview and conceptual comparison. Technical Report TR/IRIDIA/2001-13, IRIDIA, 2001.
2. Eric Bonabeau, Marco Dorigo, and Guy Theraulaz. *Swarm Intelligence. From Natural to Artificial Systems*. Studies in the Sciences of Complexity. Oxford University Press, 1999.
3. M. Dorigo and L.M. Gambardella. Ant colony system: A cooperative learning approach to the traveling salesman problem. *IEEE Transactions on Evolutionary Computation*, 1(1):53–66, 1997.
4. M. Dorigo, V. Maniezzo, and A. Colorni. Ant system: Optimization by a colony of cooperating agents. *IEEE Transactions on Systems, Man and Cybernetics - Part B*, 26(1):29–41, 1996.
5. Marco Dorigo and Gianni Di Caro. Ant algorithms for discrete optimization. *Artificial life*, 5(2):137–172, 1999.
6. Casper Joost Eyckelhof. Ant systems for dynamic problems, the TSP case - ants caught in a traffic jam. Master's thesis, University of Twente, The Netherlands, August 2001. Available throug http://www.eyckelhof.nl.
7. M. Guntsch and M. Middendorf. Pheromone modification strategies for ant algorithms applied to dynamic TSP. In *Applications of Evolutionary Computing: Proceedings of EvoWorkshops 2001*. Springer Verlag, 2001.
8. Michael Guntsch, Jurgen Branke, Martin Middendorf, and Hartmut Schmeck. ACO strategies for dynamic TSP. In Marco Dorigo et al., editor, *Abstract Proceedings of ANTS'2000*, pages 59–62, 2000.
9. Michael Guntsch, Martin Middendorf, and Hartmut Schmeck. An ant colony optimization approach to dynamic TSP. In Lee Spector et al., editor, *Proceedings of the Genetic and Evolutionary Computation Conference (GECCO-2001)*, pages 860–867, San Francisco, California, USA, 7-11 July 2001. Morgan Kaufmann.
10. T. Stützle and H. Hoos. Improvements on the ant system: Introducing MAX(MIN) ant system. In G.D. Smith, N.C. Steele, and R.F. Albrecht, editors, *Proceedings of the International Conference on Artificial Neural Networks and Genetic Algorithms*, pages 245–249. Springer-Verlag, 1997.

Anti-pheromone as a Tool for Better Exploration of Search Space

James Montgomery* and Marcus Randall

School of Information Technology, Bond University
Gold Coast, Queensland, 4229, Australia
{jmontgom,mrandall}@bond.edu.au

Abstract. Many animals use chemical substances known as pheromones
to induce behavioural changes in other members of the same species. The
use of pheromones by ants in particular has lead to the development of
a number of computational analogues of ant colony behaviour including Ant Colony Optimisation. Although many animals use a range of
pheromones in their communication, ant algorithms have typically focused on the use of just one, a substance that encourages succeeding
generations of (artificial) ants to follow the same path as previous generations. Ant algorithms for multi-objective optimisation and those employing multiple colonies have made use of more than one pheromone, but
the interactions between these different pheromones are largely simple
extensions of single criterion, single colony ant algorithms. This paper investigates an alternative form of interaction between normal pheromone
and anti-pheromone. Three variations of Ant Colony System that apply
the anti-pheromone concept in different ways are described and tested
against benchmark travelling salesman problems. The results indicate
that the use of anti-pheromone can lead to improved performance. However, if anti-pheromone is allowed too great an influence on ants' decisions, poorer performance may result.

1 Introduction

Many animal species, and insects in particular, use chemical substances called
pheromones to influence the behaviour of other animals of the same type.
Pheromones can carry many different types of information and influence behaviour in varied ways [1]. Many species of ants are known to use pheromones to
communicate to coordinate activities like the location and collection of food [2].
The success of this kind of indirect communication has lead researchers to develop a number of simulations of ant behaviour, including optimisation heuristics
such as Ant Colony Optimisation (ACO). Based on the foraging behaviour of ant
colonies, ACO has generally used a single kind of pheromone to communicate
between its (artificial) ants, as this is what biological ants do when foraging.

---

* This author is a PhD scholar supported by an Australian Postgraduate Award.

M. Dorigo et al. (Eds.): ANTS 2002, LNCS 2463, pp. 100–110, 2002.
© Springer-Verlag Berlin Heidelberg 2002

However, natural pheromonal communication often consists of a more complex interaction of a number of different pheromones [1]. Furthermore, ACO's reliance on positive feedback alone may make it difficult for it to successfully escape local optima [3,4]. Schoonderwoerd et al. [5] were some of the first to suggest that the use of an "anti-pheromone", the effect of which would be opposite to that of normal pheromone, could be a useful technique in ant algorithms for optimisation. This paper investigates ways in which the concept of an anti-pheromone can be applied to the Travelling Salesman Problem (TSP). Three variations of an Ant Colony System (ACS) that use anti-pheromone in some form are described and compared with a typical implementation of ACS.

An anti-pheromone, or any other variant of the pheromone typically used in ant algorithms, is simply a substance with a different effect to that of "normal" pheromone. Hence, a brief summary of those ant algorithms that have used more than one kind of pheromone is presented here, contrasting these ant algorithms with an approach that uses anti-pheromone. Much of the work in ant algorithms that has used more than one kind of pheromone relates to multiple colony ant systems (e.g. [3,4,6,7,8]). In most of these applications, the interaction between colonies has been relatively simple, the transfer of the best solution from one colony to update the pheromone of another colony [6,7,8].

More complex interaction has been investigated by Kawamura, Yamamoto and Ohuchi, and Kawamura et al. [3,4]. Their Multiple Ant Colony System (MACS) is highly flexible and enables pheromone from one colony to have both positive and negative effects on the behaviour of other colonies. A "negative pheromone effect" is where higher amounts of pheromone on an element actually discourage ants from choosing that element. While the MACS approach is highly flexible, it requires considerable memory and computing resources to maintain the multiple colonies, each with its own pheromone matrix, as well as to calculate the influences between colonies. The anti-pheromone ant algorithms described in this paper are simpler and require less memory and computational resources than MACS, yet can still utilise negative pheromone effects to diversify the search.

The other area in which multiple types of pheromone have been used is in multiple objective optimisation. Mariano and Morales [6] propose an Ant-Q algorithm for solving a multi-objective irrigation problem. Their algorithm, MOAQ, maintains a number of "families," one for each optimisation criterion, which communicate with each other by exchanging the best solution found by each. This is the same kind of information exchange used in multiple colony ant algorithms for single objective optimisation.

Iredi, Merkle and Middendorf [9] propose a different multi colony approach for solving bi-criterion optimisation problems. Every colony maintains two pheromone matrices, tailored to one of the optimisation criteria. Ants within a colony differ in their preference for each pheromone, so that they search different regions of the Pareto-optimal front. The idea of using different pheromones to direct the search in different areas has some merit and is a strong influence on the second anti-pheromone application we describe (see Section 3.2).

This paper is organised as follows. Section 2 has a brief overview of ACS and its governing equations. Section 3 further explains the rationale for using two kinds of pheromone and describes how we adapt ACS to make use of anti-pheromone. Section 4 shows the results of using anti-pheromone on some benchmark TSPs while Section 5 gives the conclusions of this work.

2   ACS

ACO is an umbrella term for a number of similar metaheuristics [10] including the Ant Colony System (ACS) metaheuristic [11]. A brief summary of the equations governing ACS when applied to the Travelling Salesman Problem (TSP) is provided here as it forms the basis for our anti-pheromone modifications described in the next section. The reader is referred to Dorigo and Gambardella [12] and Dorigo, Di Caro and Gambardella [10] for a more in-depth treatment of ACO.

The aim of the TSP is to find the shortest path that traverses all cities in the problem exactly once, returning to the starting city. In a TSP with N cities, the distance between each pair of cities i and j is represented by $d(i, j)$. In ACS, m ants are scattered randomly on these cities ($m \leq N$). In discrete time steps, all ants select their next city then simultaneously move to their next city. Ants deposit pheromone on each edge they visit to indicate the utility (goodness) of these edges. The accumulated strength of pheromone on edge (i, j) is denoted by $\tau(i, j)$.

Ant k located at city r chooses its next city s by applying Equations 1 and 2. Equation 1 is a greedy selection technique favouring links with the best combination of short distance and large pheromone levels. Equation 2 balances this by allowing a probabilistic selection of the next city.

$$s = \begin{cases} \arg \max_{u \in J_k(r)} \left\{ \tau(r, u)[d(r, u)]^\beta \right\} & \text{if } q \leq q_0 \\ \text{Equation 2} & \text{otherwise} \end{cases} \quad (1)$$

$$p_k(r, s) = \begin{cases} \dfrac{\tau(r,s)[d(r,s)]^\beta}{\sum_{u \in J_k(r)} \tau(r,u)[d(r,u)]^\beta} & \text{if } s \in J_k(r) \\ 0 & \text{otherwise} \end{cases} \quad (2)$$

Where:

$q \in [0, 1]$ is a uniform random number.
q_0 is the proportion of occasions when the greedy selection technique is used.
$J_k(r)$ is the set of cities yet to be visited by ant k.

The pheromone level on the selected edge (r, s) is updated according to the local updating rule in Equation 3.

$$\tau(r, s) \leftarrow (1 - \rho) \cdot \tau(r, s) + \rho \cdot \tau_0 \quad (3)$$

Where:

ρ is the local pheromone decay parameter, $0 < \rho < 1$.
τ_0 is the initial amount of pheromone deposited on each of the edges.

Upon conclusion of an iteration (i.e. once all ants have constructed a tour), global updating of the pheromone takes place. Edges that compose the best solution (over *all* iterations) are rewarded with a relatively large increase in their pheromone level. This is expressed in Equation 4.

$$\tau(r, s) \leftarrow (1 - \gamma) \cdot \tau(r, s) + \gamma \cdot \varDelta\tau(r, s) \tag{4}$$

$$\varDelta\tau(r, s) = \begin{cases} \frac{Q}{L} & \text{if } (r, s) \in \text{ globally best tour} \\ 0 & \text{otherwise.} \end{cases} \tag{5}$$

Where:

$\varDelta\tau(r, s)$ is used to reinforce the pheromone on the edges of the global best solution (see Equation 5).
L is the length of the best (shortest) tour to date while Q is a problem dependent parameter [11].
γ is the global pheromone decay parameter, $0 < \gamma < 1$.

3 Anti-pheromone Applications

As ants construct solutions they gain knowledge of which elements have high utility and which elements may, although desirable in the short-term, lead to poorer solutions in the long term. Randall and Montgomery [13] make use of this in their Accumulated Experience Ant Colony (AEAC) by weighting elements based on their long term effects on solution quality. Anti-pheromone, a substance generally laid down on the elements of poorer solutions, can have a similar effect, making known the accumulated bad experiences of ants that are otherwise lost. This is the approach taken by the first two anti-pheromone algorithms. The third algorithm takes a different approach by making normal pheromone repellent to a small number of ants, rather than depositing anti-pheromone on poorer solutions. It is included here as for those ants that see pheromone as a repellent substance, it represents an anti-pheromone.

3.1 Subtractive Anti-pheromone (SAP)

As ants construct solutions, they often identify relatively poor solutions as well as good solutions. In this version of the ACS algorithm, particular attention is paid to those poorer solutions, with pheromone being removed from those elements that make up the worst solution in each iteration. Thus, subsequent generations of ants are discouraged from using elements that have formed part of poorer solutions in the past. This constitutes the simplest way to implement anti-pheromone where the deposition of a repellent pheromone is simulated by a reduction in existing pheromone levels, as suggested by Schoonderwoerd et al. [5] in their conclusions. We refer to this application of anti-pheromone as *Subtractive Anti-Pheromone*.

The SAP algorithm is identical to ACS except for the addition of a second part to the global pheromone update, in which pheromone is removed from links that compose the worst solution in that iteration. This is described in Equation 6.

$$\tau(r, s) \leftarrow \tau(r, s) \cdot \gamma' \qquad \forall (r, s) \in v_w \tag{6}$$

Where:

γ' is the pheromone removal rate due to anti-pheromone.
v_w is the iteration worst solution.

The rate at which pheromone is removed from the elements of the iteration worst solution is controlled by the parameter γ'. The value of γ' used in the experiments is 0.5, which was found to yield the best results.

3.2   Preferential Anti-pheromone (PAP)

Iredi et al. [9] propose an ant system for solving bi-criterion optimisation problems that uses two types of pheromone, one for each criterion. Their use of a different pheromone for each optimisation criterion allows for knowledge concerning both criteria to be improved as the algorithm progresses. The second anti-pheromone application we propose, called *Preferential Anti-Pheromone*, takes a similar approach by explicitly using two types of pheromone, one for good solutions and one for poorer solutions. Ants in this version of the algorithm differ in their preference for normal pheromone versus anti-pheromone (denoted by τ') with respect to a parameter λ. The value of λ for ant k, $k = [1, m]$, is given by $\frac{k-1}{m-1}$, as in Iredi et al. [9]. Hence, instead of optimising across two objective functions, PAP allows some ants to explore apparently poorer areas of the search space while other ants focus on the solution space near the current global best solution.

Equations 1 and 2 are modified for this variant to incorporate anti-pheromone information, yielding Equations 7 and 8 respectively.

$$s = \begin{cases} \arg\max_{u \in J_k(r)} \left\{ [\lambda\tau(r, u) + (1 - \lambda)\tau'(r, u)] \cdot [d(r, u)]^\beta \right\} & \text{if } q \leq q_0 \\ \text{Equation 8} & \text{otherwise} \end{cases} \tag{7}$$

$$p_k(r, s) = \begin{cases} \dfrac{[\lambda\tau(r,s)+(1-\lambda)\tau'(r,s)]\cdot[d(r,s)]^\beta}{\sum_{u \in J_k(r)}[\lambda\tau(r,u)+(1-\lambda)\tau'(r,u)]\cdot[d(r,u)]^\beta} & \text{if } s \in J_k(r) \\ 0 & \text{otherwise} \end{cases} \tag{8}$$

Pheromone and anti-pheromone are updated equally by ants traversing links during an iteration as local updating ignores the cost of solutions produced. Thus, in addition to Equation 3 being applied, the anti-pheromone on a selected edge is updated by Equation 9.

$$\tau'(r, s) \leftarrow (1 - \rho) \cdot \tau'(r, s) + \rho \cdot \tau_0 \tag{9}$$

Upon conclusion of an iteration the global update rule in (4) is applied without modification. In addition, the global update rule for anti-pheromone given in Equation 10 is applied.

$$\tau'(r, s) \leftarrow (1 - \gamma) \cdot \tau'(r, s) + \gamma \cdot \Delta\tau'(r, s) \tag{10}$$

$$\Delta\tau'(r, s) = \begin{cases} \frac{Q}{L_w} & \text{if } (r, s) \in \text{ iteration worst tour} \\ 0 & \text{otherwise.} \end{cases} \tag{11}$$

Where:

$\Delta\tau'(r, s)$ is used to reinforce the pheromone on the edges of the iteration worst solution (see Equation 11).
L_w is the length of the worst tour from the iteration just ended.

3.3   Explorer Ants

The third anti-pheromone variant takes a different approach to the first two in that it does not associate anti-pheromone with poorer solutions. Instead, a small number of ants are chosen to behave differently from other ants by being attracted to areas with little pheromone. These *explorer ants* influence their environment by depositing pheromone in the same way as normal ants, only their preference for existing pheromone is reversed. Hence, an explorer ant finds its own pheromone undesirable. Equations 12 and 13 express how an explorer ant located at city r chooses the next city to go to s.

$$s = \begin{cases} \arg\max_{u \in J_k(r)} \left\{ [\tau_{max} - \tau(r, u)] \cdot [d(r, u)]^\beta \right\} & \text{if } q \leq q_0 \\ \text{Equation 13} & \text{otherwise} \end{cases} \tag{12}$$

$$p_k(r, s) = \begin{cases} \frac{[\tau_{max} - \tau(r,s)] \cdot [d(r,s)]^\beta}{\sum_{u \in J_k(r)} [\tau_{max} - \tau(r,u)] \cdot [d(r,u)]^\beta} & \text{if } s \in J_k(r) \\ 0 & \text{otherwise} \end{cases} \tag{13}$$

Where:

τ_{max} is the highest current level of pheromone in the system.

The *explorer ants* algorithm divides the population of ants into two groups, with a higher proportion of normal ants than explorers. We found that two explorer ants produced good results when $m = 10$. While this approach appears similar to the MACS of Kawamura et al. [4], there are some important differences. Although ants are divided into two groups they do not represent separate colonies as they share the same pheromone. Furthermore, this algorithm allows for a small number of explorer ants to be used and saves on memory requirements by using only a single pheromone matrix.

Table 1. TSP instances used in this study

Instance	Description	Optimal Cost
gr24	24 cities	1272
eil51	51 cities	426
eil76	76 cities	538
kroA100	100 cities	21282
d198	198 cities	15780
lin318	318 cities	42029
pcb442	442 cities	50778

4   Computational Experience

A control strategy (normal ACS) and the three alternative implementations were run in order to evaluate their relative performance. Table 1 describes the TSP instances [14] with which the alternative pheromone representations are tested.

The computing platform used to perform the experiments is a 550 MHz Linux machine. The computer programs are written in the C language. Each problem instance is run across 10 random seeds consisting of 3000 iterations. The ACS parameter settings used are: $\beta = -2$, $\gamma = 0.1$, $\rho = 0.1$, $m = 10$, $q_0 = 0.9$.

4.1   Results

The results are given in Table 2. The minimum ("Min"), median ("Med"), maximum ("Max") and inter-quartile range ("IQR") are used to summarise the results as they are non-normally distributed. As the results for CPU time are highly consistent for each combination of algorithm and problem, only the median CPU time (in seconds) is presented in the table.

The results for CPU time are highly consistent with the three anti-pheromone algorithms' times within 3% of those for the control. *Explorer ants* runs slightly slower than the others due to the increased computational overhead associated with evaluating the combined value of two pheromones.

To allow for statistical analysis of cost results across problems, the costs of solutions were normalised according to $\frac{c - c_{opt}}{c_{opt}}$, where c is the cost of the solution and c_{opt} is the optimal cost for its corresponding problem. As the data are non-normally distributed, the Mann-Whitney test was used to compare results.

SAP performs well on problems with less than 100 cities, producing results that are better than the control. This result is statistically significant, $p < 0.05$. On problem gr24, it found the optimal solution on every run. SAP also produces better results on eil51, and equivalent results on eil76, kroA100 and d198. However, the control performs better on problems with more than 200 cities. Across all problems there is no statistically significant difference between the two algorithms. Although in general $\gamma' = 0.5$ yields the best results for SAP on the problem instances tested, other values were investigated. It was found that

Table 2. Results for ACS control and anti-pheromone variants

Problem Instance	Algorithm	Min	Med	Max	IQR	CPU Time
gr24	Control	1272	1278	1278	6	18
	SAP	1272	1272	1272	0	18
	PAP	1272	1272	1278	0	20
	Explorer	1272	1272	1278	5	17
eil51	Control	426	430	441	7	80
	SAP	426	428	430	1	79
	PAP	426	430	436	3	83
	Explorer	426	430	439	4	78
eil76	Control	540	545	554	8	177
	SAP	539	550	558	9	176
	PAP	539	552	562	7	182
	Explorer	539	552	561	11	172
kroA100	Control	21296	21479	22178	371	307
	SAP	21319	21552	22060	382	304
	PAP	21292	21753	22754	411	315
	Explorer	21305	21515	22318	426	298
d198	Control	15948	16116	16451	151	1181
	SAP	15988	16156	16454	337	1183
	PAP	16449	16769	17182	311	1216
	Explorer	16058	16205	16425	169	1161
lin318	Control	45514	46793	47422	1031	3018
	SAP	48375	49099	50608	1026	3050
	PAP	46793	49434	50223	954	3136
	Explorer	45031	46314	48114	908	3027
pcb442	Control	60525	62420	65014	1610	5988
	SAP	62753	64596	66332	1941	5932
	PAP	61623	64156	64753	1133	6097
	Explorer	61709	63657	66663	2219	5883

increasing γ' to 0.75 improves SAP's performance on problems with more than 200 cities, bringing them closer to those achieved by the control. However, there is still a statistically significant difference between SAP with $\gamma = 0.75$ and the control on problems with more than 200 cities, $p < 0.05$.

On a per problem basis, PAP produces worse results than the control on all problems except gr24 and eil51. Analysis across all problems combined reveals a statistically significant difference between the results of the control and those of PAP, $p < 0.10$. It is possible that this poor performance is due to the local update rule in which all ants, regardless of their preference for normal pheromone versus anti-pheromone, update both pheromones by the same amount. Hence, ants with a strong preference for anti-pheromone may distract the search too much and prevent ants with a stronger preference for normal pheromone from searching near the current global best solution. A modified version of PAP was implemented in which ants locally update pheromone in proportion to their value of λ (and anti-pheromone in proportion to $(1 - \lambda)$) producing equivocal results.

Compared across all problems, no statistically significant difference exists between *explorer ants* and the control. Only on d198 is there sufficient evidence that the control performs better. *Explorer ants* performed slightly better than the control gr24 and eil51.

In general, SAP produces better solutions than PAP. Although PAP performs better than SAP on problems lin318 and pcb442, there is no statistically significant difference. On problems with less than 100 cities, SAP produces better results than *explorer ants*. However, on problems with more than 200 cities, *explorer ants* performs better. *Explorer ants* also performs better than PAP across all problems.

5   Conclusions

Although animals in nature generally use a number of pheromones in their communication, typical ant algorithms have used only one. The majority of ant algorithms that have used more than one pheromone are simple extensions of single-pheromone ant algorithms, maintaining multiple colonies and using the best solution from one colony to update the pheromone of others. More complex pheromone interactions have been used by Kawamura et al. [3,4], but these require fairly considerable computational resources to store and process multiple types of pheromone.

We have proposed three variations of the Ant Colony System that employ anti-pheromone in some form. The first, *subtractive anti-pheromone*, simulates anti-pheromone by reducing the amount of pheromone on elements of the iteration worst solution. It works well on problems with less than 200 cities. It also has a distinct advantage over multiple colony ant systems in that it stores only one kind of pheromone and is no slower than normal ACS. The second algorithm, *preferential anti-pheromone*, is less successful, producing better results than the control on only the two smallest problems. It is possible this is because ants with a strong preference for anti-pheromone distract ants with a preference for normal pheromone, or that the linear equation for deciding ants' preferences results in too few ants with a strong preference for normal pheromone. *Explorer ants*, the third anti-pheromone algorithm, changes the response to pheromone of a small number of ants, making these ants seek elements with lower pheromone levels. It can produce better solutions than the control on small problems, but produces largely equivalent results on all other problems.

Middendorf, Reischle and Schmeck [8] suggest that in multiple colony ant algorithms the amount and frequency of pheromone information exchange should be kept small. Hence, we plan to extend these algorithms so that the effects of anti-pheromone on normal pheromone are kept to a minimum. For instance, PAP could be improved by having a larger proportion of ants with a strong preference for normal pheromone as well as changing the local update rule so that ants update each kind of pheromone in proportion to their preference for that pheromone. SAP may also yield improved results if pheromone is removed from elements of poor solutions less frequently.

We have given plausible explanations for how each of the three algorithms helps to explore different areas of the search space and why this may prove beneficial. An important extension to this work is to analyse this exploration behaviour more objectively. This will involve measuring the differences in exploration between algorithms as well as the utility of this exploration.

This work is part of a wider strategy that is looking at ways of producing generic strategies to enhance ant based metaheuristics [13,15,16]. In addition to the improvements suggested above, future work will involve the extension of these anti-pheromone applications to other problems.

References

1. Vander Meer, R.K., Breed, M.D., Winston, M.L., Espelie, K.E. (eds.): Pheromone Communication in Social Insects. Ants, Wasps, Bees, and Termites. Westview Press, Boulder, Colorado (1997)
2. Heck, P.S., Ghosh, S.: A Study of Synthetic Creativity through Behavior Modeling and Simulation of an Ant Colony. IEEE Intelligent Systems **15** (2000) 58–66
3. Kawamura, H., Yamamoto, M., Ohuchi, A.: Improved Multiple Ant Colonies System for Traveling Salesman Problems. In Kozan, E., Ohuchi, A. (eds.): Operations Research/Management Science at Work. Kluwer, Boston (2002) 41–59
4. Kawamura, H., Yamamoto, M., Suzuki, K., Ohuchi, A.: Multiple Ant Colonies Algorithm Based on Colony Level Interactions. IEICE Transactions, Fundamentals **E83-A** (2000) 371–379
5. Schoonderwoerd, R., Holland, O.E., Bruten, J.L., Rothkrantz, L.J.M.: Ant-Based Load Balancing in Telecommunications Networks. Adaptive Behavior **2** (1996) 169–207
6. Mariano, C.E., Morales, E.: MOAQ: An Ant-Q Algorithm for Multiple Objective Optimization Problems. Genetic and Evolutionary Computation Conference (GECCO-99), Orlando, Florida (1999) 894–901
7. Michels, R., Middendorf, M.: An Ant System for the Shortest Common Supersequence Problem. In Corne, D., Dorigo, M., Glover, F. (eds.): New Ideas in Optimization. McGraw-Hill, London (1999) 51–61
8. Middendorf, M., Reischle, F., Schmeck, H.: Multi Colony Ant Algorithms. Parallel and Distributed Computing, Proceedings of the 15 IPDPS 2000 Workshops, Third Workshop on Biologically Inspired Solutions to Parallel Processing Problems (BioSP3), Cancun, Mexico (2000) 645–652
9. Iredi, S., Merkle, D., Middendorf, M.: Bi-Criterion Optimization with Multi Colony Ant Algorithms. Evolutionary Multi-Criterion Optimization, First International Conference (EMO'01), Zurich (2001) 359–372
10. Dorigo, M., Caro, G.D.: The Ant Colony Optimization Meta-heuristic. In Corne, D., Dorigo, M., Glover, F. (eds.): New Ideas in Optimization. McGraw-Hill, London (1999) 11–32
11. Dorigo, M., Gambardella, L.M.: Ant Colonies for the Traveling Salesman Problem. BioSystems **43** (1997) 73–81
12. Dorigo, M., Di Caro, G., Gambardella, L.M.: Ant Algorithms for Distributed Discrete Optimization. Artificial Life **5** (1999) 137-172

13. Randall, M., Montgomery, J.: The Accumulated Experience Ant Colony for the Travelling Salesman Problem. Proceedings of Inaugural Workshop on Artificial Life, Adelaide, Australia (2001) 79–87
14. Reinelt, G.: TSPLIB - A Traveling Salesman Problem Library. ORSA Journal of Computing **3** (1991) 376–384
15. Montgomery, J., Randall, M.: Alternative Pheromone Applications for Ant Colony Optimisation. Technical Report TR02-07, School of Information Technology, Bond University, Qld, Australia. Submitted to AI2002, 15th Australian Joint Conference on Artificial Intelligence, Canberra, Australia (2002)
16. Randall, M.: A General Framework for Constructive Meta-heuristics. In Kozan, E., Ohuchi, A. (eds.): Operations Research/Management Science at Work. Kluwer, Boston, MA (2002) 111–128

Applying Population Based ACO to Dynamic Optimization Problems

Michael Guntsch[1] and Martin Middendorf[2]

[1] Institute for Applied Computer Science and Formal Description Methods
University of Karlsruhe, Germany
guntsch@aifb.uni-karlsruhe.de
[2] Computer Science Group
Catholic University of Eichstätt- Ingolstadt, Germany
martin.middendorf@ku-eichstaett.de

Abstract. Population based ACO algorithms for dynamic optimization problems are studied in this paper. In the population based approach a set of solutions is transferred from one iteration of the algorithm to the next instead of transferring pheromone information as in most ACO algorithms. The set of solutions is then used to compute the pheromone information for the ants of the next iteration. The population based approach can be used to solve dynamic optimization problems when a good solution of the old instance can be modified after a change of the problem instance so that it represents a reasonable solution for the new problem instance. This is tested experimentally for a dynamic TSP and dynamic QAP problem. Moreover the behavior of different strategies for updating the population of solutions are compared.

1 Introduction

In the Ant Colony Optimization (ACO) approach a discrete combinatorial optimization problem is represented by a so called *construction graph* where feasible solutions to the problem correspond to paths through this graph [4]. Artificial ants move on the graph to generate feasible solutions. Each move from one node to the next node is based on a probabilistic decision where the probability to choose an edge is correlated to the relative amount of so called pheromone that is placed on the edge (and, when available, to additional heuristic information). In an iterative process several generations of artificial ants are changing the pheromone information since the ants that found the best solutions in a generation are allowed to add some amount of pheromone on the edges of the corresponding paths. Hence, it is the changing pheromone information that allows the algorithm to make progress so that ants in later generations usually find better solutions.

In a genetic algorithm (GA) a population of solutions (or encodings thereof) is directly transferred from one iteration to the next. Mutation and crossover operations are then used to create new solutions. The new population is obtained after selection of a subset of mostly good solutions from the old population and the newly created solutions.

M. Dorigo et al. (Eds.): ANTS 2002, LNCS 2463, pp. 111–122, 2002.
© Springer-Verlag Berlin Heidelberg 2002

A population-based ACO (P-ACO) algorithm was proposed in [7] where (as in a GA) a population of solutions (or more exactly, a population of sequences of decisions corresponding to solutions) is directly transferred to the next iteration. These solutions are then used to compute pheromone information for the ants of the new iteration. For every solution in the population some amount of pheromone is added to the corresponding edges of the construction graph (every edge has been initialized with the same nonzero initial value before).

ACO algorithms have already been used to solve dynamic optimization problems (see [2,6,8,9]). In [6,8] strategies were proposed for modifying the pheromone information in reaction to a change for a dynamic Traveling Salesperson Problem (TSP) problem. It was conjectured in [7] that P-ACO is suitable for solving dynamic optimization problems where the problem instance changes over time. In particular, this approach seems promising when dynamic changes of a problem instance are not too severe and there exists a good heuristic for modifying the solutions in the population after a change so that they become valid solutions of the new instance. Since the whole pheromone information of the P-ACO depends only on the solutions in the population it is not necessary to apply other mechanism for modifying the pheromone information (see [6,8]). This is an advantage because it will usually be faster to modify a few solutions directly than to modify the whole pheromone information of a usual ACO algorithm.

In this paper we study the behavior of P-ACO on a dynamic TSP problem and a dynamic Quadratic Assignment Problem (QAP) problem. Moreover, different strategies are investigated for updating the population of solutions. In [7] it was always the best solution of the last iteration that was added to the population and the oldest solution in the population was removed. Here we also consider removing the worst solution from the population or to make a random decision depending on the solution qualities. This is similar to the different strategies that exist for GAs to decide which individuals form the new population.

The paper is organized as follows. In Section 2 we describe ACO algorithms in general and the P-ACO with the different population update strategies. The test setup is given in Section 3. The experimental results are discussed in Section 4, and conclusions are drawn in Section 5.

2 ACO Approaches

In this section we describe the generic decision process which forms the basis for all ACO algorithms, including P-ACO. Furthermore, we discuss several methods for P-ACO to update its population and as a consequence modify the pheromone information. It is discussed how ACO is applied to the test problem classes Dynamic TSP and Dynamic QAP which will be used to compare the performance of the different strategies in Section 4.

2.1 Generic Decision Process

When constructing solutions to a problem-instance, ants proceed in an iterative fashion, making a number of local decisions which result in a global solution. For the case of TSP, an ant starts at some city and proceeds by continually choosing which city to visit next from it's current location until the tour is complete. In QAP, the ant goes to a random unassigned location and places one of the remaining facilities there, proceeding until no free locations/facilities are left.

The decisions an ant makes are probabilistic in nature and influenced by two factors: pheromone information τ_{ij} and heuristic information η_{ij}, each indicating how good it is to choose j at location i. The set of valid choices for an ant is denoted by S. With probability q_0, where $0 \leq q_0 < 1$ is a parameter of the algorithm, the ant chooses the $j \in S$ which maximizes $\tau_{ij}^{\alpha} \cdot \eta_{ij}^{\beta}$, where α and β are constants that determine the relative influence of the heuristic values and the pheromone values on the decision of the ant. With the probability of $1 - q_0$, an ant chooses according to the probability distribution over S defined by

$$\forall j \in S: \ p_{ij} = \frac{\tau_{ij}^{\alpha} \cdot \eta_{ij}^{\beta}}{\sum_{k \in S} \tau_{ik}^{\alpha} \cdot \eta_{ik}^{\beta}} \tag{1}$$

In TSP there are n cities with pairwise distances d_{ij}. The goal is to find a Hamiltonian circle which minimizes the sum of distances covered, i.e. to find a mono-cyclic permutation π of $[1:n]$ which minimizes $\sum_{i=1}^{n} d_{i\pi(i)}$. The pheromone matrix $[\tau_{ij}]$ is encoded in a city×city fashion, which means that τ_{ij} is an indication of how good it was to go from city i to city j. When initializing the pheromone matrix, all diagonal elements are therefore set to 0. The heuristic information for Formula 1 is derived from the distance between cities, i.e. $\eta_{ij} = 1/d_{ij}$. Typically, $\beta > 1$ is chosen since the η values tend to be very close to one another, resulting in a too uniform distribution of probabilities.

For QAP there are n facilities, n locations, and $n \times n$ matrices $D = [d_{ij}]$ and $F = [f_{hk}]$ where d_{ij} is the distance between locations i and j and f_{hk} is the flow between facilities h and k. The goal is to find an assignment of facilities to locations, i.e., a permutation π of $[1:n]$ such that the sum of distance-weighted flows between facilities $\sum_{i=1}^{n} \sum_{j=1}^{n} d_{\pi(i)\pi(j)} f_{ij}$ is minimized. For this problem τ_{ij} gives information about placing facility j on location i, with all pheromone values being equal at the start of the algorithm. No heuristic information is used for QAP, which is also done in other ACO algorithms for this problem[10].

After m ants have constructed a solution, the pheromone information is updated. The standard ACO approach is to evaporate all elements of the pheromone matrix by multiplication with a value smaller than 1, see ([3,5]). Afterwards, a positive update is performed by the best solution(s) by adding pheromone to the decisions which were made, i.e. τ_{ij} is increased if $\pi(i) = j$. P-ACO also performs positive updates, storing all permutations that have performed an update in the population P, but uses a different mechanism for reducing pheromone values. Instead of reducing all pheromone values, a negative update is performed when a solution leaves the population. This negative update is of the same absolute value as the previous positive update performed by it, and it is applied only to the elements which this solution had previously positively updated. Formally:

- whenever a solution π enters the population, do a positive update:
$$\forall \ i \in [1, n] : \tau_{i\pi(i)} \mapsto \tau_{i\pi(i)} + \Delta$$

– whenever a solution σ leaves the population, do a negative update:
$$\forall\, i \in [1, n] : \tau_{i\sigma(i)} \mapsto \tau_{i\sigma(i)} - \Delta$$

These updates are added to the initial pheromone value τ_{init}. Therefore, the pheromone matrix which is induced by a population P has the following values:

$$\tau_{ij} = \tau_{init} + \Delta \cdot |\{\pi \in P | \pi(i) = j\}|. \tag{2}$$

We denote the maximum possible value an element of the pheromone matrix can achieve by Formula 2 as $\tau_{max} := \tau_{init} + k \cdot \Delta$. Reciprocally, if τ_{max} is used as a parameter of the algorithm instead of Δ, we can derive $\Delta = (\tau_{max} - \tau_{init})/k$ so that with Formula 2, τ_{max} is indeed the maximum attainable value for any τ_{ij}. Note that the actual value for τ_{init} is arbitrary, as τ_{max} could simply be scaled in accordance. For reasons of clarity, we wish the row/column-sum of initial pheromone values to be 1, which means that $\tau_{init} = 1/(n-1)$ for TSP and $\tau_{init} = 1/n$ for QAP.

2.2   Population Update Strategies

Now that we have explained how a population P induces a pheromone matrix $[\tau_{ij}]$, let us consider some possibilities for managing the population, i.e. for deciding which solutions should enter the population and which should leave. A lot of work has been done on the maintenance of populations for steady state GAs, see [1], and some methods presented here are inspired by those works.

We always consider only the best solution the m ants built during the past iteration as a candidate for the population P. At the start of the P-ACO algorithm, P is empty. For the first k iterations, with k being the (ultimate) size of P, the candidate solution is automatically transferred into the population P. Starting at iteration $k+1$, we have a "full" population (k solutions) and another solution as a candidate. If the candidate should be included into the population, then another solution which is already part of the population will have to be removed from it.

Age. The easiest form of population update, which was also used in [7], is an age based strategy (called *Age*) in which the oldest solution is removed from P and the new candidate solution always enters it. This gives each solution an influence over exactly k iterations, after which it is removed from P.

Quality. Instead of having the respective best solutions of the last k iterations in the population, another obvious choice is to store the best k solutions found over all past iterations. We call this strategy *Quality*. If the candidate solution of the current iteration is better than the worst solution in P, the former replaces the latter in P. Otherwise, P does not change.

Prob. A disadvantage of the *Quality* Strategy is the possibility that after some iterations, P might consist of k copies of the best solution found so far, which will happen if the best solution is found in k separate iterations. In this case, the ants would be focused on a very small portion of the search space. To ameliorate this deficiency of the *Quality* Strategy, we introduce a further strategy, called *Prob*, which probabilistically chooses which element from P unified with the candidate solution will be removed and thus not part of P for the next iteration. Let $g(\pi)$ denote the solution value associated with a solution π. Specifically, $g(\pi) = \sum_{i=1}^{n} d_{i\pi(i)}$ for the TSP and $g(\pi) = \sum_{i=1}^{n} \sum_{j=1}^{n} d_{\pi(i)\pi(j)} f_{ij}$ for the QAP. For both problems a lower value of g is better. With $P = \{\pi_1, \ldots, \pi_k\}$ and the candidate solution denoted by π_{k+1}, we define a probability distribution over the solutions π_i, $i = 1, \ldots, k+1$:

$$p_i = \frac{x_i}{\sum_{j=1}^{k+1} x_j} \quad \text{with}$$

$$x_i = g(\pi_i) - \min_{j=1,\ldots,k+1} g(\pi_j) + avg(\pi) \quad \text{and}$$

$$avg(\pi) = \frac{1}{k+1} \sum_{j=1}^{k+1} g(\pi_j) - \min_{j=1,\ldots,k+1} g(\pi_j)$$

This provides us with the possibility of any solution from the population being removed, and the preference to remove bad solutions. Note that if the candidate solution π_{k+1} is chosen for removal, P remains unaltered.

Age&Prob. Finally, we consider a combination of two of the above strategies to form a new strategy. Specifically, we combine the *Age* and the *Prob* Strategy, using *Prob* for removal from and *Age* for insertion into the population. This is accomplished by restricting the *Prob* Strategy explained above to the elements of P, thereby always removing one element from the population. The candidate solution of the iteration is always added, according to the *Age* Strategy. This combination should exhibit the ability of the *Age* Strategy to quickly move to new parts of the search space, and the tendency of the *Prob* Strategy to hold good solutions in the population.

Elitism. In addition to these strategies, and analogously to ([5]), it is possible to introduce the concept of elitism into P-ACO. We define π_e as the elitist solution, i.e. the best solution found by the algorithm so far. If elitism is being used by P-ACO, then the population is $P = \{\pi_1, \ldots, \pi_{k-1}\} \cup \{\pi_e\}$, with the update strategy used by the algorithm only being applied to to $\{\pi_1, \ldots, \pi_{k-1}\}$. The elitist solution π_e is updated only when a better solution is found. The concept of elitism can be especially useful for the *Age* Strategy, as the ants would need to find the current best solution every k iterations otherwise in order to have it in the population.

2.3   Reacting to a Change

So far, we have only discussed how the P-ACO algorithm works on a static problem. In this subsection, we present how the algorithm reacts when a change occurs to the instance it is working on.

As has been mentioned in Section 1, the dynamical changes we consider that can occur to problem instances are the deletion and insertion of cities for TSP and deletion and insertion of locations for QAP. We will only consider changes that leave the problem size unaltered, i.e. the same number of cities/location are inserted as were deleted. For the standard ACO, these changes require a significant amount of repair to the pheromone matrix, as has been shown for TSP in ([6,8]). Also, problem-specific knowledge is required for these strategies, and it is not clear that they will work for other problem classes. However, one principal from [8] called *KeepElite* can be used for P-ACO to repair each solution of the population after a change occurred to the problem instance. For TSP, the old cities are removed, with successor and predecessor becoming neighbors in the tour, and afterwards the new cities are inserted into the tour individually in a greedy fashion at the place of least length increase. For QAP, the facilities are replaced onto the new locations individually again in a greedy fashion to cause the least increase to solution quality. The induced pheromone matrix changes accordingly.

3   Test Setup

As we have mentioned in the previous sections, the two problems used in this paper to gauge the performance of the proposed strategies are a dynamic TSP and a dynamic QAP. The algorithm used for finding good solutions is expected to cope with the insertion, deletion, or replacement of cities/locations which occur for these instances at runtime.

Our approach was to take a problem instance for TSP/QAP, randomly remove exactly half of the cities/locations, thus creating a "spare pool", and then start the algorithm, replacing c random cities/locations every t iterations between the instance the algorithm is working on and the spare pool. This procedure ensures that the optimal solution for each single instance the algorithm is trying to find is (usually) not radically different after each change. For TSP, we used rd400 from the TSPLIB [11], which is a 400 city instance, and for QAP, we chose tai150b from the QAPLIB [12], which has 150 locations. Note that for QAP, we always kept the first 75 items defined in the flow-matrix for placement on the varying locations. For severity of change and frequency of change, we chose $c \in \{1, 5, 25\}$ for TSP and $c \in \{1, 5, 15\}$ for QAP, and in both cases $t \in \{50, 200, 750\}$, creating 9 different scenarios, that is pairs (c, t), for each problem. For each scenario, 25 random instances were created, and the results averaged.

The parameters for P-ACO were chosen in accordance with the results obtained on the static variants of TSP and QAP in [7] as well as further testing. Specifically, we chose $q_0 = 0.9$ for TSP and QAP, $\tau_{max} = 1.0$ for TSP and $\tau_{max} = 5.0$ for QAP, which needs a higher τ_{max} value to direct the ants in light of a missing heuristic ($\beta = 0$). We choose $\alpha = 1$ for both TSP and QAP and $\beta = 5$ for TSP. The variable parameters of the P-ACO algorithm we explored were the population size $k \in \{1, 3, 6, 10\}$ and the replacement strategy for the population, taken from $\{Age, Quality, Prob, Prob\&Age\}$. We define the combination of a population size and a replacement strategy as a configuration. Note that for $k = 1$, Age and $Prob\&Age$ are identical. We also explored the effect of elitism by having one of the individuals in the population represent the best found solution, leaving $k - 1$ solutions to be managed by the chosen strategy. Note that for $k = 1$, using elitism for any strategy is the same as using $Quality$ without elitism.

For comparison, we also evaluated the performance of a strategy which does not repair the solutions in the population but rather restarts the algorithm each time a change occurs. For this, we used P-ACO with the best performing configuration on static TSP and QAP, which is Age with $k = 1$.

4   Results

A comparison between the population update strategies proposed in Section 2, that is Age, $Quality$, $Prob$, and $Prob\&Age$, for different population sizes k and the different scenarios described in Section 3 was conducted. This comparison is based on the best found solution quality, averaged over all 9000 iterations, of the individual configurations. The results are displayed in Figure 1 for TSP, and Figure 2 for QAP respectively. Basically, the lighter a square is in these two figures, the better the relative performance of the indicated configuration on the corresponding scenario. The best configuration always has a white square, and any configuration that performed more than 5% worse than the respective best is colored completely black. In addition to the shading, some of the squares in Figure 1 exhibit a "+" sign. This was used when employing elitism in combination with the indicated configuration led to a better solution than the omission of elitism, which for TSP performed better in the majority of cases, i.e. all those not marked with "+". A similar approach was taken in Figure 2. However, for QAP, in most cases the use of elitism led to better solutions, and hence a "-" is shown when the omission of elitism performed better.

As mentioned above, for TSP, most configurations perform better when not combined with elitism. Specifically, the best solution for each scenario was found by a configuration that did not employ elitism. It should be noted that even when the use of elitism created a better solution, the improvement was typically less than 0.5%. The cases in which configurations benefit most from using elitism are when only small changes occur to the problem, thus making it likely that the modified best solution would still represent a very good solution, and when population size is not too small and there is enough time to do exploration beside exploitation of the elite solution in the population. The two best configurations however, Age and $Prob\&Age$ with $k = 3$, exhibit best performance when not using elitism. The $Quality$ strategy performs poorly for most of the scenarios and population sizes, especially when frequent and severe changes occur. Indeed, in these cases, the $Quality$ strategy performs increasingly bad over time, with

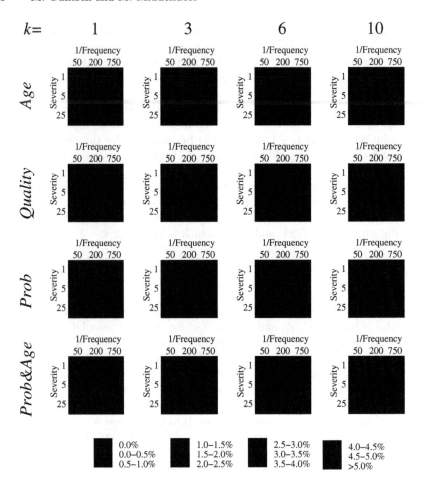

Fig. 1. Relative deviation from the respective best solution by the individual configurations for TSP is shown. A "+" indicates that the use of elitism leads to better performance; otherwise, not using elitism performed better.

the often and extensively repaired previous best solutions causing a focus on an area of the search space with bad solution quality, see Figure 3. The reason for this behavior is probably due to the combination of strong and usually good heuristic values with *Quality*'s deterministic fashion of holding only the best solutions, which effectively obstructs further exploration.

For QAP, the behavior of the individual configurations is different from TSP in some ways. Here, all configurations with $k > 1$ benefit from elitism, and even some with $k = 1$ (effectively turning them into the *Quality* strategy). Judging from the good performance of the configurations with $k = 1$ and *Quality* for larger sizes of k as well, it seems that the ants require a strong and not too variable guidance when looking for good solutions, with the quick convergence

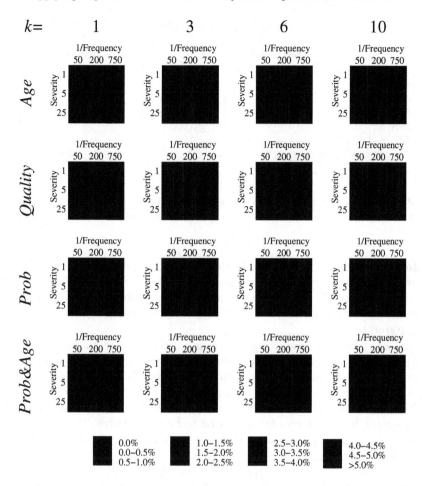

Fig. 2. Relative deviation from the respective best solution by the individual configurations for QAP is shown. A "-" indicates that not using elitism leads to better performance better; otherwise, using elitism performed better.

behavior of a small population size outweighing the better handling of changes to the problem instance by a larger population. For small populations, the difference between the individual configurations is not very large, which can be seen for 2 representative scenarios in Figure 4. Recall that these curves show integral performance, which explains why the values are still falling noticeably at the end of 9000 iterations. The only strategy which performs significantly worse than the others is *Prob*, which has neither the flexibility exhibited by the *Age* Strategy in exploring new areas of the search space nor the persistence of the *Quality* Strategy in exhaustively exploiting the surroundings of a good solution.

So far we have only examined how the dynamic strategies compare to one another. The easiest method for dealing with any dynamic problem is of course

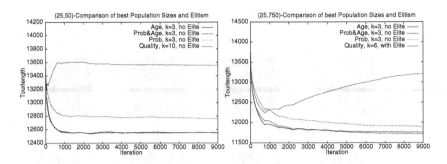

Fig. 3. Comparison of strategies in (25,50) and (25,750) scenarios for TSP (integral performance, i.e. average over all iterations from the first to the actual)

to simply restart the algorithm. If this method proves more successful than integrating the changes into an ongoing optimization process, the integration is superfluous. Figure 5 shows the integral performance for the P-ACO algorithm using the *Age* Strategy and $k = 1$ (which is one of the best configurations for using restart).

One can see that restarting performs vastly inferior to integrating the changes, for TSP as well as QAP. However, restarting starts becoming competitive in TSP when the changes to the problem become too great and/or when the change interval is long enough. For QAP, the shades are darker than for TSP, indicating that the necessary change severity and duration between changes must be even larger for restarting to be competitive.

As we have mentioned in sections 1 and 2, previous work exists for applying ACO to the dynamic TSP. A comparison for the $(c, t) = (1, 50)$ scenario is given in Figure 6. The best evaporation based ACO strategy that was studied in [8] here is the η-Strategy with $\lambda_E = 2$; for P-ACO, we use *Prob&Age* with $k = 3$. The figure shows the characteristic difference between the strategies for

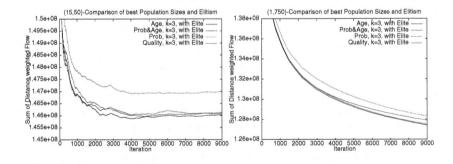

Fig. 4. Comparison of strategies in (15,50) and (1,750) scenarios for QAP (integral performance, i.e. average over all iterations from the first to the actual)

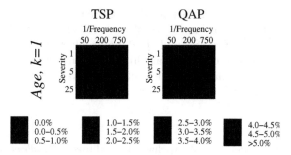

Fig. 5. Solution quality when restarting P-ACO on TSP and on QAP, compared to the best performing configuration which incorporates the changes.

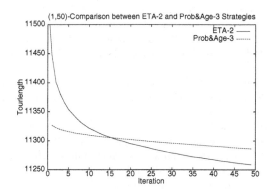

Fig. 6. η-Strategy with $\lambda_E = 2$ compared to *Prob&Age* with $k = 3$.

the evaporation based ACO and P-ACO in general. There is a comparatively large loss of solution quality by the η-strategy after a change, followed by fast improvement, while the *Prob&Age*-strategy shows small solution deterioration and slow improvement. The integral performance by the *Prob&Age*-strategy is only slightly worse than that of the η-strategy, which considering the η-strategy is TSP-specific is quite good.

5   Conclusion

We have studied the behavior of P-ACO for a dynamic TSP and a dynamic QAP problem. New strategies for updating the population of P-ACO have been described and applied with a method for repairing the solutions of the population when a change to the problem instance occurs. We compared the performance of these strategies amongst each other and with a restart of the algorithm for the problem classes TSP and QAP. With some amount of lasting guidance for the ants, either from the problem instance by providing heuristic values or by using elitism, a population size of 3 solutions seems to be the best compromise between versatility for rapid solution improvement and resilience toward changes. Aside from the fitness function for TSP and QAP, no problem specific information was used by P-ACO, which indicates that this algorithm could be employed with some success for other dynamic problem classes as well.

References

1. T. Bäck, D. B. Fogel, and Z. Michalewicz, editors. *Handbook of Evolutionary Computation*. Institute of Physics Publishing and Oxford University Press, Bristol, New York, 1997.
2. G. D. Caro and M. Dorigo. AntNet: Distributed stigmergetic control for communications networks. *Journal of Artificial Intelligence Research*, 9:317–365, 1998.
3. M. Dorigo. *Optimization, Learning and Natural Algorithms* (in Italian). PhD thesis, Dipartimento di Elettronica , Politecnico di Milano, Italy, 1992. pp. 140.
4. M. Dorigo and G. Di Caro. The ant colony optimization meta-heuristic. In D. Corne, M. Dorigo, and F. Glover, editors, *New Ideas in Optimization*, pages 11–32. McGraw-Hill, 1999.
5. M. Dorigo, V. Maniezzo, and A. Colorni. The ant system: Optimization by a colony of cooperating agents. *IEEE Trans. Systems, Man, and Cybernetics – Part B*, 26:29–41, 1996.
6. M. Guntsch and M. Middendorf. Pheromone modification strategies for ant algorithms applied to dynamic TSP. In E. B. et al., editor, *Applications of Evolutionary Computing: Proceedings of EvoWorkshops 2001*, number 2037 in Lecture Notes in Computer Science, pages 213–222. Springer Verlag, 2000.
7. M. Guntsch and M. Middendorf. A population based approach for ACO. In S. C. et al., editor, *Applications of Evolutionary Computing - EvoWorkshops 2002: Evo-COP, EvoIASP, EvoSTIM/EvoPLAN*, number 2279 in Lecture Notes in Computer Science, pages 72–81. Springer Verlag, 2002.
8. M. Guntsch, M. Middendorf, and H. Schmeck. An ant colony optimization approach to dynamic TSP. In L. S. et al., editor, *Proceedings of the Genetic and Evolutionary Computation Conference (GECCO-2001)*, pages 860–867. Morgan Kaufmann Publishers, 2001.
9. J. B. R. Schoonderwoerd, O. Holland and L. Rothkrantz. Ant-based load balancing in telecommunications networks. *Adaptive Behavior*, 5:168–207, 1996.
10. T. Stützle and M. Dorigo. ACO algorithms for the quadratic assignment problem. In D. Corne, M. Dorigo, and F. Glover, editors, *New Ideas in Optimization*, pages 33–50. McGraw-Hill, 1999.
11. http://www.iwr.uni-heidelberg.de/groups/comopt/software/tsplib95.
12. http://www.opt.math.tu-graz.ac.at/qaplib.

Cross-Entropy Guided Ant-Like Agents Finding Cyclic Paths in Scarcely Meshed Networks

Otto Wittner and Bjarne E. Helvik

Department of Telematics, Norwegian University of Science and Technology
7491 Trondheim, Norway
{wittner,bjarne}@item.ntnu.no
http://www.item.ntnu.no/~wittner

Abstract. Finding paths in networks is a well exercised activity both in theory and practice but still remains a challenge when the search domain is a dynamic communication network environment with changing traffic patterns and network topology. To enforce dependability in such network environments new routing techniques are called upon. In this paper we describe a distributed algorithm capable of finding cyclic paths in scarcely meshed networks using ant-like agents. Cyclic paths are especially interesting in the context of protection switching, and scarce meshing is typical in real world telecommunication networks. Two new next-node-selection strategies for the ant-like agents are introduced to better handle low degrees of meshing. Performance results from Monte Carlo Simulations of systems implementing the strategies are presented indicating a promising behavior of the second strategy.

1 Introduction

Finding paths in networks is a well exercised activity both in theory and practice. Still it remains a challenge especially when the search domain is a dynamic communication network environment with changing traffic patterns and network topology. The internet is such an environment, and as an increasing number of applications demanding QoS guarantees is beginning to use internet as their major communication service, efficient and dependable routing in the network becomes more important than ever.

Protection switching [1] is a well known technique for improving dependability in communication networks and commonly used in larger SDH- and ATM-networks. To enable fast recovery from link or network element failures two (or more) disjunct independent paths from source to destination are defined, one primary and one (or more) backup path. Loss of connectivity in the primary path triggers switching of traffic to the backup path. Good dependability is achieved by allocating required resources for the backup path prior to the occurrence of failures in the primary path. However maintaining the necessary mesh of backup paths in a dynamic network with a large number of active sources and destinations is a complex task [2]. Grover & al. [3,4] propose to use simple cyclic paths ("p-cycles") as a means for dependable routing in meshed networks. Protection rings are common in SDH based transport networks and guarantee protection against single link failures in the ring (assuming the ring has duplex links)[5]. All network elements on the ring can continue to communicate with each other

M. Dorigo et al. (Eds.): ANTS 2002, LNCS 2463, pp. 123–134, 2002.
© Springer-Verlag Berlin Heidelberg 2002

after a single link failure by routing all traffic over the "healthy" curve-section of the ring (Figure 1). Thus a cyclic path can provide a dependable communication service for a set

Fig. 1. Protection switching in a ring network.

of sources and destinations. Assuming p-cycles can be found, the number of necessary cycles to be maintained in a network providing a dependable communication service to a set of network elements, is likely to be far less than the number of traditional backup paths required to provide the same service.

In this paper we describe an algorithm, founded in rare event theory and cross entropy, able to find cyclic paths in networks. The fundamentals of the algorithm has previously be published in [6]. This paper enhances the original algorithm by enabling it to find cyclic paths in networks with low degrees of meshing, a common property of real world telecommunication networks. The algorithm is fully distributed with no centralized control which are two desirable properties when dependability is concerned. The algorithm can conveniently be implemented using simple (ant-like) mobile agents [7]. Section 2 introduces the foundations of the original algorithm and motivates the use of it. Section 3 describes the original next-node-selection strategy for the mobile agents as well as two new strategies. Emphasis is put upon search performance when searching for cycles in scarcely meshed networks. Section 4 presents results from Monte Carlo simulation of systems based on the strategies described in section 3. Finally section 5 summarizes, concludes and indicates future work.

2   Agent Behavior Foundations

The concept of using multiple mobile agents with a behavior inspired by foraging ants to solve routing problems in telecommunication networks was introduced by Schoonderwoerd & al. in [8] and further developed in [9,10,11]. Schoonderwoerd & al.'s work again builds on Dorigo & al.'s work on Ant Colony Optimization (ACO) [12]. The overall idea is to have a number of simple ant-like mobile agents search for paths between a given source and destination node. While moving from node to node in a network an agent leaves markings imitating the pheromone left by real ants during ant trail development. This results in nodes holding a distribution of pheromone markings pointing to their different neighbor nodes. An agent visiting a node uses the distribution of pheromone markings to select which node to visit next. A high number of markings pointing towards a node (high pheromone level) implies a high probability for an agent to continue its itinerary toward that node. Using trail marking agents together with a constant evaporation of all pheromone markings, Schoonderwoerd and Dorigo show that after a relatively short period of time the overall process converges towards having the majority of the agents following a single trail. The trail tends to be a near optimal path from the source to the destination.

2.1 The Cross Entropy Method

In [13] Rubinstein develops a search algorithm with similarities to Ant Colony Optimization [12,14]. The collection of pheromone markings is represented by a probability matrix and the agents' search for paths is a Markov Chain selection process generating sample paths in the network. ("Path" and "trail" are equivalent in this paper and will be used interchangeably.)

In a large network with a high number of feasible paths with different qualities, the event of finding an optimal path by doing a random walk (using a uniformly distributed probability matrix) is rare, i.e. the probability of finding the shortest Hamiltonian cyclic path (the Traveling Salesman Problem) in a 26 node network is $\frac{1}{25!} \approx 10^{-26}$. Thus Rubinstein develops his algorithm by founding it in rare event theory.

By importance sampling in multiple iterations Rubinstein alters the transition matrix and amplifies probabilities in Markov chains producing near optimal paths. Cross entropy (CE) is applied to ensure efficient alteration of the matrix. To speed up the process, a performance function weights the path qualities (two stage CE algorithm [15]) such that high quality paths have greater influence on the alteration of the matrix. Rubinstein's CE algorithm has 4 steps:

1. At the first iteration $t = 0$, select a start transition matrix $P_{t=0}$ (e.g. uniformly distributed).
2. Generate N paths from P_t using some selection strategy (i.e. avoid revisiting nodes, see section 3). Calculate the minimum Boltzmann temperature γ_t to fulfill average path performance constraints, i.e.

$$\min \gamma_t \text{ s.t. } h(P_t, \gamma_t) = \frac{1}{N} \sum_{k=1}^{N} H(\pi_k, \gamma_t) > \rho \tag{1}$$

where $H(\pi_k, \gamma_t) = e^{-\frac{L(\pi_k)}{\gamma_t}}$ is the performance function returning the quality of path π_k. $L(\pi_k)$ is the raw cost of path π_k (e.g. delay in a telecommunication network). $10^{-6} \leq \rho \leq 10^{-2}$ is a search focus parameter. The minimum solution for γ_t will result in a certain amplification (controlled by ρ) of high quality paths and a minimum average $h(P_t, \gamma_t) > \rho$ of all path qualities in the current batch of N paths.
3. Using γ_t from step 2 and $H(\pi_k, \gamma_t)$ for $k = 1, 2..., N$, generate a new transition matrix P_{t+1} which maximizes the "closeness" to the optimal matrix, by solving

$$\max_{P_{t+1}} \frac{1}{N} \sum_{k=1}^{N} H(\pi_k, \gamma_t) \sum_{ij \in \pi_k} \ln P_{t,ij} \tag{2}$$

where $P_{t,ij}$ is the transition probability from node i to j at iteration t. The solution of (2) is shown in [13] to be

$$P_{t+1,rs} = \frac{\sum_{k=1}^{N} I(\{r, s\} \in \pi_k) H(\pi_k, \gamma_t)}{\sum_{l=1}^{N} I(\{r\} \in \pi_l) H(\pi_l, \gamma_t)} \tag{3}$$

which will minimize the cross entropy between P_t and P_{t+1} and ensure an optimal shift in probabilities with respect to γ_t and the performance function.

4. Repeat steps 2-3 until $H(\hat{\pi}, \gamma_t) \approx H(\hat{\pi}, \gamma_{t+1})$ where $\hat{\pi}$ is the best path found.

2.2   Distributed Implementation of the Cross Entropy Method

Rubinstein's CE algorithm is centralized, synchronous and batch oriented. All results output from each step of the algorithm must be collected before the next step can be executed. In [6] a distributed and asynchronous version of Rubinstein's CE algorithm is developed. A few approximations let (3) and (1) be replaced by the autoregressive counterparts

$$P_{t+1,rs} = \frac{\sum_{k=1}^{t} I(\{r, s\} \in \pi_k) \beta^{t-k} H(\pi_k, \gamma_k)}{\sum_{l=1}^{t} I(\{r\} \in \pi_l) \beta^{t-k} H(\pi_l, \gamma_l)} \tag{4}$$

and

$$\min \gamma_t \text{ s.t. } h'_t(\gamma_t) > \rho \tag{5}$$

respectively where

$$h'_t(\gamma_t) = h'_{t-1}(\gamma_t)\beta + (1 - \beta)H(\pi_t, \gamma_t)$$

$$\approx \frac{1 - \beta}{1 - \beta^t} \sum_{k=1}^{t} \beta^{t-k} H(\pi_t, \gamma_t)$$

and $\beta < 1$, step 2 and 3 can immediately be performed when a single new path π_t is found and a new probability matrix P_{t+1} can be generated.

The distributed CE algorithm may be viewed as an algorithm where search agents evaluate a path found (and calculate γ_t by (5)) right after they reach their destination node and then immediately return to their source node backtracking along the path. During backtracking relevant probabilities in the transition matrix are updated by applying $H(\pi_t, \gamma_t)$ through (4).

The distributed CE algorithm resembles Schoonderwoerd & al.'s original system. However Schoonderwoerd's ants update probabilities during their forward search. Dorigo & al. realized early in their work on ACO that compared to other updating schemes, updating while backtracking results in significantly quicker convergence towards high quality paths. Dorigo & al.'s AntNet system [9] implements updating while backtracking, thus is more similar to the distributed CE algorithm than Schoonderwoerd & al.'s system. However none of the earlier systems implements a search focus stage (the adjustment of γ_t) as in the CE algorithms.

2.3 P-Cycles, Hamiltonian Cyclic Paths, and CE Algorithms

Grover's "p-cycles" [3] provide protection against a single link failure on any link connecting the nodes which are on the path defined by the p-cycle. This includes both on-cycle links (links traversed by the path) as well as straddling links (links not traversed but having their end nodes on the path). Intuitively a Hamiltonian cyclic path, which by definition visits all nodes once in a network, would provide a cycle potentially able to protect against any single link failure. This is also argued in [16].

The CE algorithms from both [13] and [6] show good performance when tested on optimal Hamiltonian cyclic path search problems as long as the network environment is fully meshed (all nodes have direct duplex connections). Real world telecommunication networks are seldom fully meshed. An average node degree much lager than 5 is uncommon. Finding a single Hamiltonian cyclic path in a large network with such scarce meshing can itself be considered a rare event.

In the section 3.1 we describe the selection strategy (used in CE algorithm step 2) implemented in the original CE algorithms (both [13] and [6]). They strategy struggles to find Hamiltonian cyclic paths in our 26 node test network shown in Figure 2. In section 3.2 and 3.3 we suggest two new selection strategies intended to better cope with a network topology with scarce meshing.

3 Selection Strategies

3.1 Markov Chain without Replacement

The CE algorithms in [13] and [6] implement a strict next-hop selection strategy termed *Markov Chain Without Replacement* (MCWR) in [13]. No nodes are allowed to be revisited, except for the home node when completing a Hamiltonian cyclic path.

Let

$$X_{t,r}^i(s) = I(s \notin \mathbf{V}_t^i \vee ((\mathbf{G}_{t,r} \subseteq \mathbf{V}_t^i) \wedge s = hn^i))$$

where $I(\ldots)$ is the indicator function, $\mathbf{V}_t^i$ is agent i's list of already visited nodes, $\mathbf{G}_{t,r}$ is the set of neighbor nodes to node r and hn^i is agent i's home node. Thus $X_{t,r}^i(s)$ is 1 if node s has not already been visited by agent i, or if all neighbor nodes of r have been visited by agent i and s is agent i's home node.

When MCWR is applied $P_{t,rs}$ from (4) is weighted by $X_{t,r}^i(s)$ and renormalized giving a new next-hop probability distribution

$$Q_{t,rs}^i = \frac{[I(t > D)P_{t,rs}(1 - \epsilon) + \epsilon] X_{t,r}^i(s)}{\sum_{\forall k}[I(t > D)P_{t,rs}(1 - \epsilon) + \epsilon] X_{t,r}^i(k)}$$

where D is the number of path samples required to be found to complete the initialization phase of the system (step 1). The random noise factor ϵ is set to a small value, e.g. 10^{-60}. During the initialization phase agents are forced to explore since the next-hop probability vector $\mathbf{Q}_{t,r}^i$ will have a uniform distribution over the qualified ($X_{t,r}^i(s) = 1$) neighbor nodes. See [6] for more details about the initialization phase.

If $\sum_{s \in \mathbf{G}_{t,r}} X_{t,r}^i(s) = 0$, agent i has reach a dead end and in the MCWR strategy it is terminated. When the event of finding a Hamiltonian cyclic path is rare due to scarce meshing in a network, most agents will reach such dead ends. Thus only a few "lucky" agent will be able to contribute with a path in step 2 of the CE algorithm. This will slow down the search process significantly since CE algorithms require a "smooth" search space, i.e. many suboptimal solutions should exist in addition to the optimal solutions.

3.2 Markov Chain Depth First

Instead of immediately terminating agents when dead ends are reached a "retry mechanism" can be implemented. We have tested what we call the *Markov Chain Depth First* (MCDF) strategy which allows agents to backtrack and retry searching. An MCDF-agent performs a depth first search [17] from its home node, i.e. it tries to visit nodes in such an order that when a dead end is met (a leaf node is found) all nodes have been visited only once and the home node is a neighbor of the leaf node. If a dead end is reached and either all nodes has not been visited or the home node is not a neighbor node, the agent backtracks along its path one step before continuing the search.

Let

$$X_{t,r}^{i*}(s) = I(s \notin (\mathbf{V}_t^i \cup \mathbf{D}_{t,r}^i) \vee ((\mathbf{G}_{t,r} \subseteq \mathbf{V}_t^i) \wedge s = hn^i))$$

where $\mathbf{D}_{t,r}^i$ is the set of neighbor nodes of r leading to dead ends for agent i. Thus $X_{t,r}^{i*}(s)$ is 1 if node s has not already been visited by agent i and s does not lead to a dead end, or (as for $X_{t,r}^i(s)$) if all neighbor nodes of r have been visited by agent i and s is agent i's home node.

All $\mathbf{D}_{t,r}^i$ (for $\forall_r$) are stored in a stack managed by agent i. When a fresh next node r_{+1} is chosen $\mathbf{D}_{t,r_{+1}}^i \equiv \emptyset$ is pushed onto the stack. If a dead end is reached at node r_{+1} agent i backtracks to the previously visited node r, removes (pops) $\mathbf{D}_{t,r_{+1}}^i$ from the stack and adds r_{+1} to $\mathbf{D}_{t,r}^i$ (which is now on the top of the stack).

When MCDF is applied $P_{t,rs}$ is weighted by $X_{t,r}^{i*}(s)$ in the same way $X_{t,r}^i(s)$ is for MCWR. Results in section 4 show simulation scenarios for MCDF-agents both with unlimited and limited backtracking. Unlimited backtracking implies never terminating agents but letting them search (in depth first fashion) until they find Hamiltonian cyclic paths. Limited backtracking implements a quota of backtracking steps in each agent, i.e. a certain no of "second chances" or "retries" are allow for an agent before termination.

A method with similarities to MCDF is presented in [18].

3.3 Markov Chain with Restricted Replacement

By relaxing the agent termination condition even more, we arrive at what we call the *Markov Chain with Restricted Replacement* (MCRR) strategy. The fundamental difference between this strategy and both MCWR and MCDF is a less strict condition concerning revisits to nodes. Revisits are simply allowed, but only when dead ends are reached. To ensure completion of cycles the home node is given priority when a dead end is reached and the home node is a neighbor node.

Let

$$X_{t,r}^{i**}(s) = \overline{I}(((\mathbf{G}_{t,r} \not\subseteq \mathbf{V}_t^i) \wedge s \in \mathbf{V}_t^i) \vee$$
$$((\mathbf{G}_{t,r} \subseteq \mathbf{V}_t^i) \wedge hn^i \in \mathbf{G}_{t,r} \wedge s \neq hn^i))$$

where $\overline{I}(\ldots)$ is the inverse indicator function. Thus $X_{t,r}^{i*}(s)$ is zero if an unvisited neighbor node to r exists and s has already been visited, or if all neighbor nodes have been visited and the home node is a neighbor node but s is not the home node. As for MCWR and MCDF $X_{t,r}^{i**}(s)$ weights $P_{t,rs}$.

In our simulations we consider only paths found by MCRR-agents which have visited all nodes when they return to their home node. Several of these agents will find closed acyclic paths (with loops), i.e. none Hamiltonian cyclic paths. The search space is now "smoother" and a range of suboptimal solutions exists (most none Hamiltonian). This enables step 2 in the CE algorithm to be executed with close to the same efficiently as for a fully meshed network.

However when using MCRR-agents there is no longer guaranteed that the best path found when the system converges is a Hamiltonian cyclic path. Since the agents visit all nodes, the length of acyclic closed paths where nodes have been revisited, are likely to be longer than Hamiltonian cyclic paths. Thus finding the shortest Hamiltonian cyclic path (minimization) may still be achievable.

We realize that the above statement does not hold in general since a network topology may be constructed having its shortest Hamiltonian cyclic path longer than one or more closed acyclic paths visiting all nodes. However in the context of p-cycle design closed acyclic paths may still provide protection against single link failures.

Results in section 4 are promising. MCRR outperforms both MCWR and MCDF when it comes to speed of convergence, and do in all simulation scenarios converge to a Hamiltonian cyclic path.

4   Strategy Performance

As for the simulation scenarios in [6] we have used an active network enabled version of the *Network Simulator* version 2 [19] to test the three selection strategies. Our test network topology is show in Figure 2, a 26 node network with an average number of outgoing links (degree) per node equal to 5. The low average degree implies existence of far less Hamiltonian cyclic paths compared to a fully meshed 26 node network with number of Hamiltonian cyclic paths equal to $25! \approx 10^{26}$. The network topology was generated by the *Tier 1.1* topology generator [20] with the parameter vector "1 0 0 26 0 0 9 1 1 1 1 9".

All scenarios had equal parameter settings (except for different selection strategies): $D = 19$, $\mu = 26$, $\beta = 0.998$, $\rho = 0.01$, $\rho^* = 0.95$, where μ is the number of agents operating concurrently and ρ^* the ρ reduction factor. See [6] for further descriptions of the parameters.

Table 1 and 2 compare results form simulation scenarios for the different selection strategies. Results shown are values recorded after 100, 1500 and 10 000 seconds of simulation time. Time has been chosen as the scale of progress rather than number of iterations since the time spent per iteration (per search) vary significantly between the different strategies. Simulation time is expected to be approximately proportional to real time in a network.

Columns 2, 3 and 4 in Table 1 show the total number of Hamiltonian cyclic paths found including re-discoveries. Values are averaged over 10 simulations and reported

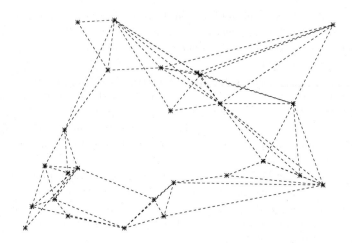

Fig. 2. Network topology used in simulation scenarios. The topology is generated by the *Tiers 1.1* topology generator [20].

Table 1. Number of paths found in simulation scenarios using the three different selection strategies. Values are averaged over 10 simulations and reported with standard deviations (prefixed by ±). Numbers in bracket are the number of agents which have finished their initialization phase.

Test	No of Hamiltonian cyclic paths found			No of none Hamiltonian cyclic paths		
scenarios	*100s*	*1500s*	*10 000s*	*100s*	*1500s*	*10 000s*
MCWR	1.0 ±0 (0)	16.0 ±2.3 (0)	109 ±8.7 (0)	16.0e3 ±52.1	246e3±213	1.6e6±533
MCDF unlimited	0.0±0 (0)	1.0 ±0.0 (0)	2.57 ±1.3 (0)	201±30.6	1.5e3±79.1	9.3e3±190
MCDF quota=10	1.0 ±0 (0)	14.0 ±4.5 (0)	85.0 ±12.4 (0)	8.8e3±24.3	134e3±156	901e3±345
MCDF quota=5	2.0 ±0.9 (0)	16.0 ±4.4 (0)	95.0 ±10.8 (0)	10.1e3±34.1	155e3±151	1.0e6±402
MCDF quota=2	1.0 ±0 (0)	14.0 ±3.7 (0)	99.0 ±12.1 (0)	11.5e3±40.9	176e3±126	1.2e6±316
MCDF quota=1	2.0 ±1.0 (0)	16.0 ±5.5 (0)	105±12.9 (0)	12.3±35.5	188e3±101	1.3e6±367
MCRR	7.0 ±2.8 (26)	**52 800±4620 (26)**	(converged)	4.9e3±56.0	**15.7e3±1580**	(converged)

Table 2. Quality of paths found in simulation scenarios using the three different selection strategies. Values are based on 10 simulation. Standard deviations are prefixed by ± and the worst of the best values are given in brackets.

Test	Best path found (Worst of best paths found)			Average path cost		
scenarios	*100s*	*1500s*	*10 000s*	*100s*	*1500s*	*10 000s*
MCWR	0.199 (0.251)	0.197 (0.214)	0.193 (0.207)	0.231±0.015	0.232±0.004	0.231±0.004
MCDF unlimited	(no paths found)	0.250 (3.273)	0.215 (0.750)	(no paths found)	0.923±1.023	254.5±670.5
MCDF quota=10	0.202 (0.371)	0.202 (0.222)	0.194 (0.207)	0.260±0.058	0.254±0.008	0.254±0.008
MCDF quota=5	0.201 (0.240)	0.201 (0.221)	0.196 (0.209)	0.228±0.014	0.240±0.011	0.243±0.008
MCDF quota=2	0.198 (0.293)	0.194 (0.224)	0.194 (0.206)	0.244±0.028	0.241±0.006	0.241±0.005
MCDF quota=1	0.204 (0.258)	0.201 (0.210)	0.193 (0.204)	0.231±0.016	0.235±0.003	0.238±0.003
MCRR	0.203 (0.230)	**0.194 (0.202)**	(converged)	0.514±0.037	**0.206±0.015**	(converged)

with standard deviations (prefixed by $\pm$). Numbers in bracket are the number of agents which have finished their initialization phase, i.e. changed search behavior from doing random walk guided only by a selection strategy to being guided both by a selection strategy and cross entropy adjusted probabilities (pheromones). Column 5, 6 and 7 show the total number of none Hamiltonian cyclic paths found, including dead ends and paths not visiting all nodes.

Column 2, 3 and 4 in Table 2 show the best of the best paths found (lowest cost) in 10 simulations with the worst of the best in brackets. And finally columns 5, 6 and 7 show path cost averaged over 10 simulations and reported with standard deviations (prefixed by $\pm$).

4.1   Markov Chain without Replacement

As expected in the MCWR scenario few Hamiltonian cyclic paths are found even after 1500 simulation seconds. The scarce meshing in the test network results in many agent reaching dead ends. The fraction of feasible to infeasible paths found is as low as $6.5 \cdot 10^{-5}$ after 1500 seconds. The paths found are of relatively good quality, but still after 10 000 seconds none of the agents have managed to collect enough paths samples to finish their initialization phases.

These results indicate the need for a different selection strategy when searching for Hamiltonian cyclic paths in networks with scarcely meshed topologies.

4.2   Markov Chain Depth First

The results for the MCDF scenarios are not promising. When unlimited backtracking is enabled even fewer path are found than for the MCWR scenario. The path quality is low because the total search time (including backtracking) is registered as path cost. This is not surprising since unlimited backtracking implies that every agent is doing an exhaustive search for Hamiltonian cyclic paths. The very reason for introducing stochastic search techniques in the first place is to avoid the need for such exhaustive searches.

When the MCDF strategy is limited by quotas of 10, 5, 2 or 1 retry the performance improves compared to unlimited backtracking but is not better than the original MCWR strategy. Also with this strategy not enough valid paths have been found after 10 000 seconds to enable any agents to complete their initialization phase. Best and average path costs are similar to the values for MCWR and no convergence is observed due to the overall low number of valid paths found.

4.3   Markov Chain with Restricted Replacement

The last row in Table 1 and 2 which presents results from MCRR scenarios, stand out from the other results. Already after 100 simulation seconds all 26 agents have completed their initialization phases, and after 1500 seconds the average path cost has converged to a value close to the best path found (see bold values). Since full convergence is experienced already around 1500 seconds, no values are given for 10 000 seconds. For all 10 simulations the best path found is a true Hamiltonian cyclic path. Figure 3 show how the ratio of Hamiltonian to none Hamiltonian cyclic paths found increases during the search process. Not long after 800 seconds most agents start finding Hamiltonian cyclic paths.

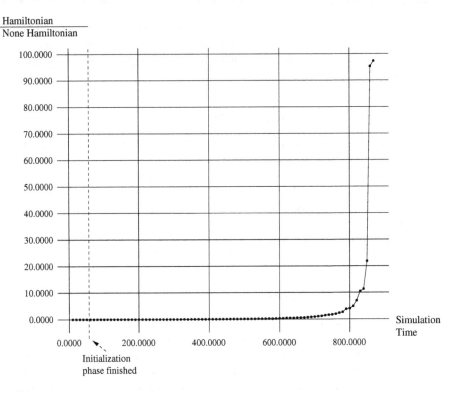

Fig. 3. Ratio of Hamiltonian to none Hamiltonian cyclic paths found when using the MCRR selection strategy.

We believe these results indicates the value of using infeasible paths (none Hamiltonian cyclic paths) as intermediate solutions during the iterative search process of the CE algorithm. The question of accepting or rejecting infeasible solutions is well known to designers of evolutionary systems [21,22]. When feasible solutions are located far from each other in the search domain a significant speedup in the search process can be achieved by letting infeasible solutions act as "stepping stones" between suboptimal and optimal feasible solutions.

5   Concluding Remarks

As an increasing number of applications demanding QoS guarantees are accessing internet, dependable routing is becoming more important than ever. This paper examines and enhances a distributed cross entropy founded algorithm designed to find cyclic paths (Hamiltonian cyclic paths) in networks. Such paths are argued to be good candidate paths when protection switching is to be implemented in meshed networks using p-cycles [16]. The algorithm presented is well suited for distributed implementation, for instance using mobile agents or active networks technology.

Previous versions of the algorithm [13,6] struggle to find Hamiltonian cyclic paths in scarcely meshed network, very much because of the strict selection strategy (Markov Chain without Replacement) in operations during the search process. In this paper we compare the original selection strategy with two new strategies. The first new strategy, *Markov Chain Depth First*, proves to be as inefficient as the original, while the second, *Markov Chain with Restricted Replacement*, outperforms both the other strategies. However, using the second strategy convergence towards feasible solutions (i.e. Hamiltonian cyclic paths) is not guaranteed. Even so results from simulation scenarios indicate a high probability of converging towards near optimal Hamiltonian cyclic paths.

More excessive parameter tuning of the algorithm is required. Also the possibility of including heuristics to speed up the convergence process even more should be investigated.

Currently a new version of the algorithm is under development which allows several species of agents compete in finding quality paths in a network. By this a set of cyclic paths providing a overall high quality p-cycle design may potentially be found.

References

1. S. Aidarous and T. Plevyak, ed., *Telecommunications Network Management into the 21st Century.* IEEE Press, 1994.
2. Otto Wittner and Bjarne E. Helvik, "Cross Entropy Guided Ant-like Agents Finding Dependable Primary/Backup Path Patterns in Networks," in *Proceedings of Congress on Evolutionary Computation (CEC2002)*, (Honolulu, Hawaii), IEEE, May 12-17th 2002.
3. W.D. Grover and D. Stamatelakis, "Cycle-oriented distributed preconfiguration: ring-like speed with mesh-like capacity for self-planning network restoration," in *Proceedings of IEEE International Conference on Communications*, vol. 1, pp. 537 –543, 7-11 June 1998.
4. D. Stamatelakis and W.D. Grover, "Rapid Span or Node Restoration in IP Networks using Virtual Protection Cycles," in *Proceedings of 3rd Canadian Conferance on Broadband Research (CCBR'99)*, (Ottawa), 7 November 1999.
5. M. Decina and T. Plevyak (editors), "Special Issue: Self-Healing Networks for SDH and ATM," *IEEE Communications Magazine*, vol. 33, September 1995.
6. Bjarne E. Helvik and Otto Wittner, "Using the Cross Entropy Method to Guide/Govern Mobile Agent's Path Finding in Networks," in *Proceedings of 3rd International Workshop on Mobile Agents for Telecommunication Applications*, Springer Verlag, August 14-16 2001.
7. Vu Anh Pham and A. Karmouch, "Mobile Software Agents: An Overview," *IEEE Communications Magazine*, vol. 36, pp. 26–37, July 1998.
8. R. Schoonderwoerd, O. Holland, J. Bruten, and L. Rothkrantz, "Ant-based Load Balancing in Telecommunications Networks," *Adaptive Behavior*, vol. 5, no. 2, pp. 169–207, 1997.
9. G. D. Caro and M. Dorigo, "AntNet: Distributed Stigmergetic Control for Communications Networks," *Journal of Artificial Intelligence Research*, vol. 9, pp. 317–365, Dec 1998.
10. T. W. B. P. F. Oppacher, "Connection Management using Adaptive Mobile Agents," in *Proceedings of 1998 International Conference on Parallel and Distributed Processing Techniques and Applications (PDAPTA'98)*, 1998.

11. J. Schuringa, "Packet Routing with Genetically Programmed Mobile Agents," in *Proceedings of SmartNet 2000*, (Wienna), September 2000.

12. Marco Dorigo and Gianni Di Caro, "Ant Algorithms for Discrete Optimization," *Artificial Life*, vol. 5, no. 3, pp. 137–172, 1999.

13. Reuven Y. Rubinstein, "The Cross-Entropy Method for Combinatorial and Continuous Optimization," *Methodology and Computing in Applied Probability*, pp. 127–190, 1999.

14. M. Zlochin, M. Birattari, N. Meuleau, and M. Dorigo, "Model-based Search for Combinatorial Optimization," IRIDIA IRIDIA/2001-15, Universite Libre de Bruxelles, Belgium, 2000.

15. Reuven Y. Rubinstein, "The Cross-Entropy and Rare Events for Maximum Cut and Bipartition Problems - Section 4.4," *Transactions on Modeling and Computer Simulation*, To appear.

16. D. Stamatelakis and W.D. Grover, "Theoretical Underpinnings for the Efficiency of Restorable Networks Using Preconfigured Cycles ("p-cycles")," *IEEE Transactions on Communications*, vol. 48, pp. 1262–1265, August 2000.

17. Kenneth A. Ross and Charles R.B. Wright, *Discrete Mathematics*. Prentice Hall, 2nd ed., 1988.

18. V. Maniezzo, "Exact and Approximate Nondeterministic Tree-Search Procedures for the Quadratic Assignment Problem," *INFORMS Journal on Computing*, vol. 11, no. 4, pp. 358–369, 1999.

19. DARPA: VINT project, "UCB/LBNL/VINT Network Simulator - ns (version 2)." http://www.isi.edu/nsnam/ns/.

20. Calvert, K.I. and Doar, M.B. and Zegura, E.W. , "Modeling Internet Topology ," *IEEE Communications Magazine*, vol. 35, pp. 160 –163, June 1997.

21. Goldberg, D., *Genetic Algorithms in Search, Optimization and MachineLearn ing*. Addison Wesley, 1998.

22. Z. Michalewicz, *Genetic algorithms + Data Stuctures = Evolution Programs*. Springer Verlag, second ed., 1996.

Insertion Based Ants for Vehicle Routing Problems with Backhauls and Time Windows

Marc Reimann, Karl Doerner, and Richard F. Hartl

Institute of Management Science, University of Vienna
Brünnerstrasse 72, A-1210 Vienna, Austria
{marc.reimann, karl.doerner, richard.hartl}@univie.ac.at
http://www.bwl.univie.ac.at/bwl/prod/index.html

Abstract. In this paper we present and analyze the application of an Ant System to the Vehicle Routing Problem with Backhauls and Time Windows (VRPBTW). At the core of the algorithm we use an Insertion procedure to construct solutions. We provide results on the learning and runtime behavior of the algorithm as well as a comparison with a custom made heuristic for the problem.

1 Introduction

Since their invention in the early 1990s by Colorni et al. (see e.g. [1]), Ant Systems have received increasing attention by researchers, leading to a wide range of applications such as the Graph Coloring Problem ([2]), the Quadratic Assignment Problem (e.g. [3]), the Travelling Salesman Problem (e.g. [4], [5]), the Vehicle Routing Problem ([6], [7]) and the Vehicle Routing Problem with Time Windows ([8]). Recently, a convergence proof for a generalized Ant System has been developed by Gutjahr ([9]).

The Ant System approach is based on the behavior of real ants searching for food. Real ants communicate with each other using an aromatic essence called pheromone, which they leave on the paths they traverse. In the absence of pheromone trails ants more or less perform a random walk. However, as soon as they sense a pheromone trail on a path in their vicinity, they are likely to follow that path, thus reinforcing this trail. More specifically, if ants at some point sense more than one pheromone trail, they will choose one of these trails with a probability related to the strenghts of the existing trails. This idea has first been applied to the TSP, where an ant located in a city chooses the next city according to the strength of the artificial trails

This leads to a construction process that resembles the Nearest Neighbor heuristic, which makes sense for the TSP. However, most of the applications of Ant Systems to other problems, also used this constructive mechanism. In order to be able to do so, the problem at hand had to be transformed into a TSP first. By doing so, structural characteristics of the problem, may disappear, thus leading to poor solutions. Moreover, for many of the problems solved with Ant Systems so far, problem specific constructive algorithms exist that exploit these structural characteristics. For example, for the classic VRP without side

M. Dorigo et al. (Eds.): ANTS 2002, LNCS 2463, pp. 135–148, 2002.

constraints, we have shown in ([7]) that the incorporation of a powerful problem specific heuristic algorithm significantly improves the performance and makes the Ant System competitive to other state-of-the-art methods, such as Tabu Search.

Building on these results, in this paper we propose an Ant System, where the constructive heuristic is a sophisticated Insertion algorithm. We apply our approach to the Vehicle Routing Problem with Backhauls and Time Windows (VRPBTW), a problem with high practical relevance. We present some preliminary results, that suggest the potential of our approach. To our best knowledge, we are not aware of existing works that deal with the application of an Ant System to the VRPBTW. The same applies to the incorporation of an Insertion algorithm within Ant Systems. These two points constitute the main contribution of our paper. While revising this paper we became aware of work done by Le Louarn et al. ([10]). In their paper, the authors use their GENI heuristic (proposed earlier in [11]) at the core of an ACO algorithm and show the potential of this approach for the TSP.

In the next section we describe the VRPBTW and review the existing literature. Section 3 deals with the Ant System algorithm, and the details of the incorporation of the Insertion algorithm. The numerical analysis in Section 4 focuses on the learning behavior of the Ant System, the effects of backhauls and the general performance of our Ant System when compared with a custom-made heuristic for the problem. Finally, we conclude in Section 5 with a summary of the paper and an outlook on future research.

2   Vehicle Routing Problems with Backhauls and Time Windows

Efficient distribution of goods is a main issue in most supply chains. The transportation process between members of the chain can be modeled as a Vehicle Routing Problem with Backhauls and Time Windows (VRPBTW). For example, the distribution of mineral water from a producer to a retailer (linehauls) may be coupled with the distribution of empty recyclable bottles from the retailer to the producer (backhauls). Both linehauls and backhauls may be constrained by possible service times at the producer and the retailers.

More formally, the VRPBTW involves the design of a set of pickup and delivery routes, originating and terminating at a depot, which services a set of customers. Each customer must be supplied exactly once by one vehicle route during her service time interval. The total demand of any route must not exceed the vehicle capacity. The total length of any route must not exceed a pre-specified bound. Additionally, it is required that, on each route, all linehauls have to be performed before all backhauls. The intuition for that is, that rearranging goods en route is costly and inefficient. The objective is to minimize the fleet size, and given a fleet size, to minimize operating costs. This problem is a generalization of the VRP, which is known to be NP-hard (cf. [12]), such that exact methods like Branch&Bound work only for relatively small problems in reasonable time.

While the VRP has received much attention from researchers in the last four decades (for surveys see [13]), the more constrained variants have only recently attracted scientific attention. The Vehicle Routing Problem with Time Windows (VRPTW) has been studied extensively in the last decade; for a recent overview of metaheuristic approaches see (e.g. [14]). The same applies for the Vehicle Routing Problem with Backhauls (VRPB, see e.g. [15]).

The VRPBTW, which combines the issues addressed separately in the works cited above, has received only very little attention. Gelinas et al.([16]) have extended a Branch&Bound algorithm developed for the VRPTW to cope with backhauling. They proposed a set of benchmark sets based on instances proposed earlier for the VRPTW, and solved problems with up to 100 customers to optimality. Their objective was to minimize travel times only. Simple construction and improvement algorithms have been proposed by Thangiah et al. ([17]), while Duhamel et al.([18]) have proposed a Tabu Search algorithm to tackle the problem. While the approach of Thangiah et al.'s considered the same objective as we do in this study, namely to minimize fleet sizes first, and then to minimize travel times as a second goal, Duhamel et al. consider the minimization of schedule times (which in addition to travel times include service times and waiting times) as the second goal.

3   Ant System Algorithms for the VRPBTW

In this section we describe our Ant System algorithm and particularly focus on the constructive heuristic used, as the basic structure of our Ant System algorithm is identical to the one proposed in Bullnheimer et al.([6]) and used in Reimann et al.([7]). The Ant System algorithm mainly consists of the iteration of three steps:

- Generation of solutions by ants according to private and pheromone information
- Application of a local search to the ants' solutions
- Update of the pheromone information

The implementation of these three steps is described below.

3.1   Generation of Solutions

As stated above, the incorporation of the Insertion algorithm as the solution generation technique within the Ant System is the main contribution of this paper. So far, in most Ant Systems solutions have been built using a Nearest Neighbor heuristic (see e.g. [6]). In Reimann et al. ([7]) we have shown for the VRP the merit of incorporating a powerful problem specific algorithm. However, the Savings algorithm used there does not perform very well for problems with time windows and/or backhauls such that we rather use a different constructive heuristic. The Insertion algorithm used in our current approach for the VRPBTW is derived from the I1 insertion algorithm proposed by Solomon ([19])

for the VRPTW. Solomon tested many different route construction algorithms and found that the I1 heuristic provided the best results.

This algorithm works as follows: Routes are constructed one by one. First, a seed customer is selected for the current route, that is, only this customer is served by the route. Sequentially other customers are inserted into this route until no more insertions are feasible with respect to time window, capacity or tour length constraints. At this point, another route is initialized with a seed customer and the insertion procedure is repeated with the remaining unrouted customers. The whole algorithm stops when all customers are assigned to routes.

In the above mentioned procedure two types of decisions have to be taken. First, a seed customer has to be determined for each route, second the attractiveness of inserting a customer into the route has to be calculated. These decisions are based on the following criteria, where we will refer to the attractiveness of a customer i as η_i:

Route initialization:

$$\eta_i = d_{0i} \qquad \forall i \in N_u,$$

where d_{0i} denotes the distance between the depot and customer i and N_u denotes the set of unrouted customers. This route initialization prefers seed customers that are far from the depot.

Customer insertion:

$$\eta_i = \max\{0, \max_{j \in R_{l_i}} [\alpha \cdot d_{0i} - \beta \cdot (d_{ji} + d_{is_j} - d_{js_j}) - (1 - \beta) \cdot (b^i_{s_j} - b_{s_j})]\}$$
$$\forall i \in N_u,$$

where s_j is the customer visited immediately after customer j in the current solution, $b^i_{s_j}$ is the actual arrival time at customer s_j, if i is inserted between customers j and s_j, while b_{s_j} is the arrival time at customer s_j before the insertion of customer i and R_{l_i} denotes the set of customers assigned to the current tour after which customer i could feasibly be inserted. Thus, this rule not only considers the detour that occurs if customer i is inserted but also the delay in service at the customer s_j to be served immediately after i. These two effects are weighted with the parameter β and compared with customer $i's$ distance to the depot, which is weighted with the parameter α. A customer being located far from the depot, that causes little detour and little delay is more likely to be chosen than a customer close to the depot and causing detour or delay.

Given these values, the procedure for each unrouted customer determines the best feasible insertion position. Afterwards, given these values η_i, customer i^* is chosen such that $\eta_{i^*} \geq \eta_i \forall i \in N_u$.

Note that, for each customer we have to check the feasibility of inserting it at any position in the current tour. While this is in principle quite tedious, in particular if tours contain a large number of customers, it can be done quite efficiently. Following Solomon we calculate for each position in the current tour the maximum possible delay at that position, that will ensure feasibility of the subour starting at that position. This calculation of the maximum possible delay has to be performed after each change to the current tour. However, by doing so, we then only have to check if the insertion of a customer at a certain position causes less delay than the maximum possible delay, in which case the insertion is feasible. [1]

To account for the fact that we have to deal with backhauls, we augment these attractiveness values in the following way:

$$\eta_i = \max\ \{0,$$
$$\max_{j \in R_{l_i}} [\alpha \cdot d_{0i} - \beta \cdot (d_{ji} + d_{is_j} - d_{js_j}) - (1 - \beta) \cdot (b^i_{s_j} - b_{s_j}) + \gamma \cdot type_i]\}$$
$$\forall i \in N_u,$$

where $type_i$ is a binary indicator variable denoting whether customer i is a linehaul ($type_i = 0$) or a backhaul customer ($type_i = 1$). The intuition is that we want to discriminate between linehaul and backhaul customers in some way.

In order to use the algorithm described above within the framework of our Ant System we need to adapt it to allow for a probabilistic choice in each decision step. This is done in the following way. First, the attractiveness for inserting each unrouted customer at its best insertion on the current tour is calculated according to the following function:

$$\eta_i = \max\ \{0,$$
$$\max_{j \in R_{l_i}} [(\alpha \cdot d_{0i} - \beta \cdot (d_{ji} + d_{is_j} - d_{js_j}) - (1-\beta) \cdot (b^i_{s_j} - b_{s_j}) + \gamma \cdot type_i) \cdot \frac{\tau_{ji} + \tau_{is_j}}{2 \cdot \tau_{js_j}}]\}$$
$$\forall i \in N_u,$$

where τ_{ji} denotes the pheromone concentration on the arc connecting locations (customers or depot) j and i. The pheromone concentration τ_{ji} contains information about how good visiting two customers i and j immediately after each other was in previous iterations. The way we use the pheromone emphasizes the effect of giving up an arc (the arc between customers j and s_j in the example above) and adding two other arcs (the arcs between customers j and i and customers i and s_j in the example above). In particular, the term $\frac{\tau_{ji} + \tau_{is_j}}{2 \cdot \tau_{js_j}}$ is larger than 1, if the average pheromone value of the arcs to be added exceeds the pheromone value of the arc to be deleted. Note, that the same pheromone utilization is done for route initialization, thus augmenting the attractiveness of initializing a route with an unrouted customer i by the search-historic information.

Then, we apply a roulette wheel selection to all unrouted customers with positive attractiveness values η_i. The decision rule used can be written as

---

[1] This procedure and the two rules for route initialization and customer insertion have been proposed by Solomon ([19]) for the VRPTW.

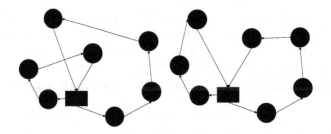

Fig. 1. An example of the application of the *Swap* operator

$$\mathcal{P}_i = \begin{cases} \dfrac{\eta_i}{\sum_{h|\eta_h>0} \eta_h} & \text{if } \eta_i > 0 \\ \\ 0 & \text{otherwise.} \end{cases} \tag{1}$$

The chosen customer i is then inserted into the current route at its best feasible insertion position.

3.2   Local Search

After an ant has constructed its solution, we apply a local search algorithm to improve the solution quality. In particular, we apply *Swap* and *Move* operators to the solution. The *Swap* operator, aims at improving the solution by exchanging a customer i with a customer j. This operator is a special case of the 4-opt operator, where the four arcs deleted are in pairs adjacent. An example of the application of this operator is given in Figure 1, where customers 3 and 4 are exchanged.

The *Move* operator tries to eject a customer i from its current position and insert it at another position. It is a special case of the 3-opt operator, where two of the three arcs deleted are adjacent. This operator is exemplified in Figure 2, where customer 5 is moved from one route to another.

Both operators have been proposed by Osman ([20]). We apply these operators until no more improvements are possible. More specifically, we first apply

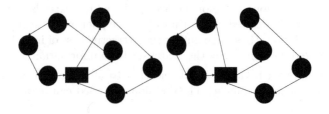

Fig. 2. An example of the application of the *Move* operator

Move and then *Swap* operators. Note, that we do not accept infeasible solutions. While Osman ([20]) proposed a more general version of these operators, where λ adjacent customers can be moved or swapped, we restrict our local search to the case where $\lambda = 1$. The reason for this is, that the operators were proposed for the classic VRP without time window and backhauling constraints. Given these additional constraints, most possible operations with $\lambda > 1$ lead to infeasible solutions. Thus, the additional computation effort to perform these more complex operations, will in general not be justified.

3.3   Pheromone Update

After all ants have constructed their solutions, the pheromone trails are updated on the basis of the solutions found by the ants. According to the rank based scheme proposed in ([5]) and ([6]), the pheromone update is as follows

$$\tau_{ij} := \rho\tau_{ij} + \sum_{\mu=1}^{p} \Delta\tau_{ij}^{\mu} + \sigma\Delta\tau_{ij}^{*} \tag{2}$$

where $0 \leq \rho \leq 1$ is the trail persistance and $\sigma = p+1$ is the number of elitists. Using this scheme two kinds of trails are laid. First, the best solution found during the process is updated as if σ ants had traversed it. The amount of pheromone laid by these elitists is $\Delta\tau_{ij}^{*} = 1/L^{*}$, where L^{*} is the objective value of the best solution found so far. Second, the p best ants of the iteration are allowed to lay pheromone on the arcs they traversed. The quantity laid by these ants depends on their rank μ as well as their solution quality L^{μ}, such that the μ-th best ant lays $\Delta\tau_{ij}^{\mu} = (p - \mu + 1)/L^{\mu}$. Arcs belonging to neither of those solutions just lose pheromone at the rate $(1 - \rho)$, which constitutes the trail evaporation.

4   Numerical Analysis

In this section we turn to the numerical analysis of our proposed approach. First we will describe the benchmark problem instances. After providing details about the parameter settings, we evaluate the influence of the pheromone information. Finally, we compare the results of our Ant System with those of Thangiah et al.'s heuristic algorithms. All our comparisons will be on the basis of the objective to first minimize fleet sizes and then minimize travel times as a second goal.

This objective was established by minimization of the following objective function:

$$L = 10000 \cdot FS + TT, \tag{3}$$

where L denotes the total costs of a solution, FS denotes the fleet size found, and TT corresponds to the total travel time (or distance). The parameter 10000 was chosen to ensure that a solution that saves a vehicle always outperforms a solution with a higher fleet size.

4.1   The Benchmark Problem Instances

The benchmark problem instances we used for our numerical tests were developed by Gelinas et al. ([16]). They used the first five problems of the r1 instances, namely r101 to r105, originally proposed by Solomon ([19]) for the vehicle routing problem with time windows (VRPTW). Each of these problems consists of 100 customers to be serviced from a central depot. The customers are located randomly around the depot. Service has to take place within a short time horizon (230 time units), and vehicle capacities are fairly loose when compared with the time window requirements at the customers. These time window requirements are varying in the data sets. The average time window length in the five instances are given in Table 1.

Table 1. Average length of the time windows

Instance	r101	r102	r103	r104	r105
Avg. Length of TW	10.0	57.4	103.0	148.3	30.0

Given these data sets Gelinas et al. randomly chose 10%, 30% and 50% of the customers to be backhaul customers with unchanged quantities, thus creating 15 different 100 customer instances.

4.2   Parameter Settings

Our aim is to use standard approaches and standard parameter settings as much as possible in order to demonstrate the benefit of intelligent combination of these approaches. However, we also performed parameter tests in order to find good combinations of these parameters for our problem. It turned out, that most of the parameters should be in the range documented in the literature.

More specifically, we analyzed the three parameters of the insertion algorithm in the ranges $\alpha \in \{0, 0.1, ..., 2\}$, $\beta \in \{0, 0.1, ..., 1\}$ and $\gamma \in \{0, 1, ..., 20\}$ on the instances described above. From this analysis we obtained the following values: $\alpha = 1$, $\beta = 1$ and $\gamma = 13$.

Note, that the parameter $\gamma = 13$ leads to a discrimination between linehaul and backhaul customers in favor of the backhaul customers. While this seems counterintuitive, it can be explained in the following way. Given the parameters α and β, the attractiveness values can become negative. However, negative attractiveness values prevent insertion. Thus, feasible solutions may be prohibited and this will generally lead to larger fleet sizes. This effect is reduced by the parameter γ, leading to tighter packed vehicles and smaller fleet sizes at the cost of increased travel times. To balance this trade-off, we chose to make the backhaul customers more attractive as they represent the minority of customers.[2]

Let n be the problem size, i.e. the number of customers to be served, then the Ant System parameters were: $m = \lceil n/2 \rceil$ ants, $\rho = 0.95$ and $p = 4$ elitists. These parameters were not extensively tested, as our experience suggests that the rank based Ant System is quite robust. However, the number of ants was reduced to $m = \lceil n/2 \rceil$ to be able to run more iterations.

For each instance we performed 10 runs of 2.5 minutes each. All runs were performed on a Pentium 3 with 900MHz. The code was implemented in C.

4.3   Evaluation of Pheromone Influence

In this section we will analyze whether our approach features pheromone learning or not. As we diverge from the standard constructive approach used in Ant Systems for Vehicle Routing, we have to check whether the utilization of the pheromone trails helps or hinders the constructive process of the ants. To do this, we compare the proposed Ant System with a stochastic implementation of the underlying algorithm, where no pheromone is used.

Table 2. Influence of the pheromone on the solution quality

Time in sec.	10% BH				30% BH				50% BH			
	ASinsert		StochInsert		ASinsert		StochInsert		ASinsert		StochInsert	
	FS	TT	FS	TT	FS	TT	FS	TT	FS	TT	FS	TT
15	17,64	1552,0	17,62	1554,3	18,34	1617,2	18,4	1629,2	18,92	1662,2	18,92	1665,6
30	17,44	1534,4	17,54	1552,0	18,24	1605,3	18,3	1628,9	18,74	1649,7	18,88	1660,1
75	17,38	1509,1	17,38	1550,9	18,08	1583,0	18,14	1624,3	18,54	1630,9	18,84	1651,7
150	17,3	1499,2	17,3	1546,7	18,02	1565,8	18,08	1623,9	18,44	1617,6	18,74	1649,0

FS...fleet size
TT...travel times

The results are shown in Table 2. We provide averaged results for the different backhaul densities after runtimes of 15, 30, 75 and 150 seconds. The table confirms that the use of the pheromone trails in decision making significantly improves solution quality. The Ant System (referred to as *ASinsert*) does find better solutions than the stochastic implementation of the Insertion algorithm (referred to as *StochInsert*). This can be seen from the last row of the table. Furthermore, the table shows that the Ant System finds good solutions faster than the stochastic Insertion algorithm. After 15 seconds the solutions of the

---

[2] However, we plan and already started to investigate other methods to solve this problem. One idea is to adjust the attractiveness values corresponding to feasible insertions, to ensure that a certain percentage of these values is positive.

Ant System are already superior to those of the stochastic Insertion algorithm, albeit the difference is small.

As more and more pheromone information is built and this matrix better reflects the differences between good and bad solutions the Ant System clearly outperforms the stochastic algorithm without pheromone information. This fact is also shown in Figure 3. We chose the case with 10% backhauls, as in this case the evolution of the fleet sizes, which are the primary goal, is similar in both approaches. So we can compare just travel times, and the figure shows that the Ant System at any point in time finds better solutions than the stochastic Insertion algorithm and moreover the difference gets larger as the number of iterations increases.

Fig. 3. Ant System ... rtion algorithm performance on problems with 10% ...

4.4 Comparison of Our Ant System with Existing Results

Let us now turn to the analysis of the solutions found with respect to absolute solution quality. As stated above, there exists one paper that studies the same objective as we do. In this paper, Thangiah et al. ([17]), propose a constructive algorithm and a number of local search descent algorithms to tackle the problem.

In Table 3 we show in detail, that is for each instance, the results of this approach together with the results of our Ant System and the stochastic Insertion algorithm. Note, that Thangiah et al. propose five different algorithms and the results in the first column of Table 3 are the best ones for each instance regardless of the version that found the solution. For our approaches we present average results over ten runs. In the rightmost two columns we report fleet sizes and travel times of the best solutions we found regardless of the algorithm that found this solution. The last row of the table reports the aggregate numbers for the approaches. Our Ant System outperforms the Thangiah's simple heuristics by approximately 2% with respect to fleet sizes, and by 2.7% with respect to travel times. However, we also see that the simple heuristics outperform our approach in three instances, namely r102 with 30% and 50% backhauls and r103 with 10% backhauls. In these instances, we did not find the same fleetsize as Thangiah's algorithms. As these instances are spread over all possible backhaul densities and our approach on average outperforms Thangiah's algorithm for each density of backhaul customers, the effect has to stem from the characteristics of the backhaul customers.

Note that we do not compare computation times for the approaches. We believe that a comparison of runtimes is meaningless in this case. First, the machines are very different, in particular ours are much faster. Second, a metaheuristic can never be expected to be as fast as a custom-made approach. Thangiah's heuristics find their solutions within less than a minute, while our approach runs for 2.5 minutes, we nevertheless believe that the results obtained by our Ant System justify the application and show the potential savings that can be achieved through the use of a metaheuristic approach. Moreover, the computation time reported for our approach refers to the total execution time of the algorithm. Of course we find good, and even better solutions than Thangiah, earlier in the search.

Table 3. Comparison of solution quality between our Ant System and other approaches

Instance	Thangiah best 5 versions		StochInsert avg. 10 runs		ASinsert avg. 10 runs		Best solutions identified by our algorithms	
	FS	TT	FS	TT	FS	TT	FS	TT
r101 10% BH	24	1842,3	22	1886,85	**22**	**1841,78**	22	1831,68
r101 30% BH	24	1928,6	**23,3**	**2000,42**	24	1931,73	23	1999,16
r101 50% BH	25	1937,6	24,1	1981,00	**24**	**1949,49**	24	1945,29
r102 10% BH	20	1654,1	**19,5**	**1695,13**	19,8	1636,20	19	1677,62
r102 30% BH	**21**	**1764,3**	22	1777,44	22	1759,18	22	1754,43
r102 50% BH	**21**	**1745,7**	22	1799,58	22	1784,15	22	1782,21
r103 10% BH	**15**	**1371,6**	16	1393,69	16	1352,23	16	1348,41
r103 30% BH	16	1477,6	16	1438,83	**16**	**1395,88**	16	1395,88
r103 50% BH	17	1543,2	17	1510,63	**17**	**1477,63**	17	1467,66
r104 10% BH	13	1220,3	12	1167,73	**11,9**	**1142,17**	11	1205,78
r104 30% BH	12	1303,5	12	1238,09	**12**	**1145,77**	12	1128,30
r104 50% BH	13	1346,6	12,6	1282,08	**12**	**1232,66**	12	1208,46
r105 10% BH	17	1553,4	17	1590,16	**16,8**	**1523,54**	16	1544,81
r105 30% BH	18	1706,7	17,1	1664,66	**16,1**	**1596,23**	16	1592,23
r105 50% BH	18	1657,4	18	1671,84	**17,2**	**1644,15**	17	1633,01
Sum	274	24052,9	270,6	24098,13	268,8	23412,79	265	23514,92

FS...fleet size
TT...travel times

Finally, we show in Figure 4 the effects of the density of backhaul customers on both fleet sizes and travel times for our Ant System. Clearly, both fleet sizes and travel times increase with the density of backhaul customers. Note however, that increasing the percentage of backhaul customers further beyond 50% will not further increase the travel times and fleet sizes. On the contrary, fleet sizes and travel times will fall again. In the extreme case of 100% backhauls we will have the same solution as in the other extreme case of 100% linehauls. Generally, with a mix of 50% linehauls and 50% backhauls, there is the smallest degree of freedom for the optimization, such that in these cases the solution quality should be worst.

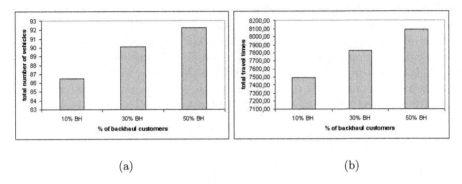

(a)   (b)

Fig. 4. Effect of backhaul customer density on fleet sizes (a) and travel times (b)

5   Conclusions and Future Research

In this paper we have proposed a promising approach for the Vehicle Routing Problem with Backhauls and Time Windows (VRPBTW). This Ant System approach is based on the well known Insertion algorithm proposed for the VRPTW by Solomon ([19]). We have proposed an approach to use pheromone information in the context of such an Insertion algorithm and shown that our algorithm benefits from this pheromone information and outperforms a custom-made heuristic proposed for the VRPBTW by Thangiah et al. ([17].)

Future research will deal with a more detailed analysis of the approach, and its application to other combinatorial problems. For the VRPBTW we will analyze the approach on larger instances. We also plan to incorporate swarm-like features by equipping each ant with its own set of parameters (of the heuristic) and adjusting these parameters in an evolutionary way during the optimization process. Furthermore, we will embed this approach in our multi-colony Ant System proposed already in Doerner et al. ([21]). This approach should help us deal better with the multiple objectives that have to be tackled in the problem.

Acknowledgments. We are grateful to two anonymous referees who provided valuable comments that improved the presentation of the paper. This work was supported by the Oesterreichische Nationalbank (OeNB) under grant #8630.

References

1. Colorni, A., Dorigo, M. and Maniezzo, V.: Distributed Optimization by Ant Colonies. In: Varela, F. and Bourgine, P. (Eds.): Proc. Europ. Conf. Artificial Life. Elsevier, Amsterdam (1991) 134–142
2. Costa, D. and Hertz, A.: Ants can colour graphs. Journal of the Operational Research Society **48**(3) (1997) 295–305
3. Stützle, T. and Dorigo, M.: ACO Algorithms for the Quadratic Assignment Problem. In: Corne, D. et al. (Eds.): New Ideas in Optimization. Mc Graw-Hill, London (1999) 33–50
4. Dorigo, M. and Gambardella, L. M.: Ant Colony System: A cooperative learning approach to the Travelling Salesman Problem. IEEE Transactions on Evolutionary Computation **1**(1) (1997) 53–66
5. Bullnheimer, B., Hartl, R. F. and Strauss, Ch.: A new rank based version of the ant system: a computational study. Central European Journal of Operations Research **7**(1) (1999) 25–38
6. Bullnheimer, B., Hartl, R. F. and Strauss, Ch.: An improved ant system algorithm for the vehicle routing problem. Annals of Operations Research **89** (1999) 319–328
7. Reimann, M., Stummer, M. and Doerner, K.: A Savings based Ant System for the Vehicle Routing Problem. to appear in: Proceedings of the Genetic and Evolutionary Computation Conference (GECCO 2002), Morgan Kaufmann, San Francisco (2002)
8. Gambardella, L. M., Taillard, E. and Agazzi, G.: MACS-VRPTW: A Multiple Ant Colony System for Vehicle Routing Problems with Time Windows. In: Corne, D. et al. (Eds.): New Ideas in Optimization. McGraw-Hill, London (1999) 63–73
9. Gutjahr, W. J.: ACO algorithms with guaranteed convergence to the optimal solution. Information Processing Letters. **82** (2002) 145–153
10. Le Louarn, F. X., Gendreau, M. and Potvin, J. Y.: GENI Ants for the Travelling Salesman Problem. CRT Research Report, Montreal (2001)
11. Gendreau, M., Hertz, A. and Laporte, G.: New Insertion and Postoptimization Procedures for the Travelling Salesman Problem. Operations Research. **40** (1992) 1086–1094
12. Garey, M. R. and Johnson, D. S.: Computers and Intractability: A Guide to the Theory of NP Completeness. W. H. Freeman & Co., New York (1979)
13. Toth, P. and Vigo, D. (Eds.): The Vehicle Routing Problem. Siam Monographs on Discrete Mathematics and Applications, Philadelphia (2002)
14. Bräysy, O. and Gendreau, M.: Metaheuristics for the Vehicle Routing Problem with Time Windows. Sintef Technical Report STF42 A01025 (2001)
15. Toth, P. and Vigo, D.: VRP with Backhauls. In Toth, P. and Vigo, D. (Eds.): The Vehicle Routing Problem. Siam Monographs on Discrete Mathematics and Applications, Philadelphia (2002) 195–224
16. Gelinas, S., Desrochers, M., Desrosiers, J. and Solomon, M. M.: A new branching strategy for time constrained routing problems with application to backhauling. Annals of Operations Research. **61** (1995) 91–109
17. Thangiah, S. R., Potvin, J. Y. and Sun, T.: Heuristic approaches to Vehicle Routing with Backhauls and Time Windows. Computers and Operations Research. **23** (1996) 1043–1057
18. Duhamel, C., Potvin, J. Y. and Rousseau, J. M.: A Tabu Search Heuristic for the Vehicle Routing Problem with Backhauls and Time Windows. Transportation Science. **31** (1997) 49–59

19. Solomon, M. M.: Algorithms for the Vehicle Routing and Scheduling Problems with Time Window Constraints. Operations Research. **35** (1987) 254–265
20. Osman, I. H.: Metastrategy simulated annealing and tabu search algorithms for the vehicle routing problem. Annals of Operations Research. **41** (1993) 421–451
21. Doerner, K. F., Hartl, R.F., and Reimann, M.: Are CompetANTS competent for problem solving - the case of a transportation problem. POM Working Paper 01/2001 (2001)

Modelling ACO: Composed Permutation Problems

Daniel Merkle[1] and Martin Middendorf[2]

[1] Institute for Applied Computer Science and Formal Description Methods
University of Karlsruhe, Germany
merkle@aifb.uni-karlsruhe.de
[2] Computer Science Group
Catholic University of Eichstätt-Ingolstadt, Germany
martin.middendorf@ku-eichstaett.de

Abstract. The behaviour of Ant Colony Optimization (ACO) algorithms is studied on optimization problems that are composed of different types of subproblems. Numerically exact results are derived using a deterministic model for ACO that is based on the average expected behaviour of the artificial ants. These computations are supplemented by test runs with an ACO algorithm on the same problem instances. It is shown that different scaling of the objective function on isomorphic subproblems has a strong influence on the optimization behaviour of ACO. Moreover, it is shown that ACOs behaviour on a subproblem depends heavily on the type of the other subproblems. This is true even when the subproblems are independent in the sense that the value of the objective function is the sum of the qualities of the solutions of the subproblems. We propose two methods for handling scaling problems (pheromone update masking and rescaling of the objective function) that can improve ACOs behaviour. Consequences of our findings for using ACO on real-world problems are pointed out.

1 Introduction

ACO is a powerful metaheuristic that is inspired by the foraging behaviour of real ants and has been applied successfully to different combinatorial optimization problems (e.g., see [2]). Unfortunately, it is difficult to analyze ACO algorithms theoretically. One reason is that they are based on sequences of random decisions of artificial ants which are usually not independent. Therefore except from convergence proofs for types of ACO algorithms with a strong elite principle [4, 5,11] not much theoretical results have been obtained so far.

A deterministic model for ACO algorithms that is based on the average expected behaviour of ants was proposed in [9,10] and used to derive numerically exact results on the dynamic optimization behaviour of ACO analytically. For small problems and for larger problems that have a simple structure (they consist of independent instances of the same small subproblem) exact results were obtained even for long runs. These results have been compared to test runs of the algorithm and it was shown that several aspects of the algorithms behaviour

M. Dorigo et al. (Eds.): ANTS 2002, LNCS 2463, pp. 149–162, 2002.

(e.g. the average solution quality over the iterations) are very well mapped by the model (see [9,10] for details). Clearly, not every aspect of ACO algorithms behaviour is reflected in the model. It should be noted that the usefulness of studying the behaviour of ACO on simple problems was also shown in [8,12].

In this paper we extend the work of [9,10] to problems that are composed of different subproblems. Numerical results are derived for the ACO model and are complemented with results of empirical tests. As a result of these investigations two methods (pheromone update masking and rescaling of the objective function) are proposed that can improve ACO behaviour. Several implications of the obtained results for the application of ACO on real-world problems are discussed.

In Section 2 we describe the simple ACO algorithm that is used in this paper. The ACO model is described in Section 3. In Section 4, we discuss how to apply the model to permutation problems that are composed of different subproblems. In Section 5, we analyze the dynamic behaviour of the ACO model and algorithm. The new methods for pheromone update masking and rescaling of the objective function are introduced in Section 6. Conclusions are given in Section 7.

2   ACO Algorithm

The test problems used in this paper are permutation problems. Given are n items $1, 2, \ldots, n$ and an $n \times n$ cost matrix $C = [c(ij)]$ with integer costs $c(i, j) \geq 0$. Let $\mathcal{P}_n$ be the set of permutations of $(1, 2, \ldots, n)$. For a permutation $\pi \in \mathcal{P}_n$ let $c(\pi) = \sum_{i=1}^{n} c(i, \pi(i))$ be the cost of the permutation. Let $\mathcal{C} := \{c(\pi) \mid \pi \in \mathcal{P}_n\}$ be the set of possible values of the cost function. The problem is to find a permutation $\pi \in \mathcal{P}_n$ of the n items that has minimal costs, i.e., a permutation with $c(\pi) = \min\{c(\pi') \mid \pi' \in \mathcal{P}_n\}$.

We give a brief description of the simple ACO algorithm that is used in this paper. Clearly, for applying ACO to real-world problems several extension and improvements have been proposed in the literature (e.g. usually ants use problem specific heuristic information or local search). The algorithm consists of several iterations where in every iteration each of m ants constructs a solution for the optimization problem. For the construction of a solution (here a permutation) every ant selects the items one after the other. For its decisions the ant uses pheromone information which stems from former ants that have found good solutions. The pheromone information, denoted by τ_{ij}, is an indicator of how good it seems to have item j at place i of the permutation. The matrix $M = (\tau_{ij})_{i,j \in [1:n]}$ of pheromone values is called the pheromone matrix.

The next item is chosen by an ant from the set $\mathcal{S}$ of items, that have not been placed so far, according to the following probability distribution that depends on the pheromone values in row i of the pheromone matrix: $p_{ij} = \tau_{ij} / \sum_{h \in \mathcal{S}} \tau_{ih}$, $j \in \mathcal{S}$ (see e.g. [3]). Note that alternative methods where the ants do not consider only the local pheromone values have also been proposed [6,8]. Before the pheromone update is done a certain percentage of the old pheromone evaporates according to the formula $\tau_{ij} = (1 - \rho) \cdot \tau_{ij}$. Parameter ρ allows to determine how strongly old pheromone influences future decisions. Then, for every item j of the best permutation found so far some amount Δ of pheromone is added to element τ_{ij}

of the pheromone matrix. The algorithm stops when some stopping criterion is met, e.g. a certain number of iterations has been done. For ease of description we assume that the sum of the pheromone values in every row and every column of the matrix is always one, i.e., $\sum_{i=1}^{n} \tau_{ij} = 1$ for $j \in [1 : n]$ and $\sum_{j=1}^{n} \tau_{ij} = 1$ for $i \in [1 : n]$ and $\Delta = \rho$.

3   Deterministic ACO Model

The deterministic ACO model that was proposed in [9,10] is described shortly in this section. In the model the pheromone update of a generation of ants is done by adding to each pheromone value the expected update value for that iteration. The effect is that individual decisions of individual ants are averaged out and a deterministic model is obtained. Clearly, since the update values in the simple ACO algorithm are always only zero or $\Delta = \rho$ the ACO model only approximates the average behaviour of the algorithm.

Given a pheromone matrix $M = (\tau_{ij})$ the probability σ_{π} to select a solution $\pi \in \mathcal{P}_n$ and the probability σ_{ij} that item i is put on place j by an ant are

$$\sigma_{\pi} = \prod_{i=1}^{n} \frac{\tau_{i,\pi(i)}}{\sum_{j=i}^{n} \tau_{i,\pi(j)}} \qquad \sigma_{ij} = \sum_{\pi \in \mathcal{P}_n} \sigma_{\pi} \cdot g(\pi, i, j) \qquad (1)$$

where $g(\pi, i, j) = 1$ if $\pi(i) = j$ and otherwise $g(\pi, i, j) = 0$. Let P be a permutation problem with corresponding cost matrix and $\mathcal{P}_{min}(P, \pi_1, \ldots, \pi_m)$ be the multiset of permutations with minimal costs from the sequence of permutations $\pi_1, \ldots, \pi_m$. In order to model a generation of m ants the average behaviour of the best of m ants has to be determined. Let $\sigma_{\pi}^{(m)}$ and $\sigma_{ij}^{(m)}$ be the the probability that the best of m ants in a generation finds permutation π, respectively selects item j for place i. Then

$$\sigma_{ij}^{(m)} = \sum_{(\pi_1, \ldots, \pi_m), \pi_i \in \mathcal{P}_n} \frac{|\{l \mid \pi_l \in \mathcal{P}_{min}(P, \pi_1, \ldots, \pi_m), \pi_l(i) = j\}|}{|\mathcal{P}_{min}(P, \pi_1, \ldots, \pi_m)|} \prod_{k=1}^{m} \sigma_{\pi_k} \qquad (2)$$

where the fraction equals the portion of permutations with $\pi(i) = j$ in the best solutions in $\pi_1, ..., \pi_m$ and $\prod_{k=1}^{m} \sigma_{\pi_k}$ is the probability that the ants have chosen $(\pi_1, ..., \pi_m)$. The pheromone update that is done in the ACO model at the end of a generation is defined by

$$\tau_{ij} = (1 - \rho) \cdot \tau_{ij} + \rho \cdot \sigma_{ij}^{(m)} \qquad (3)$$

An alternative way to compute $\sigma_{ij}^{(m)}$ is described in the following. Let $\mathcal{C}$ be the set of possible cost values for a permutation (i.e., the set of possible solution qualities). Let $\xi_x^{(m)}$ be the probability that the best of m ants in a generation finds a solution with quality $x \in \mathcal{C}$. Let $\omega_{ij}^{(x)}$ be the probability that an ant which found a solution with quality $x \in \mathcal{C}$ has selected item i for place j. Then

$$\sigma_{ij}^{(m)} = \sum_{x \in \mathcal{C}} \xi_x^{(m)} \cdot \omega_{ij}^{(x)} \qquad (4)$$

where

$$\xi_x^{(m)} = \sum_{\pi \in \mathcal{P}_n, c(\pi)=x} \sigma_\pi^{(m)}$$

and

$$w_{ij}^{(x)} = \sum_{\pi \in \mathcal{P}_n, c(\pi)=x} g(\pi, i, j) \cdot \frac{\sigma_\pi}{\sum_{\pi' \in \mathcal{P}_n, c(\pi')=x} \sigma_{\pi'}}$$

Formula (4) shows that the pheromone update in the model (see formula (3)) can be described as a weighted sum over the possible solution qualities. For each (possible) solution quality the update value is determined by the probabilities for the decisions of a single ant when it chooses between all possible solutions with that same quality. The effect of the number m of ants is only that the weight of the different qualities in this sum changes. The more ants per iteration, the higher becomes the weight of the optimal quality.

4   ACO Model for Composed Permutation Problems

Many real-world problems are composed of subproblems that are more or less independent from each other. In order to study the behaviour of ACO algorithms on such problems this is modelled in an idealized way.

Define for permutation problems P_1 and P_2 of size n a composed permutation problem $P_1 P_2$ that consists of independent instances of P_1 and P_2. As an example consider two problems P_1 and P_2 of size $n = 3$ with cost matrices C_1, respectively C_2

$$C_1 = \begin{pmatrix} 0 & 1 & 2 \\ 1 & 0 & 1 \\ 2 & 1 & 0 \end{pmatrix} \qquad C_2 = \begin{pmatrix} 0 & 2 & 4 \\ 2 & 0 & 2 \\ 4 & 2 & 0 \end{pmatrix} \tag{5}$$

The corresponding instance of the permutation problem $P_1 P_2$ has the cost matrix

$$C_1^{(2)} = \begin{pmatrix} 0 & 1 & 2 & \infty & \infty & \infty \\ 1 & 0 & 1 & \infty & \infty & \infty \\ 2 & 1 & 0 & \infty & \infty & \infty \\ \infty & \infty & \infty & 0 & 2 & 4 \\ \infty & \infty & \infty & 2 & 0 & 2 \\ \infty & \infty & \infty & 4 & 2 & 0 \end{pmatrix} \tag{6}$$

Formally, define for a permutation problem P of size n a composed permutation problem P^q of size qn such that for an instance of P with cost matrix $C = (c_{ij})_{i,j \in [1:n]}$ the corresponding instance of P^q consists of q independent of these instances of P. Let $C^{(q)} = (c_{ij})_{i,j \in [1:qn]}$ be the corresponding cost matrix of the instance of problem P^q where $c_{(l-1) \cdot n+i, (l-1) \cdot n+j} = c'_{ij}$ for $i, j \in [1 : n], l \in [1 : q]$ and $c_{ij} = \infty$ otherwise. Note, that our definition of composed permutation problems does not allow an ant to make a decision with cost ∞. We call P the elementary subproblem of P^q. Since all subproblems of P^q are the same it is called homogeneous composed permutation problem. Let P_1, P_2 be two permutation problems. Then similar as above $P_1^q P_2^r$ denotes the heterogeneous composed permutation problem that consists of q instances of P_1 and r instances of P_2.

4.1   Analyzing the ACO Model

In the following we show how the behaviour of the ACO algorithm for $m = 2$ ants on the composed permutation problem P^q can be approximated using the behaviour of the ACO model for the elementary subproblem P. Consider an instance of P with cost matrix $\mathcal{C}$. For the corresponding instance of P^q consider an arbitrary elementary subinstance — say the lth subinstance — and the quality of the solutions that 2 ants in an iteration have found on the other elementary subinstances (which form an instance of problem P^{q-1}).

Clearly, the ant that has found the better solution on subproblem P^{q-1} has better chances to end up with the best found solution on problem P^q. Since the decisions of an ant on different subinstances are independent the ant algorithm for problem P^q is modelled only on the elementary subproblem. But to model the pheromone update the difference in solution quality that the ants found on all other subinstances has to be considered. This can be modelled by giving one of the ants in an iteration a malus (i.e. a penalty value) that is added to the cost of the permutation it finds on the elementary subinstance. An ant with a malus is allowed to update only when the cost of its solution on P plus the malus is better than the solution of the other ant. For this approach it is necessary to compute the probability for the possible differences in solution quality between two ants on problem P^{q-1}. Without loss of generality assume that ant 1 does not have a malus. Let $d \geq 0$ be the malus of ant 2 and $\xi_x^{(2;d)}$ be the probability that the best of 2 ants where ant 2 has a malus d has found a solution of quality $x \in \mathcal{C}$ on P. Then for subproblem P

$$\sigma_{ij}^{(2;d)} = \sum_{x \in \mathcal{C}} \xi_x^{(2;d)} \cdot \omega_{ij}^{(x)} \tag{7}$$

Let d_{max} be the maximum difference between two solutions on the the lth subproblem. Formally, let $d_{max} := \max\{x \mid x \in \mathcal{C}\} - \min\{x \mid x \in \mathcal{C}\}$. Let d be the minimum of $d_{max} + 1$ and the difference of the cost of the permutation found by ant 2 on P^{q-1} minus the cost of the permutation found by ant 1 on P^{q-1}. Define $\phi_d^{(2)}$ as the probability that for 2 ants on problem P^{q-1} the difference of the costs of the solutions found by the 2nd best ant and the best ant is d when $d \leq d_{max}$ and when $> d$ it is the probability that this difference is $\geq d_{max} + 1$. Then for problem P^q the probability that the best of the two ants selects item j of the lth elementary subproblem at place $(l-1)n + i$ equals

$$\sigma_{(l-1)n+i,(l-1)n+j}^{(2)} = \phi_{>d_{max}}^{(2)} \cdot \sigma'_{ij} + \sum_{d=1}^{d_{max}} \phi_d^{(2)} \cdot \sigma'^{(2;d)}_{ij} + \phi_0^{(2)} \cdot \sigma'^{(2)}_{ij} \tag{8}$$

where σ' refers to probabilities for the elementary subinstance P. This equation shows that when the quality of the solutions of both ants differ by more than d_{max} then part of the probability to select the item is just the probability σ'_{ij} of a single ant to select item j at place i for the elementary subproblem P. When the solutions qualities of both ants are the same the probability $\sigma_{ij}^{\prime(2)}$ to select item j at place i equals the probability that the better of two ants on problem P selects item j at place i. All other cases correspond to the situation that one of two ants on problem P has a malus.

The larger the number q of subproblems is the larger becomes the probability $\phi_{d_{max}}^{(2)}$. An important consequence is that the (positive) effect of competition between the two ants for finding good solutions becomes weaker and a possible bias in the decisions of a single ant has more influence.

The main difficulty to determine $\sigma_{(l-1)n+i,(l-1)n+j}^{(2)}$ according to formula (8) is to compute the values $\phi_d^{(2)}$. In the following we describe how to compute $\phi_d^{(2)}$. Let ψ_d (ψ_d') be the probability that the difference of the solution quality found by the first ant minus the solution quality found by the second ant on subproblem P^{q-1} (respectively P) is d (here we do not assume that the first ant finds the better solution). The value of this difference on subproblem P^{q-1} can be described as the result of a generalized one-dimensional random walk of length $q-1$. Define I_d as the set of tuples $(k_{-d_{max}}, k_{-d_{max}+1}, \ldots, k_{d_{max}-1}, k_{d_{max}})$ with $q-1 = \sum_{i=-d_{max}}^{d_{max}} k_i$, $d = \sum_{i=-d_{max}}^{d_{max}} i \cdot k_i$ where k_i is the number of elementary subinstances of P^{q-1} where the difference between the first and the second ant is $d \in [-d_{max} : d_{max}]$. Then ψ_d can be computed as follows

$$\sum_{(k_{-d_{max}},\ldots,k_{d_{max}})\in I_d} \frac{(q-1)!}{k_{-d_{max}}! \cdot \ldots \cdot k_{d_{max}}!} \cdot (\psi_{-d_{max}}')^{k_{-d_{max}}} \cdot \ldots \cdot (\psi_{d_{max}}')^{k_{d_{max}}}$$

where the sum is over $(k_{-d_{max}}, \ldots, k_{d_{max}}) \in I_d$. Clearly, $\phi_0^{(2)} = \psi_d$ and due to symmetry, for $d \neq 0$ $\phi_d^{(2)} = 2 \cdot \psi_d$. The remarks on analysing the ACO model for the homogeneous problem can be generalized for the heterogeneous problem.

4.2   A Numerical Example

As an example consider the following elementary subproblem P_1 of the composed permutation problems P_1^2 and $P_1 P_2$ (see 5). The possible solution qualities for problem P_1 are 0, 2, and 4 and the optimal solution is to put item i on place i for $i \in [1:3]$. For problem P_2 the possible solution qualities are 0, 4, and 8. Consider the following pheromone matrix for P_1

$$M_1 = \begin{pmatrix} \tau_{11} & \tau_{12} & \tau_{13} \\ \tau_{21} & \tau_{22} & \tau_{23} \\ \tau_{31} & \tau_{32} & \tau_{33} \end{pmatrix} = \begin{pmatrix} 0.1 & 0.3 & 0.6 \\ 0.6 & 0.1 & 0.3 \\ 0.3 & 0.6 & 0.1 \end{pmatrix} \tag{9}$$

The matrix of selection probabilities on problem P_1 for a single ant and for two ants can be computed according to formula (4):

$$\begin{pmatrix} \sigma_{11} & \sigma_{12} & \sigma_{13} \\ \sigma_{21} & \sigma_{22} & \sigma_{23} \\ \sigma_{31} & \sigma_{32} & \sigma_{33} \end{pmatrix} \approx \begin{pmatrix} 0.1 & 0.3 & 0.6 \\ 0.714 & 0.111 & 0.175 \\ 0.186 & 0.589 & 0.225 \end{pmatrix} \quad \begin{pmatrix} \sigma_{11}^{(2)} & \sigma_{12}^{(2)} & \sigma_{13}^{(2)} \\ \sigma_{21}^{(2)} & \sigma_{22}^{(2)} & \sigma_{23}^{(2)} \\ \sigma_{31}^{(2)} & \sigma_{32}^{(2)} & \sigma_{33}^{(2)} \end{pmatrix} \approx \begin{pmatrix} 0.175 & 0.405 & 0.420 \\ 0.695 & 0.109 & 0.196 \\ 0.130 & 0.486 & 0.384 \end{pmatrix}$$

The matrix of selection probabilities on problem P_1 for two ants where one has a malus of 2 respectively 4 can be computed according to formula (7):

$$\begin{pmatrix} \sigma_{11}^{(2;2)} & \sigma_{12}^{(2;2)} & \sigma_{13}^{(2;2)} \\ \sigma_{21}^{(2;2)} & \sigma_{22}^{(2;2)} & \sigma_{23}^{(2;2)} \\ \sigma_{31}^{(2;2)} & \sigma_{32}^{(2;2)} & \sigma_{33}^{(2;2)} \end{pmatrix} \approx \begin{pmatrix} 0.146\ 0.351\ 0.503 \\ 0.698\ 0.118\ 0.184 \\ 0.156\ 0.531\ 0.313 \end{pmatrix}$$

$$\begin{pmatrix} \sigma_{11}^{(2;4)} & \sigma_{12}^{(2;4)} & \sigma_{13}^{(2;4)} \\ \sigma_{21}^{(2;4)} & \sigma_{22}^{(2;4)} & \sigma_{23}^{(2;4)} \\ \sigma_{31}^{(2;4)} & \sigma_{32}^{(2;4)} & \sigma_{33}^{(2;4)} \end{pmatrix} \approx \begin{pmatrix} 0.109\ 0.299\ 0.593 \\ 0.708\ 0.118\ 0.174 \\ 0.183\ 0.583\ 0.234 \end{pmatrix}$$

The influence of one subproblem on the other subproblem P_1 in P_1^2 and $P_1 P_2$ is shown in the following. The probability that the best of two ants has selected item 1 on place 1 is for subproblem P_1 of P_1^2

$$\sigma_{11}^{(2)} = \phi_0^{(2)} \cdot \sigma_{11}'^{(2)} + \phi_2^{(2)} \cdot \sigma_{11}'^{(2;2)} + \phi_4^{(2)} \cdot \sigma_{11}'^{(2;4)}$$
$$= 0.566 \cdot 0.175 + 0.398 \cdot 0.146 + 0.036 \cdot 0.109 = 0.161082$$

and for subproblem P_1 of $P_1 P_2$ it is

$$\sigma_{11}^{(2)} = \phi_0^{(2)} \cdot \sigma_{11}'^{(2)} + \phi_4^{(2)} \cdot \sigma_{11}'^{(2;4)} + \phi_{>4}^{(2)} \cdot \sigma_{11}'$$
$$= 0.566 \cdot 0.175 + 0.398 \cdot 0.109 + 0.036 \cdot 0.1 = 0.146032$$

5   Results

The dynamic behaviour of the ACO model and the ACO algorithm is investigated for $m = 2$ ants using composed permutation problems that are based on problem P_1 with cost matrix C_1 (see (5) in Section 4). The composed problems are of the form $P_1^q P_y^r$ were P_y has the cost matrix $y \cdot C_1$ (i.e. each element of C_1 is multiplied by y) for integer $y \geq 1$. Since the structure of problems P_1 and P_y is the same it is clear that the algorithm and the model show the same optimization behaviour for problems P_1^q and P_y^q. But since they differ by the scaling factor y it is interesting to see how this influences the behaviour of ACO on the composed problem $P_1^q P_y^r$. In the following the pheromone evaporation was set $\rho = 0.1$ for the model and $\rho = 0.0001$ for the algorithm (this makes the results of the algorithm more stable and allows to consider single runs, but introduces an additional difference between model and algorithm).

We are interested in composed problems because they show characteristics that can be found also in real-world applications when an optimization problem has more or less independent subproblems that differ by scaling. E.g., consider a vehicle routing problem where a vehicle that starts and ends at a depot has to visit a set of customers on a fastest tour. Assume that the customers locations are clustered because the customers are in different cities and distances between the cities are much larger than distances within a city. When the number of cities is not too large it will probably be the case that in an optimal tour all customers

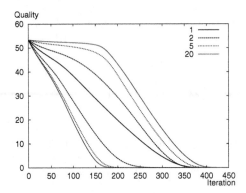

Fig. 1. Average solution quality of ACO model on subproblems P_1^{20} (upper curves) and P_y^{20} (lower curves) of $P_1^{20}P_y^{20}$ for $y \in \{1, 2, 5, 20\}$; note, for $y = 1$ there is only one curve (the middle curve); average solution quality for P_y^{20} is divided by y; initial pheromone values for all subproblems were $\tau_{ij} = 1/3$, $i, j \in [1:3]$.

in a city are visited on a subtour and the main difficulty is to find good subtours between the customers within the cities. When each city has about the same number of customers but the cities differ by size or the average driving speed possible within the cities is different (e.g. there is heavy traffic in some cities wheres other cities have only light traffic) then the problem consists of more or less independent and similar subproblems that differ mainly by a scaling factor. Clearly, it is easier to solve independent subproblems independently but this requires to detect this situation and to identify the subproblems. In practice it might often be difficult for a general optimization algorithm to do this.

5.1   The Scaling Effect

The scaling effect is shown on the test problem $P_1^{20}P_y^{20}$. Figure 1 shows the average solution quality in each iteration of the ACO model for this problem for different values of y. Note that the behaviour of the model (or the algorithm) is the same for every value of $y > 40$ for this problem because the maximal difference in solutions qualities for subproblem P_1^{20} is 40. Hence, for $y > 40$ the solution quality on subproblem P_y^{20} dominated that on subproblem P_1^{20}. The figure shows that the average solution quality on subproblem P_y^{20} converges faster to the optimal solution for larger values of y because the influence of P_1^{20} is weaker during the first iterations. For $y = 20$ the average solution quality on subproblem P_1^{20} remains nearly unimproved during the first 150 iterations because whether an ant is allowed to update depends with high probability only on the solution quality it has found on P_y^{20}.

The changes of the pheromone values over the iterations for the ACO model and algorithm are shown in figures 2 and 3 for $y = 2$ and $y = 20$, respectively. When not stated otherwise, matrix M_1 (see (6) in section 4) is the initial pheromone matrix in the rest of this paper (some effects are more visible as with the standard initial pheromone matrix). In all cases the model and the algorithm

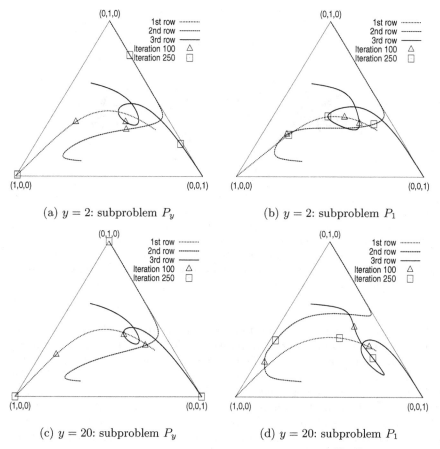

(a) $y = 2$: subproblem P_y

(b) $y = 2$: subproblem P_1

(c) $y = 20$: subproblem P_y

(d) $y = 20$: subproblem P_1

Fig. 2. Changes of pheromone values for the ACO model for $P_1^{20} P_y^{20}$ with $y = 2$ (upper part), $y = 20$ (bottom part); initial pheromone matrix M_1; left: pheromone values for subproblem P_x; right: pheromone values for subproblem P_1; all sides of the triangle have length 1 and therefore for every point within the triangle the sum of the distances from the three sides is 1; sine $\tau_{i1} + \tau_{i2} + \tau_{i3} = 1, i \in [1:3]$ for every row $i \in [1:3]$ the pheromone values in a row determine a point within in the triangle; distance from right (bottom, left) side: τ_{i1} (τ_{i2}, τ_{i3}); recall that $\tau_{i1} + \tau_{i2} + \tau_{i3} = 1, i \in [1:3]$; the state at iterations 100 and 250 is marked by small triangles (respectively boxes).

converge to the optimal solution. It was shown in [9,10] that the actual position of the selection fixed point matrix (i.e. the matrix the system would converge to when there is only one ant per generation) has a strong influence. For the first row of the pheromone matrix the pheromone values always equal the values of the selection fixed point matrix and all dynamic in this row is only due to competition. Therefore, the pheromone values for row 1 in Figure 2 approach the optimal values on an almost straight path. This is different for rows 2 and 3 (see, e.g., the loop in the curves for row 3). Comparing the behaviour of the model on subproblems P_1 and P_y it can be seen that the curves for P_y approach the

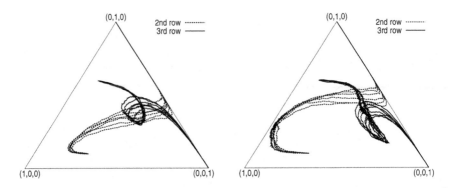

Fig. 3. Changes of pheromone values during one run of the ACO algorithm for $P_1^{20} P_y^{20}$ with $y = 20$ and initial pheromone matrix M_1; left: pheromone values for 6 of the elementary subproblems P_y; right: pheromone values for 6 of the elementary subproblems P_1; the reason that the results on only 6 of the 10 elementary subproblems are shown in each case is that the curves are hard to distinguish otherwise; see caption of Figure 2 for an explanation how to read the figure.

optimal values faster (see positions after 100 and 250 iterations) and have nearly reached the optimal values after 250 iterations. The curves for P_1 approach the optimal values much slower during the first 250 iterations. The curves for P_y differ not much for $y = 2$ and $y = 20$ (for $y = 20$ where the influence of P_y is larger the system approaches the optimal values slightly more straight and faster). In contrast to this there are big differences between the curves of P_1 for $y = 2$ and $y = 20$. For $y = 20$ the pheromone values remain more or less unchanged during iterations 100 to 250 because the system approaches the pheromone values of the selection fixed point matrix and stays there as long as competition is weak.

5.2   The Interaction of the Ants

The relative influence of pure selection (i.e., no competition), pure competition, and weak competition (where one ant has a malus) is shown in Figure 4 for an elementary subproblem of $P_1^{20} P_2^{20}$ with $y = 20$. We computed the probabilities for the possible differences d in solution qualities between the two ants on subproblems $P_1^{19} P_y^{20}$ and $P_1^{20} P_y^{19}$. Observe that the possible solution qualities for the elementary subproblems P_1 and P_y are $\{0, 2, 4\}$ and $\{0, 40, 80\}$ respectively. Figure 4 shows the probabilities $\phi_0^{(2)}$, $\phi_2^{(2)}$, $\phi_4^{(2)}$, $\phi_{>4}^{(2)}$ for $P_1^{19} P_y^{20}$ and $\phi_0^{(2)}$, $\phi_{0<d<40}^{(2)}$, $\phi_{40}^{(2)}$, $\phi_{40<d<80}^{(2)}$, $\phi_{80}^{(2)}$, $\phi_{>80}^{(2)}$ for $P_1^{20} P_y^{19}$. The case $d = 0$ corresponds to pure competition between two ants on the remaining elementary subproblem. Whereas $d > 4$ ($d > 80$) corresponds to the case where the ant without malus has always the better solution no matter how its quality on the remaining subproblem is, i.e., there is no competition on the remaining subproblem. The figure shows that there is nearly no competition during the first 200 iterations on the instances P_1 ($\phi_{>4}^{(2)} \approx 95\%$). For the instances P_y during the first 200 iterations the probability that no competition occurs ($\phi_{>80}^{(2)}$) is only about 60%. After 200

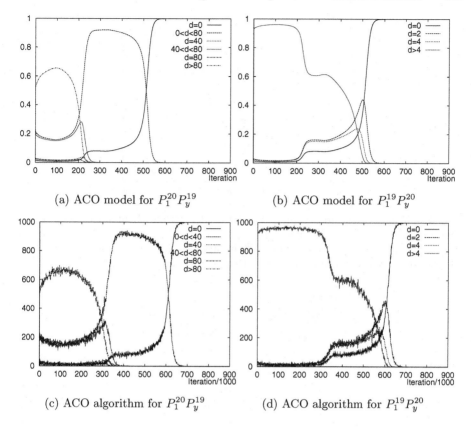

(a) ACO model for $P_1^{20} P_y^{19}$ (b) ACO model for $P_1^{19} P_y^{20}$

(c) ACO algorithm for $P_1^{20} P_y^{19}$ (d) ACO algorithm for $P_1^{19} P_y^{20}$

Fig. 4. ACO model (upper part) and ACO algorithm (bottom part) for $y = 20$ and initial pheromone matrix M_1; left: change of probabilities $\phi_0^{(2)}$, $\phi_{0<d<40}^{(2)}$, $\phi_{40}^{(2)}$, $\phi_{40<d<80}^{(2)}$, $\phi_{80}^{(2)}$, $\phi_{d>80}^{(2)}$ for $P_1^{20} P_y^{19}$; right: change of probabilities $\phi_0^{(2)}$, $\phi_2^{(2)}$, $\phi_4^{(2)}$, $\phi_{>4}^{(2)}$ for $P_1^{19} P_y^{20}$; for the ACO algorithm every point corresponds to the number of the observed differences over 1000 iterations (instead of probabilities as for the model).

iterations the pheromone matrices for instances P_x converge more and more to the optimal solution and the probability that weak or pure competition occurs for P_1 increases. At iteration 250 the pheromone matrices for instances P_y have nearly converged as can be seen from the fact that $\phi_0^{(2)} + \phi_{0<d<40}^{(2)} \approx 100\%$. At the same time the probability for pure competition ($d = 0$) is about 10% for P_1. For weak competition ($d = 2$ or 4) the probability is $\approx 30\%$ altogether. At around iteration 550 the pheromone matrices for subproblem P_1 have converged so that an ant nearly always finds the optimal solution and $\phi_0^{(2)}$ becomes $\approx 100\%$.

6   Discussion

In this section we discuss implications of our results for the application of ACO to real-world problems and propose two techniques that might improve ACO.

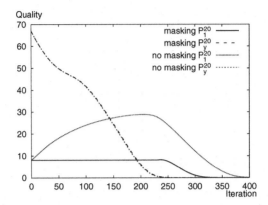

Fig. 5. Change of solution quality of ACO model with/without pheromone update masking on subproblems P_1^{20} and P_y^{20} of problem $P_1^{20}P_y^{20}$ with $y = 20$; masking was done on subproblem P_1^{20} over the first 230 iterations; initial pheromone matrix M_1 for P_y and for P_1 it is determined by $\tau_{11} = 0.9$, $\tau_{12} = 0.09$, $\tau_{21} = 0.05$, $\tau_{22} = 0.9$.

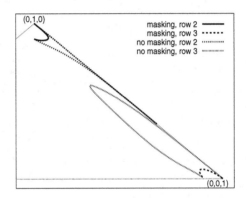

Fig. 6. Changes of pheromone values in rows 2 and 3 of elementary subproblems P_1 and P_y of problem $P_1^{20}P_y^{20}$, $y = 20$ for ACO model with/without pheromone update masking; masking was done on subproblem P_1^{20} over the first 230 iterations; initial pheromone matrix M_1 for P_y and for P_1 it is determined by $\tau_{11} = 0.9$, $\tau_{12} = 0.09$, $\tau_{21} = 0.05$, $\tau_{22} = 0.9$; left: average solution quality on P_1^{20} and P_y^{20}; Note, that only the right half of the triangle of possible values is given.

6.1   Rescaling

It was shown that scaling effects play an important rule in the optimization behaviour of ACO. E.g., the ACO model converges faster to the optimal solution for $P_1^{20}P_y^{20}$ with $y = 1$ than with $y = 20$ because in the last case the system is slow on subproblem P_1^{20}. This raises the question whether it is possible to design general rescaling techniques that can improve ACO. The basic principle is to change a problem instance by rescaling before an ACO algorithm is applied. Clearly, for real-world problems it has to be taken into account that subproblems will usually not be completely independent and it might be difficult to identify nearly independent subproblems.

6.2 Pheromone Update Masking

As shown above the dynamics of ACO is driven by competition between the ants only for parts of a problem instance. On other parts the dynamics might be driven mostly by pure selection that forces the system to approach the fixed point state. But the dynamics in the last case is not always wanted because it can bring the system in a less favorable state. The question is whether it is better to hinder the system to change the pheromone values for such parts of the problem instance as long as competition is too weak. One possibility is to mask the pheromone update in corresponding parts of the pheromone matrix, i.e. for some time the pheromone values remain the same or are only slightly changed by decreasing the evaporation rate and the amount of update. We give an example that shows that masking can improve ACOs behaviour significantly. Consider problem $P_1^{20} P_y^{20}$ with $y = 20$ where the initial pheromone matrix for P_y is M_1 and the initial pheromone matrix for P_1 is determined by $\tau_{11} = 0.9$, $\tau_{12} = 0.09$, $\tau_{21} = 0.05$, and $\tau_{22} = 0.9$. The average solution quality for the ACO model on subproblems P_1^{20} and P_y^{20} is shown in Figure 5 when using masking and without masking. Masking means here that the pheromone values in the matrices for P_1^{20} were not changed during the first 230 iterations. Without masking the average solution quality on subproblem P_1^{20} becomes increasingly worse during the first 230 iterations. This is circumvented by masking. Moreover, with masking the model converges much earlier to the optimal solution than without masking. Figure 6 shows the corresponding change of the pheromone values in rows 2 and 3 for a matrix of the elementary subproblem P_1 with masking and without masking. The second and third row of the fixed point matrix for the initial pheromone matrix are $(0, 0.1, 0.9)$, respectively $(0.1, 0.81, 0.09)$. It can be clearly seen that pheromone values move much more towards the fixed point values when no masking is used. In this example the fixed point values are fare from the optimal values which means that the system basically has to move back until it reaches the optimum.

7   Conclusion

The behaviour of ACO on a class of composed permutation problems was investigated using a deterministic model of ACO. The analytical results obtained for the model were complemented by the results of test runs of an ACO algorithm. It was shown that the behaviour of ACO is influenced by the scaling of the objective function on different subproblems and on the composition of the subproblems. Two methods were proposed for handling scaling problems (pheromone update masking and rescaling of the objective function) that can improve ACOs behaviour and possible implications for using ACO on real-worlds problems have been pointed out. Future work is to see how practical the proposed methods are for real-worlds problems.

References

1. Dorigo, M. (1992). Optimization, Learning and Natural Algorithms *(in Italian)*. PhD thesis, Dipartimento di Elettronica, Politecnico di Milano, Italy.
2. Dorigo, M., and Di Caro, G. (1999). The ant colony optimization meta-heuristic. In Corne, D., Dorigo, M., and Glover, F., editors, *New Ideas in Optimization*, 11–32, McGraw-Hill, London.
3. Dorigo, M., Maniezzo, V., and Colorni, A. (1996). The Ant System: Optimization by a Colony of Cooperating Agents. *IEEE Trans. Systems, Man, and Cybernetics – Part B*, 26:29–41.
4. Gutjahr, W. (2000). A graph-based Ant System and its convergence, *Future Generation Computer Systems*, 16:873–888.
5. Gutjahr, W. (2002). ACO algorithms with guaranteed convergence to the optimal solution. *Information Processing Letters*, 82:145–153.
6. Merkle, D., and Middendorf, M. (2000). An Ant Algorithm with a new Pheromone Evaluation Rule for Total Tardiness Problems. In Cagnoni, S., et al. (Eds.) *Real-World Applications of Evolutionary Computing*, LNCS 1803, 287–296, Springer.
7. Merkle, D., and Middendorf, M. (2001). A New Approach to Solve Permutation Scheduling Problems with Ant Colony Optimization. In Boers, E. J. W., et al. (Eds.) *Applications of Evolutionary Computing*, LNCS 2037, 213–222, Springer.
8. Merkle, D., and Middendorf, M. (2001). On the Behaviour of Ant Algorithms: Studies on Simple Problems. In *Proceedings of the 4th Metaheuristics International Conference (MIC'2001)*, Porto, 573-577.
9. Merkle, D., and Middendorf, M. (2001). Modelling the Dynamics of Ant Colony Optimization. TR 412, Institute AIFB, University of Karlsruhe. To appear in *Evolutionary Computation*.
10. Merkle, D., and Middendorf, M. (2002). Studies on the Dynamics of Ant Colony Optimization Algorithms. To appear in Proceedings of the Genetic and Evolutionary Computation Conference (GECCO-2002), New York, 2002.
11. Stützle, T. and Dorigo, M. (2000). A short convergence proof for a class of ACO algorithms. TR 2000-35, IRIDIA, Universitè Libre de Bruxelles, Belgium.
12. Stützle, T. and Dorigo, M. (2001). An Experimental Study of the Simple Ant Colony Optimization Algorithm. In Proc. 2001 WSES Int. Conference on Evolutionary Computalution Computation (EC'01), WSES-Press International, 2001.

Self-Organized Networks of Galleries in the Ant *Messor Sancta*

Jérôme Buhl[1], Jean-Louis Deneubourg[2], and Guy Theraulaz[1]

[1]Laboratoire d'Ethologie et Cognition Animale, CNRS - FRE 2382, Université Paul Sabatier
118 route de Narbonne, 31062 Toulouse Cédex, France
{buhl,theraula}@cict.fr
[2]CENOLI, Université Libre de Bruxelles
CP 231, Boulevard du Triomphe, 1050 Brussels, Belgium
jldeneub@ulb.ac.be

Abstract. In this paper we describe an individual-based model to account for the growth and morphogenesis of networks of galleries in the ant *Messor sancta*. The activity of the individuals depends only on their local perception of the immediate surroundings. Coordination between ants arises from the modifications of the environment resulting from their activity: the removal of sand pellets and a pheromone trail-laying behaviour. We show that the growth of the networks results from a self-organized process that also allows the collective adaptation of the size and shape of the network to the size of the colony.

1 Introduction

Numerous animals that live in group or societies use or create networks. Among networks, those resulting from construction behaviour such as nest structures built by social insects are probably the most fascinating but also the least understood. Several species of ants build complex networks of underground tunnels that connect chambers [1-7]. Such structures are built without any blueprint or centralization processes [8]. Ants achieve such complex features using stigmergic processes [9], in which the modification of the environment that results from individuals' activity acts as a feedback on their behaviour. When the intensity of a building stimulus changes the probability for an individual to exhibit a specific behaviour, the resulting construction is a self-organized process [10]. Self-organization leads to the emergence of large-scale patterns from a homogenous environment through many local interactions among units at the lower level [11]. It thus allows the growth of complex structures without any centralization of information at the individual level. Self-organization has been shown to be involved in numerous collective behaviours in insect colonies [10,12,13] and also in vertebrates [14-16].

Several studies report that a constant relationship exists between nest size and population within a particular species in ants [17-20]; but little is known about mechanisms involved in achieving such adaptation. A first mechanism could be a direct triggering of the digging activity according to the perception of specific cues by individuals [21], such as a chemical signal whose concentration would be correlated

M. Dorigo et al. (Eds.): ANTS 2002, LNCS 2463, pp. 163-175, 2002.

with the density of workers within the nest; however a recent study of the dynamics of nest excavation in *Lasius niger* [6,7] suggested that regulation of subterranean nest size could arise without such cues, from a combination of amplification processes and a tendency of individuals to aggregate themselves.

In this context, we studied the processes involved in the morphogenesis of galleries networks in the ant *Messor sancta* (Myrmicinae). It is a Mediterranean granivorous ant that digs large nests in sandy grounds and its colony size reaches several thousands of individuals. Nest structure is made of two parts: a central zone connecting superficial chambers to deep ones by vertical tunnels, the deepest chambers reaching several meters depth, and a peripheral zone that includes a large horizontal network of tunnels interconnecting several chambers used as seeds stocks.

The phenomenon was studied at two distinct levels: (1) at the collective level, we quantified the growth dynamics of the networks produced by 100 ants; (2) at the individual level, we quantified all the behaviours involved in the ants' activity. Finally, to understand the link between these two levels, we built a model including all the behavioural parameters drawn from experiments and we used it to study the adaptation of the resulting network for various colony sizes.

2 Experimental Result

2.1 Growth Dynamics

The general experimental set-up consisted in a sand disk of 20cm diameter and 5 mm height. It was prepared with a mould in which sand was deposited and humidified by 25 ml of sprayed water. The mould was removed, the disk covered by a glass (25 cm x 25cm) and the surface which ants had access to was restricted by a circular wall coated with fluon. 100 ants were collected from a colony and the experiment began with their random dispersal around the disk. A digital camera was placed over the arena. The duration of each experiment was 3 days, and 2 seconds of activity were recorded every 10 minutes. An image analysis software allowed us to quantify the volume of excavated sand (fig. 1).

Fig. 1. Example of tunneling pattern produced by 100 ants over 3 days. The pictures shown are the results of image analysis processes that allowed us to obtain binary images where the sand is represented by black pixels and the ground by white pixels.

Figure 2 shows the evolution of the volume of excavated sand as a function of time over 3 days in 5 experiments. In all experiments, the growth dynamics has a sigmoid shape and can be divided in 3 phases: first, an exponential growth in which the excavation speed, nil at the beginning of experiment, progressively increases; this amplification phase is followed by a short linear phase during which the excavation speed reaches its maximum value; finally, a saturation phase in which the excavation speed decreases progressively to reach zero. This suggests two kinds of feedbacks: (1) a positive feedback when the volume V of excavated sand is far from a saturation volume, V_{max} (V/V_{max} is very small): the speed dV/dt grows proportionally with V. (2) a negative feedback when V becomes closer to V_{max} ($V/V_{max} \approx 1$): dV/dt decreases and reaches a null value. This kinetics can be formalized as follows:

$$dV/dt = aV \left[1 - (V/ V_{max}) \right] \qquad (1)$$

where a is a constant. The solution of this differential equation is:

$$V(t) = V_{max} / \left[1 + ((V_{max} / V_o) - 1) . e^{-at} \right] \qquad (2)$$

where V_o is a initial non-zero volume. This relationship can be easily written in a linear form as follows:

$$Ln \left[(V_{max}/ V(t))-1 \right] = Ln \left[(V_{max} / V_o) - 1 \right] - at \qquad (3)$$

Table 1 shows the results of the linear regression test for equation (3) in each experiment and the final size of the networks. Measures were made every 20 minutes, which represents a set of 217 data points per curve. V_{max} value was estimated as the volume excavated at the end of the experiment. In all cases r^2 values were always greater than 0.9 and the parameters for linear regression test were validated ($p<0.001$).

These results show that growth dynamics is controlled by two kinds of feedbacks. Positive feedback could be the consequence of some recruitment processes, which will be confirmed later with the model. As regards negative feedback, when the saturation phase was reached, we always observed the formation of ants' aggregates. As a consequence, aggregation may be a first negative feedback involved in the saturation phase.

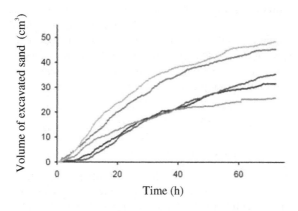

Fig. 2. Growth dynamics of networks with 100 ants over 3 days.

Table 1. Linear regression results for the logistic model (see text and equation 3 for details) and final volume and weight excavated after 3 days.

	r^2	a	Final volume (cm^3)	Final weight (g)
1	0.937	0.106 ± 0.002	33	49.04
2	0.929	0.11 ± 0.002	35.25	53.38
3	0.947	0.0986 ± 0.002	45	66.87
4	0.943	0.092 ± 0.002	25.75	38.27
5	0.923	0.0922 ± 0.002	48.25	71.7

2.2 Individual Behaviours

In this section, we describe all the individual behaviours involved in the digging activity of ants. We present the estimated mean values and probabilities to perform an action that were used in the model.

For the individual behaviour study, an artificial tunnel was created in the sand disk by adding a plastic shape 10 cm long and 3 mm wide. The number of ants was fixed to 100, and the recording time to 1 hour. The video recording was set to continuous mode.

Mean values. The distribution of walking speed was Gaussian with a mean value of 15.05 mm/s (N=100, SD=4.381, Z=1.139, p=0.149, Kolmogorov-Smirnov test). The distribution of digging time was Gaussian with a mean value of 42.04 s (N=65, SD=24.79, Z=0.7490, p=0.628, Kolmogorov-Smirnov test). The estimate weight of a pellet was 1.688 mg.

Probabilities to perform actions. Probabilities to perform actions were established from survival curves whenever it was possible. Given N_0, the total number of individuals, N the number of individuals having not yet performed an action at time t, and λ the probability to perform the action, if λ is constant, then the evolution of the population N follows equation 4; $\bullet$ _ can be estimated with a linear regression test on the logarithmic transformation of equation (4).

$$N = N_0 \, e^{-\lambda t} \tag{4}$$

$$Ln\,(N) = Ln\,(N_0) - \lambda\,t \tag{5}$$

Table 2 shows the experimental values for the probabilities to spontaneously dig the edge of the sand disc (f_s^h), to drop a pellet (P_d) and to leave the edge of the sand disk while following it (thigmotactic effect) (P_l). In each case equation (5) was tested and a linear regression validated with high r^2 values, which means that in each case the probability to perform the action is indeed constant over time and equal to the value of slope in the regression model.

Table 2. Results of the survival curves for three behaviours. r^2 values are indicated for linear regression test on equation (5) . The parameter λ represents the slope of the linear equation and corresponds to the probability to perform the action per second (see text for details).

Variable	r^2	Coefficient			
		λ	Std Err	T	p
To dig spontaneously (f_s^h)	0.985	$-2.83.10^{-2}$	0.000	-54.76	0.00
To drop a pellet (P_d)	0.971	-0.188	0.001	-17.45	0.00
To leave the edge of sand disk (P_l)	0.944	$-1.76.10^{-2}$	0.000	-24.95	0.00

The probability to dig spontaneously an anfractuosity (f_s^a) was estimated by the proportion of ants that dug after having penetrated in the artificial gallery (never dug before); its value was f_s^a=0.054 (6 ants over 111 observed). The dead-end of the tunnel represents a strong heterogeneity in the curvature of the sand in comparison with the rather straight sand walls in a tunnel or the edge of the sand disk. In this paper, we will use the term anfractuosity to refer to this kind of heterogeneity. Several species of ants are reported to be more strongly stimulated by anfractuosities [22] than by straight sand walls, and our results show that it is also the case in *Messor sancta*.

3 Model Description

The experimental analysis allowed us to quantify several parameters of the individual tunneling behaviour of ants. We developed a model entirely based on the experimental measures of individual behaviour, using a spatially-explicit individual-based simulation written in C++. In the model, ants move on a discrete 2D square-lattice preserving time and spatial scales of the experiments, with a distance of 0.8 mm between lattice cells and a time scale of 1/20 s per cycle. The environment includes 4 classes of objects: ants, ground, sand in the disk and excavated sand pellets.

3.1 Structure of the Model

Each ant can be in one of the following three internal states: free state, digging state, or transporting state. Ants can change from their present state to another as a function of the environment they meet according to the experimentally estimated probabilities.

At the beginning of the simulation, as in the experiments, ants are randomly distributed around the sand disk and are in the free state. During a cycle, each ant is randomly chosen and performs a single action one time and only one.

When an ant meets the sand disk, she can spontaneously start to dig, with two different probabilities, f_s^a and f_s^h, measured in the experiments, whether she is in presence of an anfractuosity in its immediate neighbourhood (f_s^a, high concavity in the sand wall), or not (f_s^h, straight sand wall). When an ant enters into the digging state, she stays in this state for T_d time steps and no other action can be done. When this period of time ends, the ant enters into the transporting phase. There exists another way for an ant to enter into this phase: each time a free ant occupies a place where a pellet is present, she can pick it up with a constant probability P_p. When an ant is transporting a pellet she has a constant probability P_d to spontaneously drop the pellet and return to the free state.

At the end of a cycle, each ant that has not performed an action (digging, picking or dropping a sand pellet) moves randomly to an adjacent cell, in one of the six possible directions (forward, back, 45° left or right and 90° left or right) according to a probability matrix (M_d) that favours front directions. The thigmotactic behaviour of ants is implemented by multiplying the weight of the cells adjacent to the direction blocked by sand in the matrix, so that the probability to leave the sand wall is equal to the experimental probability (P_l). Finally, an ant can move neither on a cell already occupied by two other individuals nor on a cell that contains sand and her displacement is restricted by an outside arena wall.

This basic algorithm that takes into account all the behavioural parameters that were measured in the experiments was not able to reproduce the emergence of tunnels and networks. The resulting digging dynamics was always linear until the ants have dug the whole sand disk. At this point, the main question was: how can the probabilities to dig sand evolve in order to produce networks of galleries? Several hypotheses that we won't present here, such as individual experience, have been tested, but only one additional mechanism, namely the use of trail-laying pheromones, was able to give rise to networks. The use of trail-laying pheromones is a widespread feature in ants and termites [10] and it has been shown to be involved in recruitment and digging behaviour in other ant species such as *Lasius niger* [6]. Trail-laying parameters are very hard to measure and they have been fixed by numerical exploration in the range of biological values already known from previous experiments performed on other species [6,23,24].

Two kinds of pheromone are used in the model: a trail-laying pheromone that influences ant displacement, and a digging pheromone laid at the digging site that controls the probability for an ant to dig a sand cell.

When the place in front of which an ant is located is marked with digging pheromone, the ant starts to dig with a probability given by the following sigmoid response function:

$$f_p = X^2 / (X^2 + K^2) \tag{6}$$

where X is the amount of pheromone and K a threshold constant. This class of response function provides a good approximation of the influence of pheromone on ant behaviour and it has been used in numerous works including collective recruitment and digging in ants [10,24,25]. In accordance with the experimentally measured probabilities to dig spontaneously, we determined one value of K, in presence of an anfractuosity (K_a) and a smaller value in absence of anfractuosity (K_b). In any case, each time an ant performs a digging behaviour, she will deposit a fixed amount of digging pheromone on the cell where she is located and on one cell forward according to the axis of her body.

When an ant enters the transporting phase, she will now lay down a trail pheromone at each cycle:

$$Q_t = Q_{t0} - (F \times T) \tag{7}$$

where Q_t is the number of units of pheromone laid down, Q_{t0} the initial amount of pheromone, T the number of cycles elapsed since the ant entered into the transporting phase and F the strength of the pheromone gradient. The attraction of ants by trail pheromone during their displacement is implemented by adding pheromone quantities in the matrix M_d.

At the end of a cycle, evaporation and diffusion processes occur and are applied separately on the two types of pheromones. Evaporation is expressed by the following function:

$$Q_{t+1} = Q_t - (Q_t / \mu) \tag{8}$$

Q_{t+1} is the new amount of pheromone in a cell after it has lost a fraction of it previous amount Q_t. μ is a constant representing the half-life of the pheromone.

Diffusion is implemented in the following way: a fraction of the amount of pheromone on a cell (Q_c) is diffused to each neighbouring cell that has a smaller amount of pheromone (Q_n) according to the function:

$$If \quad Q_c > Q_{n,i} \quad then \quad Q_c = Q_c - Q_c/D \quad and \quad Q_{n,i} = Q_{n,i} + Q_c/D \tag{9}$$
$$for\ each\ neighbouring\ cell\ I$$

where D is a constant that represents the diffusion coefficient of the pheromone.

Figure 3 shows a graphical representation of the model and the values of all the parameters are given in table 3.

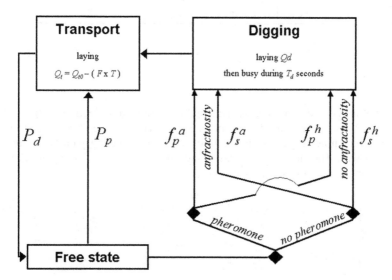

Fig. 3. Graphical representation of the model. The figure shows the three states possible for an ant and the probabilities to change from one to another (see the text for description)

Table 3. Parameters of the model.

Model parameters		Value	Origin
T_d	Time required to dig one pellet	40 s	
P_d	Probability to drop a pellet	0.188 s^{-1}	
P_p	Probability to pick up a pellet	0.102 / contact	experiments
f_s	Probability to dig spontaneously		
	-f_s^a : in presence of anfractuosity	0.054 / contact	
	-f_s^h : in absence of anfractuosity	2.83.10^{-2} s^{-1}	
f_p	Probability to dig in presence of pheromones	$fp = X^2 / (X^2 + K^2)$	
	-f_p^a : in presence of anfractuosity	$K_a = 200$	
	-f_p^h : in absence of anfractuosity	$K_b = 3500$	
Q_d	Quantity of digging pheromone per deposit	100 units	
$Q_{(t0)}$	Initial quantity of trail pheromone laid down	2000 units	estimation
F	Gradient strength of pheromone trail	5	
μ_t	Half-life parameter of the trail pheromone	45 min	
D_t	Diffusion coefficient of trail pheromone	500 000	
μ_d	Half-life parameter of digging pheromone	20 min	
D_d	Diffusion coefficient of digging pheromone	500 000	

3.2   Results

40 simulations reproducing the experimental conditions described in section 2.1 were
run with 100 ants, over a simulated time of 3 days. The volume of excavated sand was
recorded every hour. Figure 4 shows an example of tunneling pattern obtained with
the model with 100 ants over 3 days, and figure 5 shows 5 runs of the growth
dynamics obtained in the same condition.

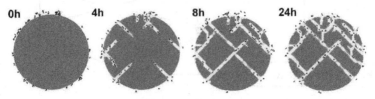

Fig. 4. Example of simulated tunneling pattern obtained with 100 ants.

The model succeeds in reproducing the logistic shape of the growth dynamics
observed in the experiments, but the time needed to reach the saturation phase was
three times shorter. This could be explained by two facts: first, the sand disk can be

dug completely in 54323 digging acts, with a sand pellet's weight of 4.298 mg, which is 2.546 times more than the experimentally measured mean weight (1.688 mg). Thus, in the model, ants excavate a volume of sand similar to the one obtained in experiments by performing around 3 times less digging acts; second, no aggregation processes were implemented in the model, so that all ants remained active during the simulation.

The linear regression described in section 2 was applied on the simulations results and the logistic model was always validated with high r^2 values (minimal r^2 value: 0.812). Comparisons were made for the mean final volume of excavated sand reached in the saturation phase in the model and the experiments for the same conditions. There was no significant difference between groups (model: 37.14 ±4.23 cm³; experiments: 37.15 ±9.35 cm³; t=-0.002; p=0.998, Student t test).

As in the experiments, the simulated dynamics appeared to be controlled by two kinds of feedbacks. A positive feedback resulting from the recruitment processes through trail and digging pheromones. Since there is no aggregation behaviour implemented in the model, the negative feedback may come from the fact that there exists some critical value of ant density in the network under which the amplification processes would fail to be efficient. At the beginning of the simulation, there is a high density of ants in the periphery of the sand disk, so that a recruitment process can take place. As the network grows, ants are more and more diluted in space, and the probability for an ant to meet a site recently dug and with sufficiently high pheromone

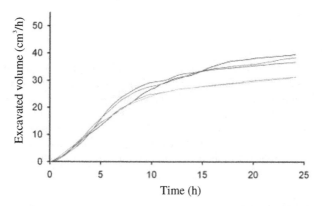

Fig. 5. Five examples of simulated growth dynamics obtained with 100 ants over the first 25 hours.

concentration becomes lower and lower. Finally, when the density of ants falls under a critical density, no recruitment can take place, since the mean time separating the visits of two different ants at the same digging site becomes too important with respect to the half-life of pheromones.

Such a mechanism of regulation of the digging activity could be a very efficient way to allow the colony to adapt the nest size to its population. If the digging activity stops when a certain density is reached, then we would observe a linear relationship between the number of ants and the total volume of excavated sand at the saturation phase.

3.3 The Influence of Colony Size

The model was used to assess the influence of the number of ants on the growth dynamics. 40 simulations were performed with 50 ants and 200 ants respectively over a simulated time of 3 days. Then, to test the predictive value of the model, we performed 5 corresponding experiments with 50 and 200 ants over 3 days.

Dynamics were logistic for all groups. The same linear regression procedure described in section 2.1 was applied to the dynamics and was in all cases validated by high r^2 values (minimal r^2 value: 0.79).

Figure 6a shows the mean volume of excavated sand as a function of the number of ants. A linear regression through origin was tested for the simulation results and validated with a high r^2 value ($r^2=0.993$; b= 0.379; t=133; p<0.001). Thus, the model predicts that the total volume of the network at saturation phase follows a linear relationship with the population size. Indeed, this relationship was confirmed by the experiments (Figure 6b). The linear regression was validated with high a r^2 values ($r^2=0.915$; b=0.487; t=12.313; p<0.001). These results show that there exists a regulation of the volume of the network according to the number of individuals. The model shows that this regulation arises in absence of any direct communication between ants; this results from a simple balance between recruitment processes and ants density. Moreover our model predicts that ants stop their activity when their mean density reaches 2.83 ±0.64 ants/cm^3. This was indeed confirmed by experiments in which the digging stopped at the same mean density value of 2.80 ±1.15 ants/cm^3.

4 Discussion

How can thousands of individuals in an ant colony coordinate their activity while digging a complex network of galleries and chambers? Ants limited cognitive abilities and their local perception of the environment as well as the dispersion of workers over many distant places prevent the colony using direct communication or centralized control to coordinate its collective activity. We have presented here a model of the growth of networks of galleries that exhibits adaptive properties. The size of the network is adapted to the size of the colony, even though each ant is not

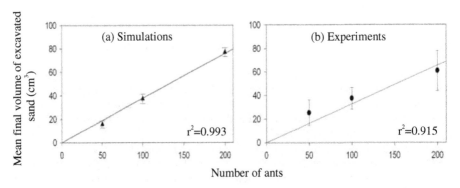

Fig. 6. Linear relationship between the final volume of excavated sand and the number of ants in the model (a) and in the experiments (b)

aware of the total number of ants in the nest. Each individual is an autonomous agent that has only a very local perception. Individuals use previous modifications in the environment by each ant as a feedback to coordinate their activity. The regulation of the network volume relies entirely on indirect communication mediated by the environment, mainly through the use of trail and digging pheromone combined with the density of ants that acts as an indirect cue. The collective regulation of the size of the galleries network is not explicitly programmed at the individual level and appeared to be an emergent property of the digging dynamics.

The study has been divided into three main steps. First, we performed an analysis at the global level through experiments with a constant population of 100 ants. We characterized the growth dynamics of the amount of excavated sand and showed that they are logistic. We then focused on the individual level and showed that several behaviours had a remarkably constant probability to occur. We finally used these experimentally measured parameters in the behavioural rules of the agents in an individual-based model. Introducing recruitment processes through pheromones, a common feature involved in several activities of ants, we showed that a group of ants is able to regulate their digging activity and stop it when a certain volume is reached, even if the individuals are still active and that nothing changed in their behavioural rules. This is consistent with the results on digging behaviour and tunnel construction in the ant *Lasius niger* [6], in which it has been shown that a volatile molecule was involved in the recruitment of ants towards a digging site.

When loaded with the experimental parameter values, the model not only leads to growth dynamics that reproduce the properties of galleries network formation, but also predicts how the growth is affected by the density of ants. We found that the final volume of excavated sand followed a linear relationship with the number of individuals. Experiments designed to test the model's predictions show that the predictions are indeed confirmed and the networks dynamics exhibit the same adaptive properties.

This coordination of digging activity in the ant *Messor sancta* emerges from a double feedback system:

(1) Positive feedbacks rapidly amplify spontaneous acts of excavation that first occur randomly and in a disorganized way all around the sand disk. Among these isolated initiation sites, some will involve several individuals so that recruitment will start. The system thus behaves in an autocatalytic way, leading the growth dynamics to its early phase of exponential growth. As it is shown in the model, this recruitment phase could result from the use of marking at the digging site with a highly volatile component and from the use of pheromone trails leading the individuals to the digging sites.

(2) Negative feedbacks progressively decrease the speed of digging and lead to its end when the volume of excavated sand reaches a critical value proportional to the number of ants. This mechanism of size adaptation of the network could result from a combination of aggregation processes and the decrease of ant density with time. This is consistent with the results of a previous study of nest size regulation in the ant *Lasius Niger* [6,7]. But the model shows that the decrease of ant's density alone could be sufficient for a colony to adapt the size of the network to the size of its population.

There exists a whole number of experimental evidence that shows that several other social phenomena such as aggregation behaviour or brood sorting in ants [25,26] result from the same kind of processes involving local amplification phenomena (aggregating where we already aggregated, dropping where we already dropped, digging where we already dug) and spatial competition between the resulting structures.

The next step will be to study the structure of the networks. An increasing number of studies suggest that the networks observed at different scales in nature could share common functional properties such as the minimization of distance between two nodes or the robustness to disconnection [27-29]. But networks in nature are not static. The origin of the functional properties of these networks has to be searched for in the processes that govern their growth. Thus a fundamental question remains to be investigated: are there common principles of growth that gives to these various networks structures the same adaptive properties?

Acknowledgments. This work was supported by the Programme Cognitique, the ATUPS program from the Université Paul Sabatier and a doctoral grant to Jerome Buhl from the French Ministry of Scientific Research.

References

1. Délye, G.:Observations sur le nid et le comportement constructeur de Messor arenarius. *Ins. Soc.*, 18, 15-20 (1971)
2. Thomé, G.: Le nid et le comportement de construction de la fourmi Messor ebenius, Forel (Hymenoptera, Formicoïdea). *Ins. Soc.*, 19, 95-103 (1972)
3. Frisch, K. von (ed.): Animal Architecture. Hutchinson, London (1975)
4. Brian, M.V. (ed.): *Social insects : Ecology and Behavioural Biology*. Chapman and Hall, London (1983)
5. Cerdan, P.: Etude de la biologie, de l'écologie et du comportement des fourmis moissonneuses du genre Messor (Hymenoptera, Formicidae) en Crau. *PhD thesis*, Univ. de Provence (Aix-Marseille I) (1989)
6. Rasse, P.: Etude sur la régulation de la taille et sur la structuration du nid souterrain de la fourmi Lasius niger. *PhD thesis*, Univ. Libre de Bruxelles (1999)
7. Rasse, P., Deneubourg, J.L.: Dynamics of Nest Excavation and Nest Size Regulation of Lasius Niger (Hymenoptera: Formicidae). *J. Insect Behav.*, 14, 433-449 (2001)
8. Theraulaz, G., Bonabeau, E. and Deneubourg, J.L.: The mechanisms and rules of coordinated building in social insects, pp. 309-330. In: Detrain, C., Deneubourg, J.L. and Pasteels, J. (eds.): *Information Processing in Social Insects*. Birkhauser Verlag, Basel (1999)
9. Grassé, P.P.: La reconstruction du nid et les coordinations interindividuelles chez Bellicositermes natalensis et Cubitermes sp. La théorie de la stigmergie : essais d'interprétation du comportement des termites constructeurs. *Ins. Soc.*, 6, 41-84 (1959)
10. Camazine, S., Deneubourg, J.L., Franks, N., Sneyd, J., Theraulaz, G., and Bonabeau, E. (eds.): *Self Organization in Biological Systems*. Princeton University Press, Princeton (2001)
11. Nicolis G. and Prigorine I. (eds.): *Self-organization in nonequilibrium systems*. Wiley, New York (1977)
12. Deneubourg, J.L. and Goss, S.: Collective patterns and decision making. *Ethol. Ecol. Evol.*, 1, 295-311 (1989)
13. Bonabeau, E.,Theraulaz, G., Deneubourg, J.-L., Aron, S. and Camazine, S.: Self-organization in social insects. *TREE*, 12, 188-193 (1997)

14. Gerard, J.F., Gonzalez, G., Guilhem, C., Le Pendu, Y., Quenette, P.Y. and Richard-Hansen, C.: Emergences collectives chez les ongulés sauvages, pp.. 171-186. In: Theraulaz, G. and Spitz, F. (eds.): *Auto-organisation et comportement*. Hermès, Paris (1997)
15. Aoki, I.: A simulation study on the schooling mechanism in fish. *Bull. Jap. Soc. Sci. Fish.* 48, 1081-1088 (1982)
16. Parrish, J. K., and EdelsteinKeshet, L.: Complexity, pattern, and evolutionary trade-offs in animal aggregation. *Science*, 284, 99-101 (1999)
17. Tschinkel, W.R.: Sociometry ans sociogenesis of colonies of the fire ant Solenopsis invicta during one annual cycle. *Ecol. Monogr.*, 63, 1-16 (1993)
18. Tschinkel, W.R.: Sociometry ans sociogenesis of colonies of the Florida harvester ant (Hymenoptera: Formicidae). *Ann. Entomol. Soc. Am.*, 92, 80-89 (1999)
19. Chrétien, L.: Organisation spatiale du matériel provenant de l'excavation du nid chez Messor barbarus et des cadavres d'ouvrières chez Lasius niger (Hymenoptera : Formicidae). *PhD thesis*, Univ. Libre de Bruxelles (1996)
20. Franks, N.R., Wilby, A., Silverman, B., Toft, C.: Self-organizing construction in ants: Sophisticated building by blind buldozing. *Anim. Behav.*, 44, 357-375 (1992)
21. Deneubourg, J.L., Franks, N.R.: Collective control without explicit coding : The case of communal nest excavation. *J. Insect. Behav.* 4, 417-432 (1995)
22. Sudd, J.H.: Specific patterns of excavation in isolated ants. Ins. Soc. 17, 253-260 (1970)
23. Jackson, B.D., Wright, and P.J. Morgan, E.D.: 3-Ethyl-2,5-dimethylpyrazine, a component of the trail pheromone of the ant Messor bouvieri. *Experientia*, 45, 487-489 (1989)
24. Edelstein-Keshet, L., Watmough J. and Ermentrout G.B.: Trail following in ants: individual properties determine population behaviour. *Behav. Ecol. Sociobiol.*, 36, 119-133 (1995)
25. Bonabeau, E., Dorigo, M. and Theraulaz, G. (eds.): *Swarm intelligence: From natural to artificial systems*. Oxford University Press, New York (1999)
26. Deneubourg, J.L., Goss, S., Franks, N., Sendova-Franks, A., Detrain, C. and Chrétien L.: The dynamics of collective sorting : Robot-Like ants and Ant-Like robots, pp. 356-363. In : Meyer, J.M. and Wilson, S.W. (eds.): *From animals to animats : Proceedings of the First International Conference on Simulation of Adaptive Behavior*. MIT Press, Cambridge, Massachusetts (1991)
27. Manrubia, S., Zanette, D.H. and Solé, R.V.: Transient dynamics and scaling phenomena in urban growth. *Fractals* 7, 1-16 (1999)
28. Watts, D.J., and Strogatz, S.H.: Collective dynamics of „small-world" networks. *Nature*, 393, 440-442 (1998)
29. Mathias, N. and Gopal, V.: Small worlds : How and why. *Phys. Rev. E.*, 63, 021117 (2001)

Solving the Homogeneous Probabilistic Traveling Salesman Problem by the ACO Metaheuristic

Leonora Bianchi[1], Luca Maria Gambardella[1], and Marco Dorigo[2]

[1] IDSIA, Strada Cantonale Galleria 2, CH-6928 Manno, Switzerland
{leonora,luca}@idsia.ch
http://www.idsia.ch
[2] IRIDIA, Université Libre de Bruxelles
CP 194/6, Avenue Franklin Roosevelt 50, 1050 Brussels, Belgium
mdorigo@ulb.ac.be
http://iridia.ulb.ac.be/~mdorigo/

Abstract. The Probabilistic Traveling Salesman Problem (PTSP) is a TSP problem in which each customer has a given probability of requiring a visit. The goal is to find an a priori tour of minimal expected length over all customers, with the strategy of visiting a random subset of customers in the same order as they appear in the a priori tour.

We propose an ant based a priori tour construction heuristic, the probabilistic Ant Colony System (pACS), which is derived from ACS, a similar heuristic previously designed for the TSP problem. We show that pACS finds better solutions than other tour construction heuristics for a wide range of homogeneous customer probabilities. We also show that for high customers probabilities ACS solutions are better than pACS solutions.

1 Introduction

Consider a routing problem through a set V of n customers. On any given instance of the problem each customer i has a known position and a probability p_i of actually requiring a visit, independently of the other customers. Finding a solution for this problem implies having a strategy to determine a tour for each random subset $S \subseteq V$, in such a way as to minimize the expected tour length. The most studied strategy is the a priori one. An a priori strategy has two components: the a priori tour and the updating method. The a priori tour is a tour visiting the complete set V of n customers; the updating method modifies the a priori tour in order to have a particular tour for each subset of customers $S \subseteq V$. A very simple example of updating method is the following: for every subset of customers, visit them *in the same order* as they appear in the a priori tour, skipping the customers that do not belong to the subset. The strategy related to this method is called the 'skipping strategy'. The problem of finding an a priori tour of minimum expected length under the skipping strategy is defined as the Probabilistic Traveling Salesman Problem (PTSP). This is an NP-hard problem [3,1], and was introduced in Jaillet's PhD thesis [8].

The PTSP approach models applications in a delivery context where a set of customers has to be visited on a regular (e.g., daily) basis, but all customers do

M. Dorigo et al. (Eds.): ANTS 2002, LNCS 2463, pp. 176–187, 2002.
© Springer-Verlag Berlin Heidelberg 2002

not always require a visit, and where re-optimizing vehicle routes from scratch every day is infeasible. In this context the delivery man would follow a standard route (i.e., an a priori tour), leaving out customers that on that day do not require a visit. The standard route of least expected length corresponds to the optimal PTSP solution.

In the literature there are a number of algorithmic and heuristic approaches used to find suboptimal solutions for the PTSP. Heuristics using a nearest neighbor criterion or savings criterion were implemented and tested by Jézéquel [9] and by Rossi-Gavioli [11]. Later, Bertsimas-Jaillet-Odoni [3] and Bertsimas-Howell [2] have further investigated some of the properties of the PTSP and have proposed some heuristics for the PTSP. These include tour construction heuristics (space filling curves and radial sort), and tour improvement heuristics (probabilistic 2-opt edge interchange local search and probabilistic 1-shift local search). More recently, Laporte-Louveaux-Mercure [10] have applied an integer L-shaped method to the PTSP and have solved to optimality instances involving up to 50 vertices.

Most of the heuristics proposed are an adaptation of a TSP heuristic to the PTSP, or even the original TSP heuristic, which in some cases gives good PTSP solutions. No application to the PTSP of nature-inspired algorithms such as ant colony optimization (ACO) [4] or genetic algorithms can be found in the literature. This paper investigates the potentialities of ACO algorithms as tour construction heuristics for the homogeneous PTSP, that is, for the PTSP where customers have the same probability of requiring a visit.

In the remainder of the paper we first introduce the PTSP objective function (section 2), then we describe the ACO algorithms which we tested (section 3), and in section 4 we report the experimental results obtained. The concluding section 5 summarizes the results obtained and indicates future directions for research on the PTSP.

2   The PTSP Objective Function

Let us consider an instance of the PTSP. We have a completely connected graph whose nodes form a set $V = \{i = 1, 2, ..., n\}$ of customers. Each customer has a probability p_i of requiring a visit, independent of the others. A solution for this instance is a tour λ over all nodes in V (an 'a priori tour'), to which is associated the expected length objective function

$$E[L_\lambda] = \sum_{S \subseteq V} p(S) L_\lambda(S) \ . \tag{1}$$

In the above expression, S is a subset of the set of nodes V, $L_\lambda(S)$ is the distance required to visit the customers in S (in the same order as they appear in the a priori tour), and $p(S)$ is the probability that all the customers in S require a visit:

$$p(S) = \prod_{i \in S} p_i \prod_{i \in V - S} (1 - p_i) \ . \tag{2}$$

Jaillet [8] showed that the evaluation of the PTSP objective function (eq.(1)) can be done in $O(n^2)$. In fact, let us consider (without loss of generality) an a priori tour $\lambda = (1, 2, \ldots, n)$; then its expected length is

$$E[L_\lambda] = \sum_{i=1}^{n} \sum_{j=i+1}^{n} d_{ij} p_i p_j \prod_{k=i+1}^{j-1} (1 - p_k) +$$

$$\sum_{i=1}^{n} \sum_{j=1}^{i-1} d_{ij} p_i p_j \prod_{k=i+1}^{n} (1 - p_k) \prod_{l=1}^{j-1} (1 - p_l) . \quad (3)$$

This expression is derived by looking at the probability for each arc of the complete graph to be used, that is, when the a priori tour is adapted by skipping a set of customers which do not require a visit. For instance, an arc (i, j) is actually used only when customers i and j do require a visit , while customers $i+1, i+2, ..., j$ do not require a visit. This event occurs with probability $p_i p_j \prod_{k=i+1}^{j-1} (1 - p_k)$ (when $j \leq n$). In the special class of PTSP instances where $p_i = p$ for all customers $i \in V$ (the homogeneous PTSP), equation (3) becomes

$$E[L_\lambda] = p^2 \sum_{r=0}^{n-2} (1 - p)^r L_\lambda^{(r)} \quad (4)$$

where $L_\lambda^{(r)} \equiv \sum_{j=1}^{n} d(j, (j + 1 + r) \bmod n)$. The $L_\lambda^{(r)}$'s have the combinatorial interpretation of being the lengths of a collection of $\gcd(n, r + 1)$ sub-tours[1] $\lambda_p^{(r)}$, obtained from tour λ by visiting one customer and skipping the next r customers. As an example, Fig. 1 shows $\lambda_p^{(0)}$ (i.e., the a priori tour), $\lambda_p^{(1)}$ and $\lambda_p^{(2)}$ for a PTSP with 8 customers.

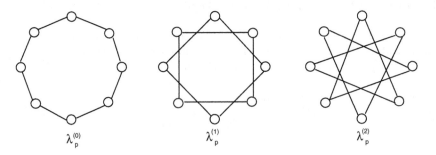

$$\lambda_p^{(0)} \qquad\qquad \lambda_p^{(1)} \qquad\qquad \lambda_p^{(2)}$$

Fig. 1. The lengths of the (sub)tours $\lambda_p^{(0)}$, $\lambda_p^{(1)}$ and $\lambda_p^{(2)}$, constitute the first three terms of the expected length for the homogeneous PTSP. From left to right, the total length of each set of (sub)tours gives the terms $L_\lambda^{(0)}$, $L_\lambda^{(1)}$ and $L_\lambda^{(2)}$ of equation (4).

---

[1] The term 'gcd' stays for 'greatest common divisor'.

3   Ant Colony Optimization

In ACO algorithms a colony of artificial ants iteratively and stochastically constructs solutions for the problem under consideration using artificial pheromone trails and heuristic information. The pheromone trails are modified by ants during the algorithm execution in order to store information about 'good' solutions. Most ACO algorithms follow the algorithmic scheme given in Fig. 2.

ACO are stochastic solution construction algorithms, which, in contrast to local search algorithms, may not find a locally optimal solution. Many of the best performing ACO algorithms improve their solutions by applying a local search algorithm after the solution construction phase. Our primary goal in this work is to analyze the PTSP tour construction capabilities of ACO, hence in this first investigation we do not use local search.

We consider a particular ACO algorithm, the probabilistic Ant Colony System, or pACS. This is an adaptation to the PTSP of the ACS algorithm [7,5], which was successfully applied to the TSP. In the following, we describe how pACS (and ACS) builds a solution and how it updates pheromone trails.

> **procedure** ACO metaheuristic for combinatorial optimization problems
>   Set parameters, initialize pheromone trails
>   **while** (termination condition not met)
>   *ConstructSolutions*
>   *ApplyLocalSearch*   *% optional*
>   *UpdateTrails*
>   **end while**

Fig. 2. High level pseudocode for the ACO metaheuristic.

3.1   Solution Construction

A feasible solution for an n-city PTSP is an a priori tour which visits all customers. Initially m ants are positioned on their starting cities chosen according to some initialization rule (e.g., randomly). Then, the solution construction phase starts (procedure *ConstructSolutions* in Fig. 2). Each ant probabilistically builds a tour by choosing the next customer to move to on the basis of two types of information, the pheromone τ and the heuristic information η. To each arc joining two customers i, j it is associated a varying quantity of pheromone τ_{ij}, and a heuristic value $\eta_{ij} = 1/d_{ij}$, which is the inverse of the distance between i and j. When an ant k is on city i, the next city is chosen as follows.

- With probability q_0, a city j that maximizes $\tau_{ij} \cdot \eta_{ij}^{\beta}$ is chosen in the set $J_k(i)$ of the cities not yet visited by ant k. Here, β is a parameter which determines the relative influence of the heuristic information.

– With probability $1 - q_0$, a city j is chosen randomly with a probability given by

$$p_k(i, j) = \begin{cases} \frac{\tau_{ij} \cdot \eta_{ij}^\beta}{\sum_{r \in J_k(i)} \tau_{ir} \cdot \eta_{ir}^\beta}, & \text{if } j \in J_k(i) \\ 0, & \text{otherwise.} \end{cases} \tag{5}$$

Hence, with probability q_0 the ant chooses the best city according to the pheromone trail and to the distance between cities, while with probability $1 - q_0$ it explores the search space in a biased way.

3.2   Pheromone Trails Update

Pheromone trails are updated in two stages. In the first stage, each ant, after it has chosen the next city to move to, applies the following local update rule:

$$\tau_{ij} \leftarrow (1 - \rho) \cdot \tau_{ij} + \rho \cdot \tau_0, \tag{6}$$

where ρ, $0 < \rho \leq 1$, and τ_0, are two parameters. The effect of the local updating rule is to make less desirable an arc which has already been chosen by an ant, so that the exploration of different tours is favored during one iteration of the algorithm.

The second stage of pheromone update occurs when all ants have terminated their tour. Pheromone is modified on those arcs belonging to the best tour since the beginning of the trial (best-so-far tour) by the following global updating rule

$$\tau_{ij} \leftarrow (1 - \alpha) \cdot \tau_{ij} + \alpha \cdot \Delta\tau_{ij}, \tag{7}$$

where

$$\Delta\tau_{ij} = ObjectiveFunc_{best}^{-1} \tag{8}$$

with $0 < \alpha \leq 1$ being the pheromone decay parameter, and $ObjectiveFunc_{best}$ is the value of the objective function of the best-so-far tour. In pACS the objective function is the PTSP expected length of the a priori tour, while in ACS the objective function is the a priori tour length.

4   Experimental Tests

4.1   Homogeneous PTSP Instances

Homogeneous PTSP instances were generated starting from TSP instances and assigning to each customer a probability p of requiring a visit, with p ranging from 0.1 to 0.9 with a 0.1 interval. We considered 21 TSP instances taken from two benchmarks. The first benchmark is the TSPLIB at http://www.iwr.uni-heidelberg.de/groups/comopt/software/TSPLIB95. From this benchmark we considered 7 symmetric instances[2] with a number of city between 30 and 200.

---

[2] The TSPLIB symmetric instances considered are oliver30, eil51, eil76, kroA100, lin105, ch150, d198.

The second benchmark is a group of instances where customers are randomly distributed on the square $[0, 10^6]$. For generating random instances we used the Instance Generator Code of the 8th DIMACS Implementation Challenge at http://www.research.att.com/~dsj/chtsp/download.html. We considered 7 uniform distributed instances and 7 clustered distributed instances from this benchmark, with a number of cities respectively equal to 50, 100, 150,..., 350.

4.2   Comparison between pACS and Other Tour Construction Heuristics

We compared pACS with two simple tour construction heuristics, the radial sort and the random best heuristic. The random best heuristic generates random tours and selects the one with the shortest expected length. Random best and pACS were run on the same machine (a Pentium Xeon, 1GB of RAM, 1.7 GHz processor) for the same CPU time ($stoptime = k \cdot n^2$ CPU seconds, with $k = 0.01$). For pACS we chose the same settings which yielded good performance in earlier studies with ACS on the TSP [5]: $m = 10$, $\beta = 2$, $q_0 = 0.98$, $\alpha = \rho = 0.1$ and $\tau_0 = 1/(n \cdot Obj)$, where n is the number of customers and Obj is the value of the objective function evaluated with the nearest neighbor heuristic [5]. For each experiment, we run 5 independent trials of pACS.

Radial sort builds a tour by sorting customers by angle with respect to the 'center of mass' of the customer spatial distribution. The 'center of mass' coordinates have been computed here by averaging over the customers coordinates. The a priori tour which radial sort builds does not depend on the customer probabilities, and a unique tour is thus used as a priori tour for all probabilities of the PTSP. Even if very simple, this heuristic is interesting for the PTSP, because of the conjecture [2] that the tour generated by radial sort is near optimal for small customer probabilities. Moreover, the combination of radial sort and the 1-shift local search have shown to be the best combination of tour construction and tour improvement heuristics in [2]. A disadvantage of radial sort is that it is only applicable to those PTSP instances where the coordinates of customers are known. In general, this is not the case for asymmetric PTSP instances, where the arc weights may have, for instance, the meaning of travel times.

The average relative performance of pACS with respect to radial sort and random best heuristics is shown in Fig. 3. The first observation is that pACS always performs better than radial sort and random best, for each probability and for each type of instance, while random best is always very poor both with respect to pACS and to radial sort. Secondly, radial sort and pACS are equivalent for small probabilities (prob = 0.1). This results supports the conjecture of near-optimality of radial sorted tours for small probability, and it is interesting that this also applies to non uniform instances, such as TSPLIB and random clustered instances.

4.3   Absolute Performance

For the PTSP instances we tested, the optimal solution is not known. Therefore, an absolute performance evaluation of a PTSP heuristic can only be done against

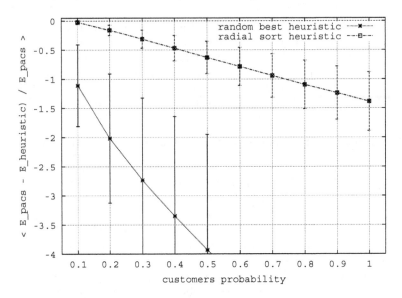

Fig. 3. Relative performance of pACS with respect to radial sort and random best heuristics. On the horizontal axis there is the customers probability. Each point is an average over 21 symmetric PTSP instances. Error bars represent average deviation, defined as $\sum_{i=1}^{n} |x_i - <x>|/n$, with $n = 21$.

some lower bound of the optimal PTSP solution, when this is available and tight enough. A lower bound to the optimal solution would give us an upper bound to the error performed by the pACS heuristic with respect to the PTSP optimum. In fact, if LB is the lower bound and $E[L_{\lambda^*}]$ is the optimal solution value, then by definition we have

$$E[L_{\lambda^*}] \geq LB . \tag{9}$$

If the solution value of pACS is $E[L_\lambda]$, then the following inequality holds for the relative error

$$\frac{E[L_\lambda] - E[L_{\lambda^*}]}{E[L_{\lambda^*}]} \leq \frac{E[L_\lambda] - LB}{LB} . \tag{10}$$

In the following we apply two different techniques for evaluating a lower bound to the optimal PTSP solution (and thus for evaluating the absolute performance of pACS). In the first case a theoretical lower bound is used while in the second case the lower bound is estimated by using Monte Carlo sampling.

Theoretical lower bound to the PTSP optimum. For the homogeneous PTSP and for instances where the optimal length L_{TSP} of the corresponding TSP is known, it is possible to use the following lower bound to the optimal expected length, as was proved in [2]

$$LB = pL_{TSP}(1 - (1 - p)^{n-1}) . \tag{11}$$

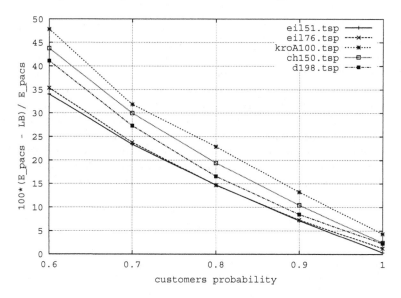

Fig. 4. Upper bound of relative percent error of pACS for 5 TSPLIB instances. Note that, for decreasing customers probability, the upper bound to the relative error becomes bigger at least partially because the lower bound to the optimum becomes less tight.

If we put this lower bound into the right side of equation (10), we obtain an upper bound of the relative error of pACS. Fig. 4 shows the absolute performance of pACS, evaluated with this method, for a few TSPLIB instances. From the figure we see that, for example, pACS finds a solution within 15% of the optimum for a homogeneous PTSP with customers probability 0.9. This technique for evaluating the absolute performance of a PTSP heuristic is rigorous, but has the limitation that the lower bound for small probabilities is not tight, so that it produces big overestimates of the error. The following technique is more flexible and gives better estimates of the error, even if, as we will see, it also has some limitations.

Estimated a posteriori optimum. The expected tour length under re optimization is defined as the average of the lengths of the optimal TSP solution to each subset of customers, and it is also called a posteriori optimum, since it is the value obtained by solving a TSP problem once the set of customers requiring a visit on a certain day is known. The a posteriori optimum is a lower bound on the optimal PTSP solution, because the length induced by the PTSP a priori tour on a subset of customers cannot be smaller than the optimal TSP solutions for that subset of customers.

The exact evaluation of the a posteriori optimum is impractical, because it requires the solution of 2^n instances of the TSP to optimality. The technique proposed in [2], consists in making two approximations. First, only a random sample of the 2^n subsets of customers is selected, by means of a stratified Monte Carlo sampling (see [2] for a detailed description). Second, each random sample of customers S is solved to near optimality as a TSP by choosing the best of $|S|/\gamma$ random tours (γ is a parameter) and applying to it the 3-opt local search.

This technique can be applied easily only to small instances (say, up to 100 customers). Otherwise, care must be taken in order to avoid overflow when generating the stratified samples from a set of 2^n subsets of customers. We report average results for 10 random uniform and clustered instances in Fig. 5. In our tests we used $\gamma = 0.5$ and about 400 samples. From the figure we see that pACS is within 8% of the optimum when applied to uniform random instances, while it is within 14% of the optimum if the instances are clustered.

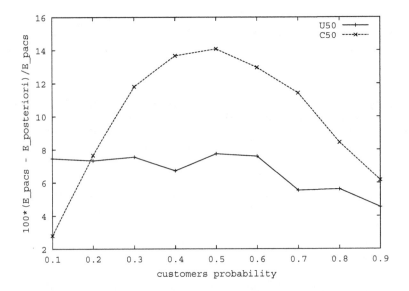

Fig. 5. Relative percent error with respect to the estimated a posteriori optimum for random uniform instances (U50) and for random clustered instances (C50) of 50 customers. Each point is an average over 10 random instances.

4.4   Comparison between pACS and ACS

In some cases an a priori tour found by a TSP heuristic can also be a good solution for the PTSP. An example of this is the a priori tour found by radial sort for small probabilities, as discussed in section 4.2. In this section we address the question of whether an a priori tour found by the ACS heuristic is also good for the PTSP, or at least as good a the solution found by pACS. In order to assess the relative performance of ACS versus pACS independently of the details of the settings, the two algorithms were run with the same parameters. The choice of

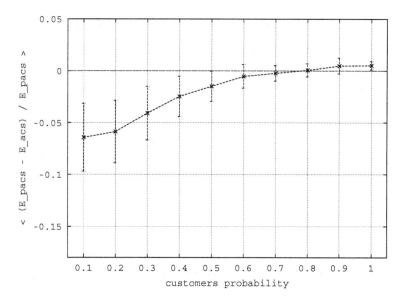

Fig. 6. Relative performance of pACS versus ACS for the homogeneous PTSP. The vertical axis represents $(E[L_\lambda(pACS)] - E[L_\lambda(ACS)])/E[L_\lambda(pACS)]$. On the horizontal axis there is the customer probability p. Each point of the graph is an average over 21 symmetric homogeneous PTSP instances. Note that for $p = 1$ ACS outperforms pACS, since for a fixed CPU stopping time ACS makes more iterations.

this parameter setting is the simplest among many other possibilities for comparing ACS and pACS. In fact, it would also be useful to tune pACS parameters, not only to achieve a better performance, but also to see how much they differ from ACS parameters (which are tuned on the TSP). Fig. 6 summarizes the results obtained. The figure shows the relative performance of pACS versus ACS, averaged over the 21 tested symmetric PTSP instances. From the figure we see that for small enough probabilities pACS outperforms ACS. Nevertheless, for all the problems we tested there is a range of probabilities $[p_0, 1]$ for which ACS outperforms pACS. The critical probability p_0 at which this happens depends on the problem.

The reason why pACS does not always perform better than ACS is clear if we consider two aspects. The first is the time complexity (speed) of ACS versus pACS. In both algorithms one iteration (i.e., one cycle through the *while* condition of Fig. 2) is $O(n^2)$ [6], but the constant of proportionality is bigger in pACS than in ACS. To see this one should consider the procedure *Update Trail* of Fig. 2, where the best-so-far tour must be evaluated in order to choose the arcs on which pheromone is to be updated. The evaluation of the best-so-far tour requires $O(n)$ time in ACS and $O(n^2)$ time in pACS. ACS is thus faster and always performs more iterations than pACS for a fixed CPU time.

The second reason why ACS performs better than pACS for high probabilities is that the length of an a priori tour (ACS objective function) may be considered as an $O(n)$ approximation to the $O(n^2)$ expected length (pACS objective function). In general, the worse the approximation, the worse will be the solution quality of ACS versus pACS. The quality of the approximation depends on the set of customer probabilities p_i. In the homogeneous PTSP, where customer probability is p for all customers, it is easy to see the relation between the two objective functions. For a given a priori tour λ of length L_λ we have

$$\Delta = L_\lambda - E[L_\lambda] = (1 - p^2)L_\lambda - \sum_{r=1}^{n-2}(1 - p)^r L_\lambda^{(r)}, \tag{12}$$

which implies

$$\Delta \sim O(q) \tag{13}$$

for $q \to 0$, with $q = 1 - p$. Therefore, the higher the probability, the better is the a priori tour length L_λ as an approximation for the expected tour length $E[L_\lambda]$.

5   Conclusions and Future Work

In this paper we investigated the potentialities of pACS, a particular ACO algorithm, for the homogeneous PTSP. We showed that the pACS algorithm is a promising tour construction heuristic for the PTSP. We compared pACS with other tour construction heuristics and we provided an estimation of the absolute error with respect to the optimal PTSP solution for some instances. We also compared pACS to ACS, and we showed that for customers probability close to 1, the ACS heuristic is a better alternative than pACS.

In this paper the ACO metaheuristic was applied without a local search for improving the a priori tour. The study of an efficient local search for the PTSP, which should greatly improve the solution quality of pACS and of any tour construction heuristic in general, is an important direction of research. At present we are investigating the heterogeneous PTSP, for different probability configurations of customers. This is an interesting issue, since it is closer to a real-world problem than the homogeneous PTSP.

Acknowledgments. This research has been partially supported by the Swiss National Science Foundation project titled "On-line fleet management", grant 16R10FM, and by the "Metaheuristics Network", a Research Training Network funded by the Improving Human Potential programme of the CEC, grant HPRN-CT-1999-00106. The information provided in this paper is the sole responsibility of the authors and does not reflect the Community's opinion. The Community is not responsible for any use that might be made of data appearing in this publication. Marco Dorigo acknowledges support from the Belgian FNRS, of which he is a Senior Research Associate.

References

1. D. J. Bertsimas. *Probabilistic Combinatorial Optimization Problems*. PhD thesis, MIT, Cambridge, MA, 1988.
2. D. J. Bertsimas and L. Howell. Further results on the probabilistic traveling salesman problem. *European Journal of Operational Research*, 65:68–95, 1993.
3. D. J. Bertsimas, P. Jaillet, and A. Odoni. A priori optimization. *Operations Research*, 38:1019–1033, 1990.
4. M. Dorigo, G. Di Caro, and L. M. Gambardella. Ant algorithms for discrete optimization. *Artificial Life*, 5(2):137–172, 1999.
5. M. Dorigo and L. M. Gambardella. Ant Colony System: A cooperative learning approach to the traveling salesman problem. *IEEE Transactions on Evolutionary Computation*, 1(1):53–66, 1997.
6. M. Dorigo, V. Maniezzo, and A. Colorni. The Ant System: Optimization by a colony of cooperating agents. *IEEE Transactions on Systems, Man, and Cybernetics – Part B*, 26(1):29–41, 1996.
7. L. M. Gambardella and M. Dorigo. Solving symmetric and asymmetric TSPs by ant colonies. In *Proceedings of the 1996 IEEE International Conference on Evolutionary Computation (ICEC'96)*, pages 622–627. IEEE Press, Piscataway, NJ, 1996.
8. P. Jaillet. *Probabilistic Traveling Salesman Problems*. PhD thesis, MIT, Cambridge, MA, 1985.
9. A. Jézéquel. *Probabilistic Vehicle Routing Problems*. Master's thesis, MIT, Cambridge, MA, 1985.
10. G. Laporte, F. Louveaux, and H. Mercure. An exact solution for the a priori optimization of the probabilistic traveling salesman problem. *Operations Research*, 42:543–549, 1994.
11. F. A. Rossi and I. Gavioli. *Aspects of Heuristic Methods in the Probabilistic Traveling Salesman Problem*, pages 214–227. World Scientific, Singapore, 1987.

Toward the Formal Foundation of Ant Programming

Mauro Birattari, Gianni Di Caro, and Marco Dorigo

IRIDIA, Université Libre de Bruxelles
CP 194/6, Av. Franklin D. Roosevelt 50, 1050 Brussels, Belgium
{mbiro,gdicaro,mdorigo}@ulb.ac.be
http://iridia.ulb.ac.be

Abstract. This paper develops the formal framework of *ant programming* with the goal of gaining a deeper understanding on *ant colony optimization* and, more in general, on the principles underlying the use of an iterated Monte Carlo approach for the multi-stage solution of combinatorial optimization problems. *Ant programming* searches for the optimal policy of a multi-stage decision problem to which the original combinatorial problem is reduced. In order to describe *ant programming* we adopt on the one hand concepts of optimal control, and on the other hand the *ant* metaphor suggested by *ant colony optimization*. In this context, a critical analysis is given of notions such as state, representation, and sequential decision process under incomplete information.

1 Introduction

In the last decade, a number of algorithms inspired by the foraging behavior of ant colonies have been introduced for the approximate solution of combinatorial optimization problems (see [8,10,9] for extensive reviews). The framework of *ant colony optimization* [8,10] gave recently a first unifying description of (most of) these algorithms. Loosely speaking, *ant colony optimization* presents the following features. A graph is defined in a way that each solution of the combinatorial problem corresponds to at least one path on the graph itself. The weights associated to the edges are such that the cost of a path equals the cost of the associated solution. In this sense, the goal of *ant colony optimization* is to find a path of minimum cost. To this end, a number of paths are incrementally generated in a Monte Carlo fashion, and the observed costs are used to bias the generation of further paths. This process is iterated with the aim of gathering information on the graph and of eventually producing a path of minimum cost. In *ant colony optimization*, the above described algorithm is visualized in terms of a metaphor in which the generation of a path is represented as the walk of an *ant* that, at each node, stochastically selects the following one on the basis of local information called *pheromone trail* [1]. In turn, the *pheromone trail* is modified by the *ants* in order to bias the generation of future paths toward better solutions.

The very possibility of obtaining better solutions by exploiting memory about solutions generated so far is the basic assumption of *ant colony optimization*. A further implicit assumption concerns what this memory should consist in. In spite of the key role played in all implementations of *ant colony optimization*, this assumption was never critically discussed before: The formal definition of *ant colony optimization* [8] envisages, for each optimization problem, a unique way of defining the memory. To clarify

M. Dorigo et al. (Eds.): ANTS 2002, LNCS 2463, pp. 188–201, 2002.

this issue, let us consider an optimization problem whose solutions are expressed by *ant colony optimization* as a sequence of components. *Ant colony optimization* generates solutions in the form of paths in the space of such components. Memory is kept of all the observed transitions between components. A degree of desirability is associated to each transition depending on the quality of the solutions in which it occurred so far. While a new solution is being incrementally generated, a component y is included with a probability that is proportional to the desirability of the transition between the last component included and y itself. Even if it seems natural that memory should be associated with pairs of solution components, as assumed by *ant colony optimization*, in this paper we maintain that such an assumption is just a matter of choice. Indeed, this is only one of the possible *representations* of the solution generation process that can be adopted for framing information about solutions previously observed. As it will be clear in the following, this representation is neither optimal nor the most natural, provided that a correct analysis of the problem at hand is given. Our analysis will be based on a clear understanding of the concept of state of the process of incremental solution construction.

In this paper, we propose a novel formal description of the combinatorial optimization problems to which *ant colony optimization* applies, and we analyze the implication of adopting a generic solution strategy based on the incremental Monte Carlo construction of solutions biased by a memory. This paper is an abridged version of a previously unpublished work by the same authors [4] and complements other recent theoretical analysis [15,21,12]. The paper introduces *ant programming* as an abstract class of algorithms which presents the characterizing features of *ant colony optimization* but which is more amenable to theoretical analysis for what concerns the concepts of representation and state. In particular, *ant programming* bridges the terminological gap between *ant colony optimization* and the fields of optimal control [3] and reinforcement learning [17]. Accordingly, the name *ant programming* was chosen for its assonance with *dynamic programming*, with which *ant programming* has in common the stress on the concept of state and the related idea of reformulating an optimization problem as a multi-stage decision problem and then searching for a good (hopefully optimal) decision *policy* for the latter. Both in dynamic programming and in *ant programming*, such a reformulation is not trivial and requires an *ad hoc* analysis of the optimization problem under consideration. These concepts, being among the main issues in this research, will be discussed in detail in the rest of the paper: Section 2 shows how to reformulate a discrete optimization problem into a discrete-time optimal control problem and then into a shortest path problem. Section 3 introduces the concepts of *graph of the representation*, *phantasma* and *sequential decision process* under incomplete information. Section 4 introduces and discusses the *ant programming* abstract class of algorithms. Section 5 discusses the main issues and describes the future developments of our research.

2 Discrete Optimization, Optimal Control, and Shortest Paths

Let us consider a discrete optimization problem defined by a finite set S of feasible solutions and by a cost function J. The set S is:

$$S = \{s_1, s_2, \ldots, s_N\}, \quad N \in \mathbb{N}, \quad N < \infty, \tag{1}$$

where each solution s_i is a n_i-tuple

$$s_i = (s_i^0, s_i^1, \ldots, s_i^{n_i-1}), \quad n_i \in \mathbb{N}, \quad n_i \leq n < \infty, \tag{2}$$

with $n = \max n_i$, and $s_i^j \in Y$, where Y is a finite set of *components*. The cost function $J : S \to \mathbb{R}$ assigns a cost to each feasible solution s_i. The optimization problem is therefore the problem of finding the element $\bar{s} \in S$ which minimizes the function J:

$$\bar{s} = \arg\min_{s \in S} J(s). \tag{3}$$

Being the set S finite, the minimum of J on S indeed exists. If such minimum is attained for more than one element of S, it is a matter of indifference which one is considered.

A feasible solution in S can be built incrementally starting from the 0-tuple $x_0 = ()$, and adding one-at-a-time a component. The generic iteration can be described as:

$$x_j = (u_0, \ldots, u_{j-1}) \to x_{j+1} = (u_0, \ldots, u_{j-1}, u_j), \quad \text{with } u_j \in Y, \tag{4}$$

where x_j is a *partial solution* of length j. A partial solution x_j is called *feasible* if it can be completed into a feasible solution $s_i \in S$, that is, if at least one feasible solution $s_i \in S$ exists, of which x_j is the initial sub-tuple of length j. It is understood that a process generating a sequence of feasible partial solutions necessarily ends up into a feasible solution. For each feasible partial solution x_j, we define the set $U(x_j) \in Y$ of all the possible new components u_j that can be appended to x_j giving in turn a feasible (partial) solution x_{j+1}.

Now, the set X of all feasible tuples x_j is finite since both the set S and the length of each feasible solution s_i are finite. Moreover, it can be shown that $S \subset X$, since all the solutions s_i are composed by a finite number of components, all belonging to Y.

Since a feasible solution can be obtained incrementally, the original optimization problem can be reformulated as a *multi-stage decision process* in which the optimal solution $\bar{s}$ is obtained by a sequence of decisions concerning the set Y of the components. Such a way of proceeding results particularly natural when the cost $J(s_i)$ of a solution s_i is expressed as a sum of contributions c_{j+1}, each related to the fact that a particular component u_j is included in the solution s_i itself after a sequence of components described by the tuple x_j. Formally, a function $\mathcal{C} : X \setminus \{x_0\} \to \mathbb{R}$ must be conveniently defined, which associates a cost c_{j+1} to each tuple x_{j+1}.[1]

The finite-horizon multi-stage decision process described above can be thoroughly seen as a deterministic *discrete-time optimal control problem* [5]. The tuple x_j can be seen as the *state* at time $t = j$ of a discrete-time dynamic system whose state-transition application is such that the state at time $t+1$ is obtained by appending the current control action $u_t \in U(x_t)$ to the state x_t:

$$\begin{cases} x_{t+1} = [x_t, u_t], \\ y_{t+1} = u_t, \end{cases} \tag{5}$$

The set of the feasible actions, given the current state, is a subset of the range of the output: $U(x_t) \subset Y$.

---

[1] Given rule (4), the tuple x_{j+1} determines uniquely the tuple x_j and the component u_j, and is in turn determined uniquely by them. Therefore, the function $\mathcal{C}$ could be equivalently defined as a function mapping on the real line an ordered pair $\langle x_j, u_j \rangle$, a transition $\langle x_j, x_{j+1} \rangle$, or even the triplet $\langle x_j, u_j, x_{j+1} \rangle$.

Now, let $\mathcal{U}$ be the set of all the admissible control sequences that bring the system from the initial state x_0 to a terminal state belonging to S: The generic element of $\mathcal{U}$, $u = \langle u_0, u_1, \ldots, u_{\tau-1} \rangle$, is such that the corresponding state trajectory, which is unique, is $\langle x_0, x_1, \ldots, x_\tau \rangle$, with $x_\tau \in S$, and $u_t \in U(x_t)$, for $0 \le t < \tau$. In this sense, the dynamic system defines a mapping $\mathcal{S} : \mathcal{U} \to S$ which assigns to each admissible control sequence $u \in \mathcal{U}$ a final state $s = \mathcal{S}(u) \in S$.

The problem of optimal control consists in finding the sequence $\bar{u} \in \mathcal{U}$ for which the sum J of the costs c_t, incurred along the state trajectory, is minimized:

$$\bar{u} = \arg \min_{u \in \mathcal{U}} J\big(\mathcal{S}(u)\big), \tag{6}$$

where with "arg min" we denote the element of $\mathcal{U}$ for which the minimum of the composed function $J \circ S$ is attained. If such a minimum is attained for more than one element of $\mathcal{U}$, it is a matter of indifference which one is considered.

It is apparent that the solution of the problem of optimal control stated in (6) is equivalent to the solution of the original optimization problem (3), and that the optimal sequence of control actions $\bar{u}$ for the optimal control problem determines uniquely the optimal solution $\bar{s}$ of the original optimization problem. Since the set X is discrete and finite, together with all the sets $U(x_t)$, for all $x_t \in X$, and since trajectories have a fixed maximum length n, all the possible state trajectories of the system (5) can be conveniently represented through a weighted and oriented graph with a finite number of nodes. Let $\mathcal{G}(X, U)$ be such a graph, where X is the set of nodes and U is the set of edges, and let $C : U \to \mathbb{R}$ be a function that associates a weight to each edge. In terms of system (5), each node of the graph $\mathcal{G}(X, U)$ represents a state x_t of the system. The set $U \subset X \times X$ is the set of the edges $\langle x_t, x_{t+1} \rangle$. Each of the edges departing from a given node x_t represents one of the actions $u_t \in U(x_t)$, feasible when the system is in state x_t. Finally, the function C is defined in terms of the function $\mathcal{C}$. Namely, $c_{t+1} = C(\langle x_t, x_{t+1} \rangle) = \mathcal{C}(x_{t+1})$ is the cost of the edge $\langle x_t, x_{t+1} \rangle$. Furthermore, on the graph $\mathcal{G}(X, U)$ we can single out the initial state x_0, as the only state with no incoming edges, and the set S of the terminal nodes from which no edges depart. In terms of the graph $\mathcal{G}(X, U)$ and of the function C, the optimal control problem (6) can be stated as the problem of finding the *path of minimal cost* from the initial node x_0 to any of the terminal nodes in S.

As already mentioned in Section 1, the solution strategy of *ant colony optimization* is based on the iterated generation of multiple paths on a graph that encodes the optimization problem under consideration. As it will be defined in the following, this graph is obtained as a transformation of the graph $\mathcal{G}$ consisting in an aggregation of nodes. In previous works on *ant colony optimization*, the graph resulting from such a transformation was the only graph taken into consideration explicitly. In this paper, we move the focus on the original graph $\mathcal{G}$ and on the properties of the transformation.

3 Markov and Non-Markov Representations

Consistently with the optimal control literature, we have called *state* each node of the graph $\mathcal{G}(X, U)$ and, by extension, we call *state graph* the graph $\mathcal{G}$ itself. In the following, the properties of the state graph will be discussed in the perspective of the solution of problem (3), and in relation to the solution strategy of *ant colony optimization*. The ant metaphor will be used to visualize abstract concepts. In particular, we will picture the state evolution of system (5), and therefore the incremental construction of a solution, as the walk of an *ant* on the state graph $\mathcal{G}$. In the following, the state x_t at time t will be called interchangeably the "partial solution," the "state of the system," or, by extension, the "state of the ant."

The state of a stochastic or deterministic dynamic system can be informally thought of as the piece of information that gives the most predictive description possible of the system at a given time instant.[2] Since what is known in the literature as *Markov property* is related precisely to the concept of state, it is clear that the state, when correctly conceived, is *always* a state in the Markov sense: When described in terms of its state, any discrete-time system is *intrinsically* Markov.[3] It is therefore of dubious utility to state the Markov property with respect to a dynamic system *tout court*. Of much greater significance, it is to assert the Markov property of a *representation*. Informally, we call a representation the structure in which an agent[4] frames experience: an agent refers to a representation for describing the state of the system, for possibly keeping memory of observed trajectories, and for performing predictions or control actions. In the limit, a representation might bear the same information as the state. In this case the Markov property holds for such representation. In the more general case, a representation is of non-Markov type, that is, it gives less information than the state. Being non-Markov is therefore a characteristic of the interaction system-agent and is related to the fact that the

---

[2] A detailed analysis of the concept of state in the context of ant colony optimization can be found in [4]. A general analysis of the concept of state is given in the classical literature on linear system theory [20], dynamic programming [2], and optimal control [3].

[3] For a discrete Markov decision process, the following holds by definition: $P(x_{t+1}|x^t, u^t) = P(x_{t+1}|x_t, u_t)$, where $x^t = (x_t, x_{t-1}, x_{t-2}, \dots)$ and $u^t = (u_t, u_{t-1}, u_{t-2}, \dots)$ indicate the past history of x and u, respectively. Now, let us consider a time-varying system whose state dynamic is given by $x_{t+1} = f_t(x_t, u_t, \xi_t)$ where x_t and u_t are respectively state and input at time t, and the state disturbance $\xi_t \sim P(\xi)$ is a white noise independent of the state and the input in the following sense: $P(\xi_t|x^t, u^t) = P(\xi_t)$. Clearly, x_{t+1} is a random variable whose distribution is $P(x_{t+1}|x_t, u_t) = P(\Xi_{x_{t+1}})$ where $\Xi_{x_{t+1}} = \{\xi : f_t(x_t, u_t, \xi) = x_{t+1}\}$ is the set of the values ξ that, for the given x_t and u_t, map to x_{t+1}, and $P(\Xi_{x_{t+1}})$ indicates the probability of observing a ξ belonging to such a set. The Markov property holds when the above introduced time-varying system is seen as a decision process. In particular:

$$P(x_{t+1}|x^t, u^t) = \sum_{\Xi_{x_{t+1}}} P(\xi|x^t, u^t) = \sum_{\Xi_{x_{t+1}}} P(\xi) = P(\Xi_{x_{t+1}}) = P(x_{t+1}|x_t, u_t).$$

The treatment given above assumes that ξ is a discrete variable. Thought the property holds also for continuous ξ, the proof for the general case involves a more complex notation and goes beyond the scope of this footnote.

Conversely, any discrete Markov decision process is a state description of a system in the Kalman sense. It is straightforward to verify that any $x_{t+1} \sim P(x_{t+1}|x_t, u_t)$ can be written in the form $x_{t+1} = f_t(x_t, u_t, \xi_t)$ where the dependence on time t accounts for the fact that in the definition of the Markov property the distributions at different temporal instants need not be the same. Since $x_{t+1} = f_t(x_t, u_t, \xi_t)$ is the classical form in which the state dynamic of a generic time-varying system can be given, the assertion is proved.

[4] By *agent* we mean any entity acting on or observing purposely the system at hand.

agent describes the system in terms of a representation that brings less information than a state description. In general, such a shortcoming of the representation can be ascribed to the inability of the agent to obtain information on the system, or to the deliberate choice of reducing the amount of information to be handled. In this second case, we are facing a *quality-complexity dilemma*.

In the context of *ant colony optimization*, as pointed out in Section 1, the basic assumption is that better solutions can be obtained by exploiting memory about previously generated ones. In this context, the discussion proposed above entails two major issues. First, for most combinatorial optimization problems of interest for which the state space grows exponentially with the size of the problem itself, it is clear that it is infeasible to gather and use memory about solutions in terms of a state description: it is very unlikely that a trajectory has exploitable superpositions with previously generated ones. Therefore, in *ant colony optimization* it is necessary to refer to a representation that reduces the information retained about the current state. This determines some sort of *aliasing* of distinct states which induces a criterion for *generalizing* previous experience. Second, as it will be made clear in the following, since a generic representation is non-Markov, it is not possible to generate feasible solutions on the basis of the sole representation. Therefore, it is necessary to refer to a state description in order to insure that a feasible solution be generated. These two issues, taken together, force to devise a strategy for the incremental generation of solution that on the one hand refers to a state description for guaranteeing feasibility, and on the other hand refers to a representation for optimizing the quality of the generated solution. The characteristics of the representation to be adopted reflect the design choice regarding the trade-off associated with the *quality-complexity dilemma*. *Ant programming* makes explicit the necessity to refer both to a representation and to a state description. Every step in the incremental construction of a solution consists of two sub-steps: first, a set of feasible candidate actions is defined on the basis of information pertaining to the state description; second, one of such candidates is selected on the basis of its desirability expressed in terms of the representation. In this sense, *Ant programming* introduces the categories needed for understanding some mechanisms already adopted in *ant colony optimization* such as, for instance, keeping and updating at each step the list of the components whose inclusion into the solution under construction would make the latter unfeasible. Such a list implicitly brings information about the state of the solution construction process.

For the class of problems discussed in this paper, a formal definition of a representation can be given with reference to the state graph $\mathcal{G}(X, U)$. We define the *representation graph* as the graph $\mathcal{G}_r(Z_r, U_r)$, where Z_r is the set of the nodes and U_r is the set of the edges. Furthermore, we call *generating function of the representation* the function $r : X \rightarrow Z_r$ that maps the set X of the states onto the set Z_r. The function r associates therefore to every elements of X an element in Z_r: every element $z_t \in Z_r$ has *at least* one preimage in X, but generally the preimage is not unique. The notation $r^{-1}(\{z_t\}) = \{x_\tau | r(x_\tau) = z_t\}$ indicates the set of states x_τ whose image under r is z_t. The function r induces an equivalence relation on X: Two states x_i and x_j are *equivalent* according to the representation defined by r, if and only if $r(x_i) = r(x_j)$. In this sense, a representation can be seen as a *partition* of the set X. In the following, we will call each $z_t \in Z_r$ a *phantasma*, adopting the term used by Aristotle with the meaning of *mental image.*[5] With such a term we want to stress that, from the point of view of an agent that observes the system through the representation r, z_t plays the role of the *phenomenal perception*, that is, what is retained about the system at time t for optimization purposes.[6]

Thanks to the notion of *phantasma*, we can give a precise interpretation to the concept of representation in the context of the control problem (6). As we pointed out before, the state evolution of the system (5) can be described as the walk of an *ant* on $\mathcal{G}(X, U)$. Let us assume now that the *ant* visits in sequence the nodes $x_0, x_1, \ldots, x_n$. The same sequence, under the representation induced by r, appears as a sequence $z_0, z_1, \ldots, z_n$ where for each i, with $0 \leq i \leq n$, z_i is the *phantasma* of the state x_i, that is, $z_i = r(x_i)$. In the *ant* metaphor, we say that the *ant*, though moving on the state graph $\mathcal{G}(X, U)$, *represents* its movement on the representation graph $\mathcal{G}_r(Z_r, U_r)$. In control theory, the process that carries the state into what we call a *phantasma*, is related to the concept of *state-space reduction*.[7]

In the same spirit of the definition of the set Z_r, also the set of the edges U_r can be defined in terms of the generating function r. The set $U_r \subset Z_r \times Z_r$ is the set of the edges $\langle z_i, z_j \rangle$ for which an edge $\langle x_i, x_j \rangle \in U$ exists on the state graph such that x_i and x_j are the preimages under r of z_i and z_j, respectively. Formally:

$$U_r = \Big\{ \langle z_i, z_j \rangle \mid \exists \langle x_i, x_j \rangle \in U : z_i = r(x_i),\, z_j = r(x_j) \Big\}.$$

When the system is described through a generic representation r, the subset $U_r(t) \subset U_r$ of the admissible control actions at time t cannot usually be described in terms of the *phantasma* z_t alone, but needs for its definition the knowledge of the underlying state x_t. In other words, for the generic generating function r, the *phantasma* z_t does not bring the same information as the state x_t and therefore the corresponding representation is non-Markov. The adoption of a non-Markov representation is by no means free from complications. While on the graph $\mathcal{G}$ every (partial) path is a (partial) feasible solution and *vice versa*, on $\mathcal{G}_r$ this property does not hold anymore. As far as the construction of feasible solutions is concerned, $\mathcal{G}$ is not therefore superseded by $\mathcal{G}_r$: As anticipated before, the graph $\mathcal{G}_r$ and the information stored on it are used for optimizing the construction of a solution while the graph $\mathcal{G}$ is used for guaranteeing feasibility. In any case, because of the loss of topological information induced by the transformation from $\mathcal{G}$ to $\mathcal{G}_r$ and since the optimization process is based on $\mathcal{G}_r$, in the general case only sub-optimal solutions will be obtained.

The parallel of the weight function C of $\mathcal{G}$ for the graph $\mathcal{G}_r$ cannot be defined in a straightforward manner for a generic r. Moreover, it results more useful to define the

---

[5] Aristotle (384–322 BC) *De Anima*: "The soul never thinks without a mental image."

[6] As an example, let us consider the case in which the set Z_r coincides with the set of solution components Y and $r : [x_t, u_t] \mapsto u_t$. This is the typical transformation adopted in the applications of *ant colony optimization* to the traveling salesman problem and to other combinatorial optimization problems. For this reason, such a transformation will be denoted in the following as r_{aco}.

[7] Yet, the result of a *state-space reduction* does not have a standard name in control theory and the various terms used always bring a direct reference to the concept of state: e.g. *reduced state*. It is just in order to underline the important qualitative difference between the properties of the state and those of the result of a *state-space reduction*, that we introduce here the term *phantasma* to denote the latter.

weights of the edges of the graph $\mathcal{G}_r(Z_r, U_r)$ so that they describe the quantity that in *ant colony optimization* is called *pheromone trail*. The function $T : U_r \to \mathbb{R}$ will be used in the process of selecting an action by an *ant* when perceiving a given *phantasma*, and will be iteratively modified in order to improve the quality of the solutions generated. The definition of the function T will be given in Section 4.

4   Ant Programming

In this section we introduce *ant programming* as a new class of algorithms that deal with the optimization problems (3) under the form described by (6). *Ant programming* is inspired by *ant colony optimization*, and from the latter it inherits the essential features, the terminology and the underlying philosophy. The aim of this section is mostly speculative: we do not describe a specific algorithm, but rather a class of algorithms, in the sense that we define a general resolution strategy and an algorithmic structure where some components are functionally specified but left uninstantiated.

4.1   The Three Phases of Ant Programming

Two are the essential features of *ant programming*. The first is the incremental Monte Carlo generation of complete paths over the state graph $\mathcal{G}$, on the basis of desirability information provided by the function T associated with the representation graph $\mathcal{G}_r$. The second is the update of the desirability information in $\mathcal{G}_r$ on the basis of the cost of the generated solutions and the use of such information to bias subsequent generations. These two features are described in terms of the three *phases* that, when properly iterated, constitute *ant programming*: At each iteration, a new set of *ants*, hereafter called a *cohort*, is considered. Each *ant* in the *cohort* undergoes a *forward* phase that determines the generation of a path, and a *backward* phase that states how the costs experienced along such a path should influence the generation of future paths. Finally, each iteration is concluded by a *merge* phase that combines the contribution of all the *ants* of the *cohort*. The three phases *forward*, *backward*, and *merge* are in turn characterized by the three *operators* π, ν, and σ respectively.

The forward phase. Using the terminology of *ant colony optimization* and in the light of the formalization given in Section 3, *ant programming* metaphorically describes each Monte Carlo run as the walk of an *ant* over the graph $\mathcal{G}(X, U)$, where at each node a random experiment determines the following node. In the ant metaphor, the random experiment is depicted as a *decision* taken by the *ant* on the basis of a probabilistic policy parameterized in terms of the function T, usually called the *pheromone trail*, defined on the set of edges of the graph $\mathcal{G}_r(Z_r, U_r)$.

The *forward* phase can be described as follows: Let us suppose that after t decision steps the partial solution built so far is $(u_0, \ldots, u_{t-1})$. The state of the solution generation process is therefore $x_t = (u_0, \ldots, u_{t-1})$. In the ant metaphor, this fact is visualized as an *ant* being in the node x_t of $\mathcal{G}(X, U)$. The *ant* perceives the state x_t in terms of the *phantasma* $z_t = r(x_t)$. In the general case, it is not possible to express the set $U_r(t)$ of admissible actions available to the *ant* when in z_t only in terms of z_t itself, and of the information given by $\mathcal{G}_r$. The set $U_r(t)$ of the admissible actions at time t is indeed:

$$U_r(t) = U_r(z_t|x_t) = \left\{ \langle z_t, z_{t+1} \rangle \in U_r \mid z_t = r(x_t), \exists u \in U(x_t) : z_{t+1} = r([x_t, u]) \right\}.$$

The decision of the *ant* consists in the selection of one element from the set $U_r(z_t|x_t)$ of the available transitions, as described at the level of the graph $\mathcal{G}_r$. Once an element, say $\langle z_t, z_{t+1} \rangle$, is selected, the partial solution is transformed according to Eq. 4 and Eq. 5: $x_{t+1} = [x_t, u_t] = (u_0, \ldots, u_{t-1}, u_t)$, where $x_{t+1} \in r^{-1}(\{z_{t+1}\})$ is one of the preimages of the *phantasma* z_{t+1}. In terms of the metaphor, this state transition is described as a movement of the *ant* to the node x_{t+1} of $\mathcal{G}$ which in turn is perceived by the *ant* as a movement to the *phantasma* $z_{t+1} = r(x_t)$ on $\mathcal{G}_r$.

The decision among the elements of $U_r(z_t|x_t)$ is taken according to the first operator of *ant programming*: the *stochastic policy* π. Given the current *phantasma* and the set of admissible actions $U_r(z_t|x_t)$, the policy selects an element of $U_r(z_t|x_t)$ as the outcome of a random experiment whose parameters are defined by the weights $T(\langle z_t, z_{t+1} \rangle)$ associated with the edges $U_r(z_t|x_t)$ of the graph $\mathcal{G}_r(Z_r, U_r)$. Accordingly we will adopt the following notation to denote the stochastic policy:

$$\pi\left(z_t, U_r(z_t|x_t); T|_{U_r(z_t|x_t)}\right). \tag{7}$$

With the notation $T|_{U_r(z_t|x_t)}$ we want to suggest that, when in z_t, the full knowledge of the function T is not strictly needed to select an element of the set $U_r(z_t|x_t)$. Indeed it is sufficient to know the restriction of T to the subset $U_r(z_t|x_t)$ of the domain U_r.[8] The function T plays the role of parameter of the policy π: changing T will change the policy itself.

In relation to the definition of the policy π, it is worth noticing here how the decision process uses the information contained in the two graphs $\mathcal{G}$ and $\mathcal{G}_r$: The decision is taken on the basis of information pertaining to the graph $\mathcal{G}_r$, restricted by the knowledge of the actual state x_t which in turn is a piece of information pertaining to the graph $\mathcal{G}$.

Given the abstract definition (7) of the policy π, the *forward* phase can be defined as the sequence of steps that take one *ant* from the initial state x_0, to a solution, say $s = x_\tau$, of the original combinatorial problem (3). Each of such steps is composed by three operations: first define, on the basis of the current state x_t, the set $U_r(z_t|x_t)$ of the available transitions; second select a transition on $\mathcal{G}_r$; and third move on $\mathcal{G}$ from the current node x_t to the neighboring node x_{t+1}. Formally, the single *forward* step is described as:

$$\begin{aligned} \langle z_t, z'_{t+1} \rangle &= \pi\left(z_t, U_r(z_t|x_t); T|_{U_r(z_t|x_t)}\right); \\ x_{t+1} &= \mathcal{F}\left(x_t, \langle z_t, z'_{t+1} \rangle\right); \\ z_{t+1} &= r(x_{t+1}), \end{aligned} \tag{8}$$

where the operator π is the stochastic policy that indicates the transition to be executed as seen on the graph $\mathcal{G}_r$, and where with the operator $\mathcal{F}$ we denote the operation of

---

[8] This fact is the expression of one of the feature of *ant programming*, namely the *locality* of the information needed by the *ant* in order to take each elementary decision. Such a feature plays and important role in the implementation, allowing a *distribution* of the information on the graph of the representation $\mathcal{G}_r$.

selecting one preimage x_{t+1} of z'_{t+1} and moving to it on the graph $\mathcal{G}$ from the current state x_t. Such a movement on $\mathcal{G}$ will be indeed "perceived" by the *ant* as a movement to the *phantasma* $z_{t+1} = r(x_{t+1}) = z'_{t+1}$, as requested by the policy π.

The backward phase. The ultimate goal of *ant programming* is to find a policy $\bar{\pi}$, not necessarily stochastic, such that a sequence of decisions taken according to $\bar{\pi}$ leads an *ant* to define the solution $\bar{s}$ which minimizes the cost function J of the original optimization problem (3).

Since the generic policy (7) is described parametrically in terms of the function T, that is, in terms of the weights associated to the edges of the graph $\mathcal{G}_r$, a search in the space of the policies amounts to a search in the space of the possible weights of the graph $\mathcal{G}_r$ itself. From a conceptual point of view, the function T is to be related to Hamilton's *principal function* of the calculus of variations, and to the *cost-to-go* and *value function* of dynamic programming and reinforcement learning. More precisely, the function T can be closely related to the function that in the reinforcement learning literature is known as "*state*-action value function," and that is customarily denoted by the letter Q. In fact, $T(\langle z_t, z_{t+1} \rangle)$ determines, as to (7), the probability of selecting the action "go to *phantasma* z_{t+1}" when the current *phantasma* is z_t. It therefore associates to the *phantasma*-action pair, a number which represents the *desirability* of performing such an action in the given *phantasma*. In this respect, it is clear the similarity with the role of the function Q in reinforcement learning.[9] The value of $T(\langle z_t, z_{t+1} \rangle)$ is generally given as a *statistic* of the observed cost of paths containing the transition $\langle z_t, z_{t+1} \rangle$. It therefore brings information on the quality of the solution that can be obtained by "going to z_{t+1}" when in z_t. Also in this respect, it can be stated a parallel with the function Q which indeed informs on the long-term cost of a given action, provided that future actions are selected optimally. In *ant programming*, as generally in reinforcement learning, the search in the space of the policies is performed through some form of *generalized policy iteration* [17]. Starting from some arbitrary initial policy, *ant programming* iteratively generates a number of paths in order to *evaluate* the current policy and then *improves* it on the basis of the result of the evaluation. At each iteration, therefore, a *cohort* of *ants* is considered, each generating a solution through a *forward* phase. Once the solution is completed, each *ant* traces back its path proposing at each visited *phantasma* an update of the local values of the function T on the basis of the costs experienced in the forward movement. This phase is denoted in the terminology of *ant programming* as the *backward* phase of the given *ant*. The actual new value of T is obtained by some combination of the values proposed by the *ants* of the *cohort*. This phase is denoted as the *merge* phase.

Let us now see in detail the *backward* phase for a given single *ant*. Let us consider a complete path $x = \langle x_0, x_1, \ldots, x_\tau \rangle$ over the graph $\mathcal{G}$. If $z = \langle z_0, z_1, \ldots, z_\tau \rangle$ is the complete forward path as seen under r, and $c = \langle c_1, \ldots, c_\tau \rangle$ is the experienced sequence of costs, then the single step of the *backward* phase is:

---

[9] An important difference is precisely that the function Q supposes a direct knowledge of the state, while T refers to the *phantasma*. In reinforcement learning, the situation in which more states are not perceived as distinct is termed *perceptual aliasing* [19].

$$z_t = \mathcal{B}(z_{t+1}, z),$$
$$T'(\langle z_t, z_{t+1}\rangle) = \nu(c, T), \tag{9}$$

where the operator $\mathcal{B}$ indicates a single step backward on $\mathcal{G}_r$, along the forward trajectory z. The operator ν is the key element of the *backward* phase. It has the role of proposing a new value for the weight associated to each visited edge $\langle z_t, z_{t+1}\rangle$, on the basis of the sequence of costs experimented during the *forward* phase, and of the current values of the function T. Hence, in our pictorial description of *ant programming*, this phase is pictured through an *ant* that "traces back" its forward path and leaves on such a path some information. >From a logical point of view, the different strategies for propagating the information gathered along a path are to be related to the different *update* strategies in reinforcement learning. In particular, to propose values of T' only for the visited transitions and on the basis of the cost of the associated solution, is equivalent to what in reinforcement learning is called *Monte Carlo update* [17]. On the other hand, it is equivalent to a *Q-learning update* [18] to propose a value of T' for a visited transition on the basis of the experienced cost for the transition itself and of the minimum of the current values that T assumes on the edges departing from the node to which the considered transition leads. The details of the definition of the *backward* phase, and in particular of the operator ν are not given as part of the description of *ant programming* and are left uninstantiated.

The merge phase. In the same spirit, we leave here undefined in its details also the *merge* phase which combines the different functions T' proposed by the individual *ants* of the same *cohort*. At this level of our description it will be sufficient to note that, for every transition $\langle z_t, z_{t+1}\rangle \in U_r$, the actual new value of $T(\langle z_t, z_{t+1}\rangle)$ will be some linear or nonlinear function of the current value of $T(\langle z_t, z_{t+1}\rangle)$, and of the different $T'_j(\langle z_t, z_{t+1}\rangle)$, where j is the index ranging over the *ants* of the *cohort*. The *merge* phase will be therefore characterized by the operator σ:

$$T(\langle z_t, z_{t+1}\rangle) = \sigma\big(T(\langle z_t, z_{t+1}\rangle), T'_1(\langle z_t, z_{t+1}\rangle), T'_2(\langle z_t, z_{t+1}\rangle), \dots\big). \tag{10}$$

Different possible instances of the operators ν and σ will be discussed in a future work.

4.2   The Algorithm and the Metaphor

The abstract definition of *ant programming* was given in previous sections in terms of the operators π, ν, and σ. In order to define an instance of the *ant programming* class, such operators need to be instantiated. Together with the operators π, ν, and σ, the other key element in the definition of an instance of the class, is the generating function r that defines the relation between the state graph $\mathcal{G}$ and the representation $\mathcal{G}_r$. We will therefore denote an instance of *ant programming* with the 4-tuple $\mathcal{I} = \langle r, \pi, \nu, \sigma\rangle$. Indeed, other elements are to be instantiated as, for example, the number of ants composing a *cohort* and the way of initializing the function T. Anyway, such elements are either less relevant, or are to be defined as a more or less direct consequence of the definition of $\mathcal{I}$.

In particular, the 4-tuple $\mathcal{I}$ gives an operative definition of the function T. As seen in the previous sections, the generating function r, together with the graph $\mathcal{G}$, gives the topology of the graph $\mathcal{G}_r$ and determines therefore the domain of the function T. The operator π defines how the values of T are used in the decision process, while the operators ν and σ define how the function T is to be modified on the basis of the quality of the solutions obtained. According to the pictorial description of *ant programming*, the function T is called *pheromone trail* and defines the policy π followed by the *ant* during the forward walk. Once a solution s is completed, the *ant* traces back its forward path and *deposits its pheromone* to update the function T. The role of the *pheromone trails* T is therefore to make available the information gathered on a particular path by one *ant* belonging to one given *cohort*, to other *ants* of a future *cohort*; it is therefore a form of *inter-cohort* communication mediated by the graph $\mathcal{G}_r$. >From the terminology adopted in the studies on social insects [13], it is customary to refer to such indirect communication with the term *stigmergy* [7].

At this point, having defined the 4-tuple $\mathcal{I}$, we have completed the definition of the elements that are necessary to handle the complexity of the combinatorial problem (3) in the spirit of the solution strategy originally suggested by *ant colony optimization*.

5   Discussion and Future Work

Future work will concentrate on the analysis of *ant programming* and on the properties of its possible instances. In particular, it is of paramount importance to gain a full understanding of the impact of the choice of r, the *generating function of the representation*, on the resulting algorithms. Such a function associates a *phantasma* to the current state and therefore can be informally thought of as the "lens" under which the process of incremental construction of a solution is seen. In this sense, "the *ant* never thinks without a *phantasma*" and, as far as the decision process is concerned, this is to be understood as "the *ant* takes decisions on the basis of the *phantasma*." The generating function determines therefore the information on the basis of which decisions will be taken. At the extreme, the generating function might be a one-to-one mapping. In this case, only one state is associated to a *phantasma*, and *vice versa*. As a consequence, the state graph $\mathcal{G}$ and the representation graph $\mathcal{G}_r$ have the same topological structure and, therefore, the representation enjoys the Markov property. Accordingly, we refer to this extreme instance of the *ant programming* class with the name of *Markov ants*. *Markov ants* face directly the exponential explosion of the number of edges of the graph $\mathcal{G}$. Nevertheless, since r is a one-to-one mapping, no two states are *aliased* in the representation. As a consequence, the policy that according to (7) selects the action on the basis of the current *phantasma*, indeed implicitly bases the choice on the actual underlying state. >From this fact, different appealing properties follow. It can be shown, for instance, that an optimal policy exists, and that it is deterministic. The performance of *Markov ants* can be improved if the pheromone trails T and the operator ν are designed in such a way that the Markov property of the representation is fully exploited. This can be done by defining T as a costs-to-go function, and by allowing the operator ν to *bootstrap* [17]. In this way *Markov ants* would reduce to an algorithm of the *temporal difference* class [17]. Anyway, *Markov ants* are not meant to be implemented. The focus of *ant programming* is indeed on problems whose Markov representation is computationally intractable and, in such situations, *Markov ants* are ruled out by their very own nature. Still, *Markov ants* remain of great theoretical interest.

Another class of instances of *ant programming* is of much greater practical interest. These instances are characterized by the function r_{aco}, as in Footnote 6, that associates a *phantasma* with one and only one of the possible solution components. The function r_{aco} generates the representation used in almost all the implementations of *ant colony optimization* since the first "template" instance developed by Marco Dorigo and colleagues [11,6] back in 1991. Accordingly, we call *Marco's ants* the instances of this class. Thanks to the concepts introduced in this paper, it becomes apparent that the representation graph generated by r_{aco} is much more compact than the state graph. In order to compensate this drastic loss of information, most of the instances of *ant colony optimization* adopt some additional device both to guarantee the feasibility and to improve the quality of the solutions being built. As far as feasibility is concerned, all instances of *ant colony optimization* use an implicit description of the state graph usually in the form of a list of components already included into the solution under construction. As far as quality is concerned, two major approaches have been followed. In the first approach, some additional *a priori* knowledge about the problem at hand, has been combined to the estimate of the function T for the definition of the decision policy. In the second approach, local optimization procedures, *ad hoc* tailored on the problem at hand, have been used in order to improve the quality of the solutions generated by the *ants*. Some of the resulting implementations have been shown to be comparable to or better than state-of-the-art techniques on several NP-hard problems. Moreover, under "reasonable" assumptions on the characteristics of the other components of the algorithm, *ant colony optimization* has been proved to asymptotically converge in probability to the optimal solution [14,16].

Future developments of this work will analyze in detail the properties of the two above mentioned instances: *Markov ants* and *Marco's ants*. Further, it will be of great practical interest to evaluate the possibility of designing other instances of *ant programming* that, on the one hand, keep an eye on the practical implementation, as *Marco's ants* do, and that, on the other, try to preserve as much as possible the properties of a state-space representation, going therefore in the direction of *Markov ants*.

Acknowledgments. Mauro Birattari acknowledges support from the Metaheuristics Network, a Research and Training Network funded by the Commission of the European Communities under the Improving Human Potential programme, contract number HPRN-CT-1999-00106. The work of Gianni Di Caro has been supported by a Marie Curie Fellowship of the European Community programme Improving the Human Research Potential under contract number HPMF-CT-2000-00987. The information provided is the sole responsibility of the authors and does not reflect the Community's opinion. The Community is not responsible for any use that might be made of data appearing in this publication. Marco Dorigo acknowledges support from the Belgian FNRS, of which he is a Senior Research Associate.

References

1. R. Beckers, J. L. Deneubourg, and S. Goss. Trails and U-turns in the selection of the shortest path by the ant Lasius Niger. *Journal of Theoretical Biology*, 159:397–415, 1992.
2. R. Bellman. *Dynamic Programming*. Princeton University Press, Princeton, NJ, USA, 1957.
3. D. P. Bertsekas. *Dynamic Programming and Optimal Control*. Athena Scientific, Belmont, MA, USA, 1995. Vols. I and II.

4. M. Birattari, G. Di Caro, and M. Dorigo. For a formal foundation of the Ant Programming approach to combinatorial optimization. Part 1: The problem, the representation, and the general solution strategy. Technical Report TR-H-301, ATR Human Information Processing Research Laboratories, Kyoto, Japan, 2000.

5. V. Boltyanskii. *Optimal Control of Discrete Systems*. John Wiley & Sons, New York, NY, USA, 1978.

6. M. Dorigo. *Optimization, Learning and Natural Algorithms* (in Italian). PhD thesis, Dipartimento di Elettronica, Politecnico di Milano, Milan, Italy, 1992.

7. M. Dorigo, E. Bonabeau, and G. Theraulaz. Ant algorithms and stigmergy. *Future Generation Computer Systems*, 16(8):851–871, 2000.

8. M. Dorigo and G. Di Caro. The ant colony optimization meta-heuristic. In D. Corne, M. Dorigo, and F. Glover, editors, *New Ideas in Optimization*, pages 11–32. McGraw-Hill, New York, NY, USA, 1999.

9. M. Dorigo, G. Di Caro, and T. Stützle (Editors). Special issue on "Ant Algorithms". *Future Generation Computer Systems*, 16(8), 2000.

10. M. Dorigo, G. Di Caro, and L. M. Gambardella. Ant algorithms for distributed discrete optimization. *Artificial Life*, 5(2):137–172, 1999.

11. M. Dorigo, V. Maniezzo, and A. Colorni. The ant system: An autocatalytic optimizing process. Technical Report 91-016 Revised, Dipartimento di Elettronica, Politecnico di Milano, Milan, Italy, 1991.

12. M. Dorigo, M. Zlochin, N. Meuleau, and M. Birattari. Updating ACO pheromones using stochastic gradient ascent and cross-entropy methods. In S. Cagnoni, J. Gottlieb, E. Hart, M. Middendorf, and R. Raidl, editors, *EvoCOP 2002: Applications of Evolutionary Computing*, volume 2279 of *Lecture Notes in Computer Science*, pages 21–30. Springer-Verlag, Heidelberg, Germany, 2002.

13. P. P. Grassé. La reconstruction du nid et les coordinations interindividuelles chez *bellicositermes natalensis et cubitermes sp.* La théorie de la stigmergie: essai d'interprétation du comportement des termites constructeurs. *Insectes Sociaux*, 6:41–81, 1959.

14. W. Gutjahr. A graph-based ant system and its convergence. Special issue on Ant Algorithms, *Future Generation Computer Systems*, 16(8):873–888, 2000.

15. N. Meuleau and M. Dorigo. Ant colony optimization and stochastic gradient descent. *Artificial Life*, 8(2):103–121, 2002.

16. T. Stützle and M. Dorigo. A short convergence proof for a class of ACO algorithms. *IEEE Transactions on Evolutionary Computation*, 6(4), 2002, in press.

17. R. S. Sutton and A. G. Barto. *Reinforcement Learning. An Introduction*. MIT Press, Cambridge, MA, USA, 1998.

18. C. J. C. H. Watkins. *Learning from Delayed Rewards*. PhD thesis, King's College, Cambridge, United Kingdom, 1989.

19. S. D. Whitehead and D. H. Ballard. Learning to perceive and act. *Machine Learning*, 7(1):45–83, 1991.

20. L. Zadeh and C. Desoer. *Linear System Theory*. McGraw-Hill, New York, NY, USA, 1963.

21. M. Zlochin, M. Birattari, N. Meuleau, and M. Dorigo. Model-base search for combinatorial optimization. Technical Report TR/IRIDIA/2001-15, IRIDIA, Université Libre de Bruxelles, Brussels, Belgium, 2001.

Towards Building Terrain-Covering Ant Robots

Jonas Svennebring and Sven Koenig

College of Computing
Georgia Institute of Technology
Atlanta, GA 30312-0280, USA
{jonas,skoenig}@cc.gatech.edu

Abstract. We study robots that leave trails in the terrain to cover closed terrain efficiently and robustly. Such ant robots have so far been studied only theoretically for gross ant robot simplifications in unrealistic settings. In this article, we design ant robots for a realistic robot simulation environment. We discuss how large the markers should be that form the trail, how frequently they should be dropped, how large the sensed floor area below the ant robots should be, how the ant robots should move depending on where the markers are in the sensed area, and how the markers should be deleted to avoid saturating the floor with markers. We then report experiments that we have performed to understand the behavior of the resulting ant robots better, including their efficiency and robustness in situations where they are failing, they are moved without realizing this, and markers are deleted. Finally, we report the results of a large-scale experiment where ten ant robots covered an area of 25 by 25 meters repeatedly over 85 hours.

1 Introduction

Robotics researchers study ants for two reasons, namely to learn more about them [5] and to build better robots [3]. Researchers that are interested in the second objective often build ant robots that use trails for navigation, similar to the pheromone trails used by many ants [1]. Some researchers have imitated nature closely [20] while others only got inspiration from it. The trails have been be either real [23,24,25] or virtual [4,21,27]. An ant robot that follows a real trail arrives at its destination without having to know its exact coordinates, which completely eliminates solving difficult and time-consuming localization problems. It needs only simple sensors, namely sensors that are able to sense the trails, which are artificial landmarks that can be carefully designed to simplify sensing. In this article, we are interested in building teams of ant robots that robustly cover closed terrain once or repeatedly (that is, where every location is once or repeatedly swept by the body of a robot) which is important for mine sweeping, surveillance, surface inspection, or guarding terrain [7]. Repeated coverage is important for all of these tasks, even mine sweeping since mine sensors can fail to detect mines during the first coverage [8]. The main problem that robots have to overcome is that they never know exactly where they are in the terrain. The currently popular POMDP-based robot architectures [16] attempt

M. Dorigo et al. (Eds.): ANTS 2002, LNCS 2463, pp. 202–215, 2002.

time step 0

0	0	0		0	0
0	0	0		0	0
0	0			0	0
0	0	0	0	0	0
0	0	0	0	0	0

time step 1

3	0	0		0	0
0	0	0		0	0
0	0			0	0
0	0	0	0	0	0
0	0	0	0	0	0

time step 2

3	1	0		0	0
2	0	0		0	0
0	0			0	0
0	0	0	0	0	0
0	0	0	0	0	0

time step 3

3	1	0		0	0
2	2	0		0	0
1	0			0	0
0	0	0	0	0	0
0	0	0	0	0	0

time step 4

3	1	0		0	0
2	2	1		0	0
1	2			0	0
0	0	0	0	0	0
0	0	0	0	0	0

time step 5

3	1	1		0	0
2	2	1		0	0
1	2			0	0
0	2	0	0	0	0
0	0	0	0	0	0

time step 6

3	1	1		0	0
2	2	2		0	0
1	2			0	0
0	2	1	0	0	0
0	1	0	0	0	0

time step 7

3	1	2		0	0
2	2	2		0	0
1	2			0	0
0	2	1	0	0	0
1	1	1	0	0	0

Fig. 1. Previous Work: Node Counting.

to overcome this problem by providing robots with the best possible location estimates [26]. However, this approach is complicated and can be brittle for robots that are small and cheap and thus have extremely noisy actuators and sensors. We are exploring ant robots as an alternative to this approach. They can navigate robustly in closed terrain without knowing where they are. Like trail-following ant robots, they only have to leave markers in the terrain and sense the markers in their neighborhood. Different from trail-following ant robots, however, they need to move away from the trails rather then follow them. Teams of ant robots have the potential to cover closed terrain faster and be more fault tolerant than single ant robots. For example, teams of ant robots robustly cover terrain even if they do not communicate with each other except via the trails, do not have any memory, do not know the terrain, cannot maintain maps of the terrain, nor plan complete paths. In particular, teams of ant robots cover terrain even if the ant robots are moved without realizing this (say, by people running into them and pushing them accidentally to a different location), some of the ant robots fail, or some trails are destroyed. The trails also coordinate the ant robots implicitly and allow them to cover terrain faster than without any communication.

In earlier work, we and other researchers have demonstrated the advantages of covering terrain with ant robots but only for gross ant robot simplifications in unrealistic settings [18,30]. These approaches are unsuitable for implementations on real robots. In this article, we study more realistic approaches for covering terrain with ant robots, using much more realistic robot simulations than has been done before. We demonstrate experimentally that large teams of our ant robots cover large terrains repeatedly over long periods of time. Our goal is to build real ant robots based on our design, using the Pebble robots from IS Robotics.

2   Related Work

Real-time search methods [10,19] are able to cover graphs repeatedly with a small cover time [11,12,13,14,15,17,28,29,32], which suggests that they can be used to build ant robots that cover terrain [18,30]. This has been studied so far by imposing a regular four-connected grid over the terrain. The ant robots can

Fig. 2. Pebbles Robot.

move to each of the four neighboring cells of their current cell provided that the destination cell is traversable. They maintain a number for each cell (initially: zero) and always move from the current cell to the neighboring cell with the smallest number. Various real-time search methods have been developed [13, 19,28]. They differ only in how they update the numbers. *Node counting* [22] is probably the simplest real-time search method. Its numbers correspond to how often the cells have been visited by ant robots. Each ant robot always increases the number of its current cell by one directly before it moves and then moves to the neighboring cell that has been visited the least number of times by all ant robots. Figure 1 demonstrates the behavior of three ant robots that all use node counting. White cells are empty and gray cells are blocked. For simplicity, we make the (unrealistic) assumption in the figure that the ant robots move in a given sequential order and that several ant robots can be in the same cell at the same time. If a cell contains an ant robot, one of its corners is marked. Different corners represent different ant robots. In previous work, we have shown that teams of ant robots that each use node counting cover all cells repeatedly, analyzed the resulting cover time, and studied other properties of the resulting behavior experimentally [18]. However, node counting is unsuitable for implementations on real robots. Its unrealistic assumptions include that the ant robots only move in discrete steps, that several ant robots can be in the same cell at the same time, that the ant robots have markers of a large number of different intensities available, that they can mark cells uniformly, and so on. Methods for ant coverage in continuous spaces [31] assume no actuator or sensor noise or largely depend on random motion. Different from these more theoretical approaches, we investigate in this article how to design ant robots that perform well in realistic robot simulation environments.

3   Methodology

We use a realistic robot simulation environment to study how to build ant robots that use variations of node counting to cover closed terrain repeatedly. Figure 2 shows the target of our design, the Pebbles robot from IS Robotics. We simulated the Pebbles robot in the TeamBots simulator, a realistic multi-robot simulator [6]. Figures 6 and 7 show snapshots of the simulator. The ant robots use a schema-based navigation strategy with three behaviors that are active at the same time, namely our navigation behavior, a collision avoidance behavior, and random noise. Each behavior produces its own recommendation for how the robot should move, in form of a vector. The robot then moves in the direction of the weighted average of all vectors [2]. We optimized the weights by hand for the physical characteristics of the simulated Pebbles robot. The test terrain has a size of 10 by 10 meters and is discretized into cells with a precision of 1 by 1 centimeters. We say that the first coverage of the terrain is completed when every cell has been swept at least once by the body of a robot. In general, we define an additional coverage of the terrain to be completed when every cell has been swept at least once by the body of a robot after the previous coverage was completed. Unless stated otherwise, we report a cover time that has been averaged over five runs each, together with its corresponding 95 percent confidence interval.

In the following, we first report simulations that we performed to design the ant robots. We then report simulations that we performed to understand the behavior of the resulting ant robots better, including their efficiency and robustness in situations where they are failing, they are moved without realizing this, and markers are deleted. In this context, we also demonstrate that a large team of our ant robots indeed covers a large terrain repeatedly over long periods of time without ant robots getting stuck.

4   Design Decisions

Our overall design decision was to design ant robots that create trails by dropping markers, for example, dripping drops of a chemical substance. We assume that each cell either does not contain the chemical or is saturated by it. If a drop of the chemical drips onto a cell that is already saturated by the chemical, it might spill to a randomly selected cell within a ten centimeter range but has no effect otherwise. We need to decide how large the markers should be, how frequently they should be dropped, how large the sensed floor area below the ant robots should be, how the ant robots should move depending on where the markers are in the sensed area, and how the markers should be deleted to avoid saturating the floor with markers. All of these design decisions are interrelated.

4.1   Marker Size

We would like to keep the markers as small as possible. This way, we avoid having to refill expensive marker material frequently. Fortunately, smaller marker sizes result in faster coverage. Consequently, we use the smallest marker size in our experiments that can still be detected reliably, that is, markers of size 1 by 1 centimeters.

4.2   Drop Frequency

We would like to drop markers as infrequently as possible, for the same reason why we would like to keep them as small as possible. Fortunately, even drop frequencies of only one marker every three seconds result in an acceptable cover time. The optimal drop frequency turns out to be about 6.6 markers per second or one marker every 15 centimeters traveled if the ant robot moves at its maximum speed of one meter per second. These results can be explained as follows: Dropping no markers results in a random walk, which is inefficient. Higher drop frequencies result in a more intelligent cover behavior and thus speed up coverage. A small drop frequency is sufficient as long as the overall number of markers in the sensor field is small since then the influence of each marker on the movement of the ant robots is high. If the drop frequency is too high, trails can become barriers that are time consuming to cross, which increases the cover time. To understand this phenomenon, assume that a trail separates two parts of a room that does not contain any other trails. This trail is hard to cross for ant robots because they get repelled from it. The trail even gets reinforced every time an ant robot approaches it without crossing. The emerging barrier can only be crossed easily once the marker density in the part of the room that contains the ant robot has become sufficiently high. Based on these results, we use constant a drop frequency of 6.6 markers per second in our experiments.

4.3   Sensor Field

We would like the sensor field to be as small as possible to keep the cost of the sensors small. However, larger sensor fields result in faster coverage since they increase the information that the ant robots have available. Thus, we use a sensor field of 50 by 50 centimeters in our experiments, the size of the Pebbles robot.

4.4   Robot Movement

The navigation method specifies the heading of the ant robot at each point in time. We would like the ant robots to avoid turning too much. This makes them fast and makes it easier for them to cross trails (because they cannot quickly turn away when they approach them), which prevents barriers from forming and decreases the cover time. This is consistent with previous results that also suggest that motion changes should be avoided unless they are absolutely necessary [9]. We studied two different navigation methods: the vectorsum method and the discrete-4 method.

– The *vectorsum method* associates a vector with every marker in the sensor field. The vector points from the marker to the center of the sensor field. The vectorsum method then sums up the vectors, normalizes the resulting vector, and adds it to a vector that points forward with a magnitude that is proportional to the speed of the ant robot. The vectorsum method then recommends to move in the direction of the resulting vector. Figure 3 (left)

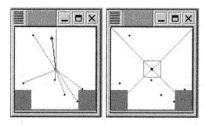

Fig. 3. Vectorsum (left) and Discrete-4 and Discrete-8 Navigation Methods (right).

shows an example. The window corresponds to the sensor field. The dots in the window correspond to the markers, and the arrow corresponds to the movement recommendation of the vectorsum method.

- The *discrete-4 method* partitions the sensor field as shown in Figure 3 (right). The square in the center of the sensor field remains unused to increase stability. The discrete-4 method then recommends to move in the direction (forward, backward, left, or right) that corresponds to the partition of the visual field that contains the fewest markers.[1]

The paths are straighter and coverage tends to be more intelligent the more directions a navigation method can recommend. This is an advantage of the vectorsum method, that can recommend arbitrary directions, over the discrete-4 method, that can recommend only four different directions. However, the straighter the paths, the more likely the ant robot will follow a trail if it is pushed onto it. This is a disadvantage of the vectorsum method over the discrete-4 method. We therefore decided to explore the discrete-8 method, a hybrid between the two navigation methods. The *discrete-8 method* partitions the sensor field exactly like the discrete-4 method but can recommend to move in eight different directions. It behaves as follows:

- It recommends to move forward (backward) if the fewest markers are in the front (back) partition and the left and right partitions have approximately the same number of markers.
- It recommends to move left (right) if the fewest markers are in the left (right) partition.

---

[1] In case several partitions tie for the fewest markers, the discrete-4 method uses the following tie-breaking strategy: If the left and right partitions have approximately the same number of markers, it recommends to move backward if the back partition has fewer markers than the front partition, otherwise it recommends to move forward. In all other cases, it recommends to move forward if the front partition ties for the fewest markers, recommends to move left if the left partition ties for the fewest markers but not the front partition, and otherwise recommends to move right. For example, the discrete-4 method recommends to move forward in the situation shown in Figure 3 (right).

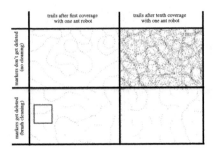

Fig. 4. Trails (with and without Cleaning).

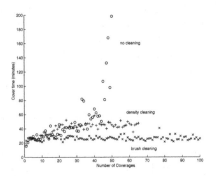

Fig. 5. Cover Time as Function of Number of Coverages.

- It recommends to move forward/left (forward/right) if the fewest markers are in the front partition and the left (right) partition has fewer markers than the right (left) partition.
- It recommends to move backward/left (backward/right) if the fewest markers are in the back partition and the left (right) partition has fewer markers than the right (left) partition.

It turns out that the discrete-8 method results in a smaller cover time than both of the other navigation methods. We therefore use the discrete-8 method in our experiments.

4.5   Cleaning Method

The top row of Figure 4 shows that the terrain gets saturated with markers over time. (The square in the lower left corner corresponds to the simulated robot.) The graph "no cleaning" in Figure 5 shows how this causes the cover time to increase for a single ant robot. For example, a 95-percent confidence interval for the cover time is 15.4 ± 1.5 minutes for the first coverage but 30.6 ± 4.0 minutes for the fifth coverage. Eventually, the terrain is saturated with markers and the behavior of the ant robot degrades to a random walk, which is inefficient. The cover time for the 50th coverage is already on the order of several hours. This is an important difference to the earlier gross ant robot simplifications in unrealistic settings described in the section on "Related Work," where the cover time does not increase over time. Thus, the trails now need to either evaporate or get deleted to keep the cover time small in the long run. Evaporating trails are problematic because the evaporation rate needs to get optimized for each application, for example, the size of the terrain. Thus, we propose to use long-lasting trails, such as trails of fluorescence or phosphorescence chemicals, that the ant robots delete themselves. This is different from previous work on trail-following ant robots, that used short-lasting trails whose lifetime is too short for covering terrain (such as heat trails [23], alcohol trails [25] and odor trails [24]).

We tested two simple methods for deleting markers. *Density cleaning* deletes a marker if there are two markers next to each other. *Brush cleaning*, on the other hand, deletes all markers in the cleaning area. It is an important design decision where to place the cleaning area to avoid that ant robots delete all markers that they have just placed or more subtle problems that result from them perceiving areas with deleted markers as not having been covered yet. This can result in ant robots following other ant robots or ant robots moving back to areas that were just covered. We therefore let the ant robots delete markers on either side below their belly, whereas they add markers in the center below their belly. Figure 3 shows the cleaning area in gray.

Brush cleaning is similar to vacuum cleaning and thus easier to implement than density cleaning but more prone to the problems just described. Figure 5 shows how the cover time increases with the number of coverages without cleaning, with density cleaning, and with brush cleaning. Density cleaning initially behaves like no cleaning because it only deletes markers if their density is so large that markers are next to each other. Since deleting markers removes information that is especially important when markers are still sparse, the cover time without cleaning and with density cleaning is smaller than the one with brush cleaning during the first five coverages. However, the cover time with density cleaning and brush cleaning is much smaller than the cover time without cleaning in the long run. The cover time with density cleaning is 45.1 ± 8.5 minutes in the long run, whereas the one with brush cleaning is only 26.5 ± 2.9 minutes. We therefore use brush cleaning in our experiments. The bottom row of Figure 4 shows that this prevents the terrain from getting saturated with markers.

5   Large-Scale Experiment

Based on the results reported in the previous section, we decided to use markers of size 1 by 1 centimeters, a drop frequency of 6.6 markers per second, a sensor field of 50 by 50 centimeters, the discrete-8 navigation method, and brush cleaning. The line in Figure 6 shows the beginning of the resulting trajectory of a single ant robot that conforms to these design decisions. We placed ten of these ant robots into an area of 25 by 25 meters that resembled a factory floor with two production lines and a number of office rooms, as shown in Figure 7. The

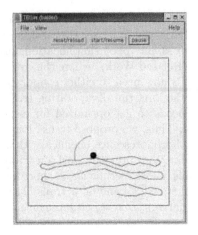

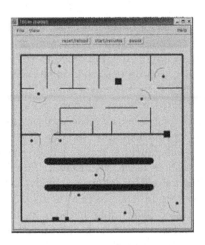

Fig. 6. Sample Cover Behavior.   **Fig. 7.** Large-Scale Experiment.

ant robots were started in three groups but quickly distributed evenly across the factory floor. They covered the factory floor 35 times during 80 hours, with an average cover time of about 146.9 ± 15.2 minutes. This experiment shows that a large team of our ant robots covers a large terrain repeatedly over long periods of time without ant robots getting stuck. The resulting cover time is much smaller than the cover time that results when the ant robots move randomly, which we simulated by making the ant robots not drop trails (but leaving everything else unchanged). With this change, the ant robots have problems to traverse small passages and thus no longer cover the terrain in a reasonable amount of time. However, our experiment also shows that the cover time can be improved further since a single ant robot that moves on an optimal trajectory can already cover the terrain in about 20 minutes if it moves at its maximum speed of one meter per second, although this ideal case cannot be achieved in practice since ant robots cannot follow trajectories precisely and at full speed.

6   Experimental Evaluation

We now report experiments that we performed to understand the behavior of our ant robots better, including their efficiency and robustness in situations where they are failing, they are moved without realizing this, and markers are deleted.

6.1   Scaling with Terrain Size

Figure 8 (left) shows how the cover time increases with the size of the terrain. The cover time is a linear function of the size of the terrain, and the variance of the cover time increases with the size of the terrain. This is an important result because some of our earlier theoretical results for gross ant robot simplifications

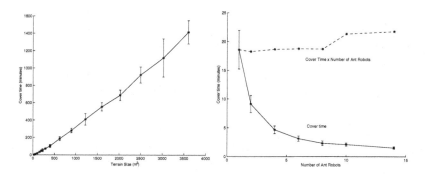

Fig. 8. Cover Time as Function of Terrain Size (left) and Number of Ant Robots (right).

in the unrealistic settings already described in the section on "Related Work" suggested that the worst-case cover time might not scale well with the size of the terrain and could potentially even increase exponentially [18].

6.2   Scaling with Number of Ant Robots

Figure 8 (right) shows how the cover time decreases with the number of ant robots. The cover time is inversely proportional in the number of ant robots. This result is similar to our earlier theoretical results for gross ant robot simplifications in unrealistic settings [18] except that increasing the number of ant robots has an additional advantage here: Trails can become barriers that are time consuming to cross. The more ant robots there are, the less this is an issue since it becomes more likely that there are ant robots on both sides of the barrier. Furthermore, the ant robots are able to cross barriers more easily since they get repelled from each other which might push them across barriers.

6.3   Coping with Error Conditions

One of the attractive properties of ant robots is that they cover closed terrain robustly even in situations where they are failing, they are moved without realizing this, and markers are deleted. We demonstrated these properties earlier for gross ant robot simplifications in unrealistic settings [18] and show in the following that our ant robots share these advantages.

Failing Ant Robots. We first measure the cover time when ant robots fail. This is important because ant robots can malfunction. Every six seconds in the experiment, every functional ant robot failed with a given probability and then remained motionless for an amount of time that was uniformly distributed between 0 and 2.8 minutes. Figure 9 (left) shows how the cover time increases with the failure probability, averaged over 50 runs each. The cover time is a linear

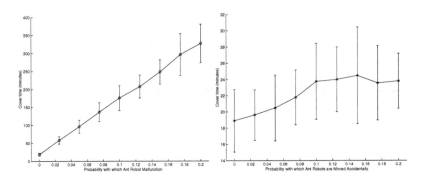

Fig. 9. Cover Time as Function of Probability with which Ant Robots Malfunction (left) and are Moved (right).

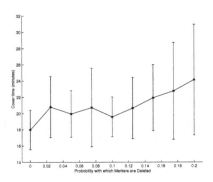

Fig. 10. Cover Time as Function of Probability with which Markers are Deleted.

function of the failure probability, and the variance of the cover time increases with the failure probability. Thus, our ant robots continue to cover the terrain and the cover time increases gracefully with the probability of failing.

Moving Ant Robots. We now measure the cover time when ant robots are accidentally moved without realizing this. This is important because people or other ant robots can easily run into them and accidentally push them to a different location. Every six seconds in the experiment, every ant robot was moved with a given probability in a random direction for a random distance (but not more than 25 centimeters). Figure 9 (right) shows how the cover time increases with the movement probability, averaged over 50 runs each. The cover time remains small even if the probability of being moved is 0.2, which is unrealistically high. Thus, our ant robots continue to cover the terrain and the cover time increases gracefully with the probability of being moved.

Deleting Markers. Finally, we measure the cover time when markers are accidentally deleted. This is important because markers can get destroyed due to wind, dust, rain, humans, or other ant robots. We test two different scenarios. In the first scenario, randomly selected individual markers are deleted. This could happen, for example, due to wind and dust. In the second scenario, all markers in randomly selected areas are deleted. This more systematic destruction of markers could happen, for example, due to humans walking around.

- In the first scenario, during each time step in the experiment, one randomly selected marker was deleted with a given probability. Our ant robots continue to cover the terrain and the cover time does not change significantly. This is so because the number of markers in the simulated area was in the thousands. Thus, deleting a small number of markers uniformly across the area decreases the overall marker density somewhat but has no large impact.
- In the second scenario, during each time step in the experiment, all markers in one randomly selected area of size 20 by 20 centimeters were deleted with a given probability. Figure 10 shows how the cover time increases with the erasure probability, averaged over 50 runs each. Our ant robots continue to cover the terrain and the cover time increases gracefully with the probability of areas being deleted.

7   Conclusions

In this article, we have described how to build ant robots that leave trails in the terrain to cover closed terrain efficiently and robustly. The trails convey information to the ant robots, which allows them to cover terrain more intelligently and faster than with random walks. The ant robots do not even need to be localized, which completely eliminates solving difficult and time-consuming localization problems. We reported the results of a large-scale experiment in a realistic robot simulation environment where ten ant robots covered an area of 25 by 25 meters repeatedly over 85 hours to demonstrate that our ant robots indeed cover closed terrain repeatedly over long periods of time without getting stuck. We also showed experimentally that our ant robots robustly cover terrain even if they are moved without realizing this (say, by people running into them), some ant robots fail, and some markers are destroyed. Our next step is to build ant robots in hardware, using the Pebbles robots from IS robotics. We are also working on schemes that decrease the cover time of our ant robots even further while continuing to let them cover closed terrain robustly without knowing where they are.

Acknowledgments. We thank Ashwin Ram for making his hardware available to us. The Intelligent Decision-Making Group is partly supported by NSF awards to Sven Koenig under contracts IIS-9984827, IIS-0098807, and ITR/AP-0113881 as well as an IBM faculty partnership award. The views and conclusions contained in this document are those of the authors and should not be interpreted as representing the official policies, either expressed or implied, of the sponsoring organizations, agencies, companies or the U.S. government.

References

1. F. Adler and D. Gordon. Information collection and spread by networks of patrolling ants. *The American Naturalist*, 140(3):373–400, 1992.
2. R. Arkin. Motor-schema based mobile robot navigation. *International Journal of Robotics Research*, 8(4):92–112, 1989.
3. R. Arkin. *Behavior-Based Robotics*. MIT Press, 1998.
4. T. Balch and R. Arkin. Avoiding the past: A simple, but effective strategy for reactive navigation. In *International Conference on Robotics and Automation*, pages 678–685, 1993.
5. T. Balch, Z. Khan, and M. Veloso. Automatically tracking and analyzing the behavior of live insect colonies. In *Proceedings of the International Conference on Autonomous Agents*, pages 521–528, 2001.
6. T. Balch and A. Ram. Integrating robotics research with JavaBots. In *Proceedings of the AAAI Spring Symposium*, 1998.
7. H. Choset. Coverage for robotics - a survey of recent results. *Annals of Mathematics and Artificial Intelligence*, 31:113–126, 2001.
8. D. Gage. Randomized search strategies with imperfect sensors. *Proceedings of SPIE Mobile Robots VIII*, 2058:270–279, 1993.
9. W. Huang. Optimal line-sweep-based decompositions for coverage algorithms. In *Proceedings of the International Conference on Robotics and Automation*, pages 27–32, 2001.
10. T. Ishida. *Real-Time Search for Learning Autonomous Agents*. Kluwer Academic Publishers, 1997.
11. S. Koenig. Exploring unknown environments with real-time search or reinforcement learning. In *Advances in Neural Information Processing Systems 12*, volume 11, pages 1003–1009, 1999.
12. S. Koenig and R.G. Simmons. Exploration with and without a map. In *Proceedings of the AAAI Spring Symposium on Learning Action Models*, pages 28–32, 1993. Available as AAAI Technical Report WS-93-06.
13. S. Koenig and R.G. Simmons. Real-time search in non-deterministic domains. In *Proceedings of the International Joint Conference on Artificial Intelligence*, pages 1660–1667, 1995.
14. S. Koenig and R.G. Simmons. Easy and hard testbeds for real-time search algorithms. In *Proceedings of the National Conference on Artificial Intelligence*, pages 279–285, 1996.
15. S. Koenig and R.G. Simmons. Solving robot navigation problems with initial pose uncertainty using real-time heuristic search. In *Proceedings of the International Conference on Artificial Intelligence Planning Systems*, pages 145–153, 1998.
16. S. Koenig and R.G. Simmons. Xavier: A robot navigation architecture based on partially observable Markov decision process models. In D. Kortenkamp, R. Bonasso, and R. Murphy, editors, *Artificial Intelligence Based Mobile Robotics: Case Studies of Successful Robot Systems*, pages 91–122. MIT Press, 1998.
17. S. Koenig and B. Szymanski. Value-update rules for real-time search. In *Proceedings of the National Conference on Artificial Intelligence*, pages 718–724, 1999.
18. S. Koenig, B. Szymanski, and Y. Liu. Efficient and inefficient ant coverage methods. *Annals of Mathematics and Artificial Intelligence – Special Issue on Ant Robotics*, 31:41–76, 2001.
19. R. Korf. Real-time heuristic search. *Artificial Intelligence*, 42(2-3):189–211, 1990.

20. D. Lambrinos, R. Möller, R. Labhart, R. Pfeifer, and R. Wehner. A mobile robot employing insect strategies for navigation. *Robotics and Autonomous Systems*, 30:39–64, 2000.

21. D. Payton, M. Daily, B. Hoff, M. Howard, and C. Lee. Autonomy-oriented computation in pheromone robotics. In *Proceedings of the Autonomous Agents Workshop on Autonomy Oriented Computation*, 2001.

22. A. Pirzadeh and W. Snyder. A unified solution to coverage and search in explored and unexplored terrains using indirect control. In *Proceedings of the International Conference on Robotics and Automation*, pages 2113–2119, 1990.

23. R. Russell. Heat trails as short-lived navigational markers for mobile robots. In *Proceedings of the International Conference on Robotics and Automation*, pages 3534–3539, 1997.

24. R. Russell, D. Thiel, and A. Mackay-Sim. Sensing odour trails for mobile robot navigation. In *Proceedings of the International Conference on Robotics and Automation*, pages 2672–2677, 1994.

25. R. Sharpe and B. Webb. Simulated and situated models of chemical trail following in ants. In *Proceedings of the International Conference on Simulation of Adaptive Behavior*, pages 195–204, 1998.

26. S. Thrun. Probabilistic algorithms in robotics. *Artificial Intelligence Magazine*, 21(4):93–109, 2000.

27. R. Vaughan, K. Stoey, G. Sukhatme, and M. Mataric. Whistling in the dark: Cooperative trail following in uncertainty localization space. In *Proceedings of the International Conference on Autonomous Agents*, pages 187–194, 2000.

28. I. Wagner, M. Lindenbaum, and A. Bruckstein. On-line graph searching by a smell-oriented vertex process. In S. Koenig, A. Blum, T. Ishida, and R. Korf, editors, *Proceedings of the AAAI Workshop on On-Line Search*, pages 122–125, 1997. Available as AAAI Technical Report WS-97-10.

29. I. Wagner, M. Lindenbaum, and A. Bruckstein. Efficiently searching a dynamic graph by a smell-oriented vertex process. *Annals of Mathematics and Artificial Intelligence*, 24:211–223, 1998.

30. I. Wagner, M. Lindenbaum, and A. Bruckstein. Distributed covering by ant-robots using evaporating traces. *IEEE Transactions on Robotics and Automation*, 15(5):918–933, 1999.

31. I. Wagner, M. Lindenbaum, and A. Bruckstein. MAC vs. PC: Determinism and randomness as complementary approaches to robotic exploration of continuous unknown domains. *International Journal of Robotics Research*, 19(1):12–31, 2000.

32. V. Yanovski, I. Wagner, and A. Bruckstein. Edge ant walk for patrolling networks. In *Proceedings of the International Workshop on Ant Algorithms*, 2000.

A New Ant Colony Algorithm Using the Heterarchical Concept Aimed at Optimization of Multiminima Continuous Functions

Johann Dréo and Patrick Siarry

Laboratoire d'Étude et de Recherche en Instrumentation Signaux et Systèmes (L.E.R.I.S.S.), Université de Paris XII Val-de-Marne
61 avenue du Général de Gaulle, 94010 Créteil, France
{dreo,siarry}@univ-paris12.fr

Abstract. Ant colony algorithms are a class of metaheuristics which are inspired from the behaviour of real ants. The original idea consisted in simulating the trail communication, therefore these algorithms are considered as a form of adaptive memory programming. A new formalization is proposed for the design of ant colony algorithms, introducing the biological notions of heterarchy and communication channels.

1 Introduction

Having recently given raise to a new metaheuristic method, the ant colony metaphor proved to be a successful approach to solve "difficult" optimization problems. The first algorithm inspired from the ant colony functioning is the "ant system" (Colorni & al. 1991), which has been applied to many combinatorial problems. Until now, there are few adaptations of such algorithms to continuous optimization problems. The first algorithm designed for continuous function optimization was *CACO* (for *Continuous Ant Colony Optimization*) (Bilchev & al. 1995) which is close to the original metaphor, but unfortunately, the use of two different processes inside the *CACO* algorithm leads to a delicate setting of parameters.

All of these algorithms use a particular trait of real ants behaviour: the pheromonal trail laying. Indeed, ants colonies are often viewed as distributed systems capable of solving complex problems by the way of *stigmergy*, which is a form of indirect communication mediated by modifications of environment. But the trail-laying behaviour is also a part of the *recruitment* process, as the recruitment is defined by biologists as "a special form of assembly in which members of a society are directed to some point in space where work is required" (Hölldobler & Wilson 1990, p.642). According to this definition, another optimization method has been developed for continuous optimization: the *API* algorithm (Monmarché & al. 2000); but *API* makes a poor use of memory that generally characterizes ant colony systems (Taillard & al. 1998).

Our point of view is that ant colony metaphor can be defined as a model using trail communication or, more widely, as a recruitment process. According

M. Dorigo et al. (Eds.): ANTS 2002, LNCS 2463, pp. 216–221, 2002.

to this idea, pheromonal trail laying may not be the only way to comprehend the ant colony metaphor for optimization problems (Bonabeau & al. 1997). Our recent research on modelling ants behaviour (Dréo 2001) has shown that the importance of inter-individuals communication may be underestimated, and that including it into an ant colony system may improve performances by accelerating the diffusion of information.

We were interested by these various ways of approaching the same idea and we thought that it would be useful to gather them in a single formalism. Indeed, there is another interesting way of approaching the social insects behaviour : the notion of *dense heterarchy*, developed in Sect. 2. An *heterarchical algorithm* takes advantages from a flow of information passing through a population of agents. These informations are exchanged using *communication channels* (see Sect. 3) and permit that a form of description of the objective function emerges from the system. In Sect. 4, we present an experimental result. Conclusion makes up the last section.

2   The Notion of Dense Heterarchy

2.1   A Biological Definition

This notion was firstly introduced by Wilson in 1988 to describe the information flow inside an ant colony (Wilson & Hölldobler 1988). In this idea, the two communication channels that we have evoked in the introduction are present, the stigmergic channel as well as the direct communication between individuals. One of the important issues here is that informations *flow* through the colony, each of the ants can communicate with any other (which makes the heterarchy being *dense*). This process constructs a kind of densely connected network which is not set up in a *hierarchical* but in a *heterarchical* manner (cf. Fig. 1).

Fig. 1. In a dense heterarchy, unlike hierarchy, any individual can communicate with any other one

Such a system has emergent properties, indeed if each agent operates with elementary rules and a limited accuracy, the whole population organized in heterarchy with mass communication can show an emergent pattern. These findings are well known in the study of self-organization (Camazine & al. 2000). To summarize, the notion of dense heterarchy describes the way the ant colony handles the informations that it receives from the environment. Each ant can communicate with any other at any time and the informations flow through the colony.

2.2 Heterarchical Algorithms

We propose here a simple formalization in order to apply the notion of dense heterarchy to an optimization problem. The main concept for implementing such an *heterarchical algorithm* is the idea of *communication channel* (cf. Fig. 2). These channels transmit an information, here the localization of a food source, and have some properties, like stigmergy and memory, which can be combined in the same channel, so that a large variety of different channels can be built.

Fig. 2. The informations pass from a part of the population to another one through a communication channel that has some specific properties

For example, some basic properties of the channels are listed below:

1. *Scope*: the way the information goes through the population. A sub-group of the population (from one to n agents) can exchange informations with another group of agents.
2. *Memory*: the way the information persists in the system. The information can be stored during some period of time or be transitory.
3. *Integrity*: the way the information is evolving in the system. The information can be modified, by one or more agents, by an external process, or not.

3   The Communication Channels

3.1 The Trail Channel

The first version of our algorithm is trying to be as close as possible to a continuous version of the original *ACO* (*Ant Colony Optimization*) algorithm (Colorni & al. 1991) which was designed for combinatorial problems. Due to this assumption, it has somewhat the same design as the local search part of the *CACO* algorithm (Bilchev & al. 1995), which was also inspired by the first Ant System.

This implementation uses only one communication channel which is inspired from the trail laying behaviour of ants. Here, each ant can deposit a certain amount of *pheromone* as a spot on the search space, proportionally to the amelioration of the objective function she founds on her way. These pheromonal spots can be perceived by all the members of the population, and diffuse into the environment. The ants are attracted by each spot according to its distance and to the amount of pheromone it contains. Each artificial ant has a "range" parameter that is normally distributed over the population. Each ant draws a random distance and then jumps of this length in the direction of her weighted gravity center, some noise modifying the final position.

To summarize this behaviour under the vision of the heterarchical concept, this stigmergic communication channel shows the following properties, underlined in Sect. 2.2:

1. *Scope*: When one ant lays a pheromonal spot, all the ants can subsequently perceive it.
2. *Memory*: The information persists in the system during a certain period of time, independently of the agents.
3. *Integrity*: The information is modified by time, to reproduce the pheromone evaporation.

There is some similarity between this algorithm and the "path-relinking" algorithm introduced by Glover (Glover & Laguna 1997), as the ants are moving through a set of informative points. We can also notice that the behavior of the algorithm reminds the *Particle Swarm Optimization* (*PSO* Kennedy & al. 1995) as the agents tend to gather in the same place, the whole population evolving as a swarm.

3.2   Using the Direct Inter-individual Communication

As we have said in Sect. 1, some biological works have led us to take an interest out of the common vision of the "trail" ant colony optimization. Indeed, we have implemented another communication channel which possesses the properties of the direct inter-individual interactions that can be observed in societies of some social insects as ants.

In concrete terms, each ant can send "messages" to another one. An ant receiving a message stores it in a stack with other incoming messages. In a second time, a message is read randomly in the stack. Here, the information sent is the position of the sender – an artificial ant – and the value of the objective function. The receiver compares the sender's value with its own value and decides if it moves near the sender's position. But if the receiver's value is better than the sender's one, then the receiver sends a message to another ant randomly chosen, and suppresses the read message. One can notice that the system needs to be "activated", so that an important parameter is the number of messages initially set.

This communication channel shows the following properties:

1. *Scope*: When an ant sends a message, only one ant can perceive it.
2. *Memory*: The information persists in the system during a certain period of time, under the form of ants memory.
3. *Integrity*: The informations stored are static.

4   Experimental Result

To illustrate the behaviour of an heterarchical algorithm, we have applied it on the B_2 function, which has two dimensions and some local minima. The two channels play complementary parts. Indeed, the direct channel leads to a form of intensification, because it gives more importance to the best points without taking into account the previously encountered regions. On the contrary, the trail channel –with its memory property– permits to perform a kind of diversification, by taking into account the previously evaluated points. Globally, the ants gather at the global optimum. However some of them are kept during some iterations near local optima by some pheromonal spots still persisting within the search space, afterwards evaporation and direct communications prevent them from being trapped. The way the trail channel permits a form of diversification is pointed out with more difficulties. The ants are moving around a gravity center that is the global optimum, but continue to explore the search space, as they are attracted by local minima.

One interesting issue is the way the two channels works in *synergy*. Indeed, on the simple B_2 function, the direct channel seems to be more appropriate as it allows the algorithm to converge more rapidly (Fig. 3). But if we take a look to the variation of the standard deviation for the algorithm using the two communication channels (Fig. 3), we notice that there is a kind of *periodicity*. This means that the population tends to gather near a value at one time and then tends to disperse. In other words, there is an alternation of short intensification phases (low deviation) and diversification phases (high deviation). This behaviour of the algorithm seems to be an emergent pattern, which cannot be observed when only one of the two channels is used. Thus the algorithm regulates itself the way the two channels are working, until a stable state is found.

Fig. 3. Average and standard deviation of the objective function values in the population at different steps

5   Conclusion

We argue that the heterarchical concept can be interesting to design new ant colony algorithms, in particular aimed at the optimization of continuous multiminima functions. Such a biological concept was not exploited until now for the design of optimization algorithms, using only stigmergic processes. We propose to extend the ant colony metaphor to take into account several communication processes. According to our first tests it seems that an heterarchical algorithm implementing two complementary communication channels shows interesting emergent properties, like a self-management of the relative influence of the two channels.

Regarding the future, one important issue consists in improving the automatic tuning of parameters, before testing the algorithm through a large set of analytical test functions, and comparing its efficiency to that of competing metaheuristics.

References

G. Bilchev and I.C. Parmee. The Ant Colony Metaphor for Searching Continuous Design Spaces. *Lecture Notes in Computer Science*, 993:25–39, 1995.

E. Bonabeau, A. Sobkowski, G. Theraulaz, and J.-L. Deneubourg. Adaptive Task Allocation Inspired by a Model of Division of Labor in Social Insects. *BCEC*, pages 36–45, 1997.

S. Camazine, J.L. Deneubourg, N. Franks, J. Sneyd, G. Theraulaz, and E. Bonabeau. *Self-Organization in Biological Systems*. 2000.

A. Colorni, M. Dorigo, and V. Maniezzo. Distributed Optimization by Ant Colonies. In Elsevier Publishing, editor, *Proceedings of ECAL'91 - European Conference on Artificial Life*, pages 134–142, 1991.

J. Dréo. Modélisation de la mobilisation chez les fourmis. Mémoire de dea, Université Paris7 & Université Libre de Bruxelles, 2001.

Glover F. and M. Laguna. *Tabu Search*. Kluwer Academic Publishers, 1997.

B. Hölldobler and E.O. Wilson. *The Ants*. Springer Verlag, 1990.

J. Kennedy and R. C. Eberhart. Particle swarm optimization. In *Proc. IEEE Int. Conf. on Neural Networks*, volume IV, pages 1942–1948, Piscataway, NJ: IEEE Service Center, 1995.

N. Monmarché, G. Venturini, and M. Slimane. On how Pachycondyla apicalis ants suggest a new search algorithm. *Future Generation Computer Systems*, 16:937–946, 2000.

E.D. Taillard, L. Gambardella, M. Gendreau, and J-Y. Potvin. Adaptive Memory Programming: A Unified View of Metaheuristics. In *EURO XVI Conference Tutorial and Research Reviews booklet*, Brussels, 1998. EURO.

E.O. Wilson and B. Hölldobler. Dense Heterarchy and mass communication as the basis of organization in ant colonies. *Trend in Ecology and Evolution*, 3:65–68, 1988.

An Ant-Based Framework for Very Strongly Constrained Problems

Vittorio Maniezzo and Matteo Milandri

Department of Computer Science, University of Bologna, Italy
{maniezzo,milan}@csr.unibo.it

Abstract. Metaheuristics in general and ant-based systems in particular have shown remarkable success in solving combinatorial optimization problems. However, a few problems exist for which the best performing heuristic algorithm is not a metaheuristic. These few are often characterized by a very highly constrained search space. This is a situation in which it is not possible to define any efficient neighborhood, thus no local search is available. The paradigmatic case is the set partitioning problem, a problem for which standard Integer Programming solvers outperform metaheuristics. This paper presents an extended ant framework improving the effectiveness of ant-based systems to such problems. Computational results are presented both on standard set partitioning problem instances and on vertical fragmentation problem instances. This last is a real world problem arising in data warehouse logical design.

1   Introduction and Motivation

The effectiveness of different metaheuristic algorithms is often leveled by the use of problem-specific local search routines, which can themselves ensure a good quality result solution. Any algorithm incorporating them will gain a solution quality assurance, whereas algorithms such as genetic algorithms or ACO systems, which could work without, often get good performance only by incorporating local search. There exist however problems, and classes of problems, for which local search is of limited effectiveness, if of any at all. Among them a prominent role is played by very strongly constrained problems. These are problems for which efficient polynomial neighborhoods contain few solutions, or none at all, and local search is of very limited use. Probably, the most significant of such problems is the set partitioning (SP) problem, the problem of partitioning a given set into mutually independent subsets while minimizing a cost function defined as the sum of the costs associated to each of the eligible subsets. Its importance derives from the fact that many actual situations can be modeled as SP, and in fact many combinatorial optimization problems (crew scheduling, vehicle routing, project scheduling, warehouse location to name a few) can be modeled as SP with maybe some additional constraints. The limited utilization of these models reflects the limited effectiveness of heuristic and metaheuristic algorithms on large-scale SP instances.

This paper proposes an ant-based algorithm for solving real world problems whose mathematical model extends SP. In order to do so, since the basic ACO

M. Dorigo et al. (Eds.): ANTS 2002, LNCS 2463, pp. 222–227, 2002.

framework is ineffective, it is necessary to modify it. Modifications of the basic framework aimed at increasing effectiveness, accepting higher computational costs, intersect another main research issue for ACO systems: the notion of ant *state*. ACO systems have in the trail laying mechanism their characteristic feature. Trails are meant to bias on the basis of past experience ants' decisions about how to complete the partial solution they have constructed. Trail should thus be defined for each partial solution. This is obviously infeasible; fortunately however there exist a large number of problems for which the next decision can be effectively based only on the last one, instead of on the whole of the partial solution. A typical case is the TSP, where it is sufficient to consider the last node visited by the ant instead of the whole path so far followed. A large number of problems where ACO systems proved effective are of this last type. SP is not such. To deal with it, it is necessary to extend the ants state by considering other elements besides the last decision taken.

1.1   The Set Partitioning Problem

The Set Partitioning problem can be modeled as follows. Let $x_i, i = 1, \ldots, n$ be a binary variable denoting whether or not the $i-th$ subset is part of the solution and let $A = [a_{ij}], i = 1, \ldots, n, j = 1, \ldots, m$, be a 0-1 coefficient matrix whose columns correspond to the subset and whose rows to the set elements: $a_{ij} = 1$ means that the $j-th$ element is a member of the $i-th$ subset. Each subset j has an associated cost c_j. Furthermore, let I be the index set of all subsets and J the index set of all elements. A mathematical formulation of SP is the following.

$$(SP) \quad z(SP) = Min \quad \sum_{i \in I} c_i x_i \tag{1}$$

$$s.t. \quad \sum_{i \in I} a_{ij} x_i = 1 \quad j \in J \tag{2}$$

$$x_i \in \{0, 1\} \quad i \in I \tag{3}$$

The linear programming relaxation of formulation SP usually gives a good lower bound to the problem. Another good lower bound can be obtained by a Lagrangean relaxation of constraints (2), that is by associating a Lagrangean penalty $\lambda_j, j = 1, \ldots, m$ to each constraint (2) and thereby penalizing the relaxed constraint in the objective function of the resulting problem LSP. Currently, no ACO algorithm has been proposed for SP, and a direct implementation of the basic ACO framework is incapable of obtaining feasible solutions for many standard testset instances. The best performing metaheuristic for SP is a genetic algorithm due to Chu and Beasley [2]. Effective heuristics were proposed by Atamtürk and others [1].

Our interest in SP was risen by the need of solving a problem arising in the context of logical design of data warehouses (DW), namely the Vertical Fragmentation Problem (VFP). DWs are foremost systems for improving the support given to company internal decision processes and data analysis procedures. A

DW enables executives to retrieve summary data, derived by those present in operational information systems. An essential feature of a successful DW is a fast query response. DW design follows three main phases: conceptual, logical and physical design. VFP is a part of logical design and has the objective of minimizing the query response time by reducing the number of disk pages to be accessed. VFP problem details are rather intricate, and we refer the interested reader to [3] and [5] for problem details. While the specific formulation elements are not important for this paper, we point out that a substantial number of constraints are actually SP constraints, thus it is possible to solve problem VFP only if an effective means for solving SP is available. The only heuristic for VFP was presented in [5], where the authors describe a preliminary design of an ACO system. The algorithm in this paper builds on that experience.

2   An Ant-Based Heuristic for the SP

The approach advocated by this paper includes in a standard ACO algorithm elements taken from bounded enumeration procedures, which are tree search procedures where the number of nodes which can be expanded at each tree level is limited from above by a parametric constraint k. Greater values of k entail a bigger search space to explore, k unlimited makes the algorithm become a complete enumeration procedure. In the case of SP, each level of the search tree is associated with one constraint. Level 0 corresponds to empty solution, level 1 considers the first constraint, and so on. The order according to which constraints are considered in successive tree expansions greatly affects the algorithm performance. There are two main differences between standard ACO and the algorithm presented in the following, which will be denoted BE-ANT for short: i) the trail laying policy and ii) a synchronization step among ants after each expansion.

Trail is not laid directly on the components which build up a solution (the subsets) but on the couplings (component/element), that is (variable / constraint). A high trail value τ_{ij} indicates that a particular variable $i \in I$ demonstrated to be a good choice for covering constraint $j \in J$. Thus the same variable i may have different desirability, depending on the particular choice (which variable to use for covering constraint j) an ant has to take.

A *synchronization* step is implemented after each ant expands its partial solution, in order to define the different possible further expansions.

Essentially, the number of ants is equal to the value of parameter k. At each level, one ant ant_i is assigned to each of the k branches to be expanded and computes its possible expansion set E_i. Then all expansion sets are united, ordered by non-decreasing cost and the k of them, if so many exist, are chosen for further expansion.

There are three cases that make a branch non eligible for further expansion: 1) The partial solution associated with the branch is infeasible; 2) The partial solution associated with the branch has a cost already greater than that of the best one already found, thus it cannot lead to an improvement of the best solution found, and 3) The partial solution associated with the branch is not among the k ones chosen for the expansion of the level. A branch which can be expanded is a *valid* branch. Algorithm BE-ANT for SP is as follows.

1. i=1. The first constraint to be covered is chosen. The set E of feasible expansions is initialized with all feasible expansions of the empty solution.

2. The value $\alpha \cdot \tau_{ij} + (1 - \alpha)\eta_i$ is computed for each $j \in E$, where α, τ and η have the usual ACO interpretation. The partial solutions are accordingly ordered.

3. k valid branches in E are selected, by means of the usual ant probability selection formula, and one ant is assigned to each of them.

4. $i = i + 1$. Each ant j decides the constraint to cover at level i and accordingly computes the expansions E_j of its current partial solution. $E = \bigcup_j E_j$.

5. If $i < m$ goto Step 2.

6. If (end_condition) then Stop else update trails, goto Step 1.

To completely specify the algorithm, it is necessary to define how to compute the η and the τ values. These are problem-specific elements, which in the case of the SP we implemented as follows. The desirability η_i of a column $i \in I$ can be set equal either to the sum of the dual variable values of the still uncovered constraints which are covered by column i or simply to the number of still uncovered constraints which are covered by column i (we tested both possibilities). Thus, columns which cover the most difficult / the greater number of uncovered constraints are preferred. Trails are updated by means of the formula introduced in [4]. Notice however that trails are laid on the coupling (i, j), that is, we explicitly increase or decrease trails stating how proficient it was to cover constraint j with variable i.

As a last comment to the implementation of BE-ANT for SP it is worth saying that the lower bound to the SP was computed by means of the Lagrangean relaxation of the problem and that BE-ANT was intertwined with a subgradient optmization routine, so that step 6. of BE-ANT actually was:

6. If (end_condition) then Stop else update trails, perform one subgradient iteration, goto Step 1.

The same procedure introduced above was used to solve the VFP. The essential difference between VFP and SP is that VFP, besides the SP constraints, must also satisfy other constraints, namely Set packing constraints, a knapsack constraint and a number of subprolems-linking constraints. To deal with this, we separated VFP into two related subproblems: the SP subproblem and the subproblem defined by sep packing and knapsack constraints. The remaining constraints were relaxed in a Lagrangean fashion to obtain a Lagrangean decomposition of VFP. This yielded a lower bound. The upper bound was found by means of BE-ANT.

3   Computational Results

The algorithms described in the previous sections were implemented in Visual C++ v. 6.0 and run on a 677 Pentium III machine with 256 Mb of RAM. Tests were made on two benchmark sets. The first one consists of instances, downloaded from OR-LIB, which were used in [2] for validating the genetic algorithm. Due to space constraints (6 pages) we do not report the whole problem set: we selected a representative subset of difficult problems (several instances in [2] are acknowledgely very easy). The second testset consists of the VFP instances used in [3].

Table 1. Computational results on SP instances

probl.	LP Opt	IP Opt	BE-ANT	CPU sec.	BCGA
NW01	114852.0	114852	n.f.	-	n.f.
NW02	105444.0	105444	115929	6117.35	108816
NW03	24447.0	24492	24492	950.54	24492
NW04	16310.7	16862	16978	1033.63	16862
NW06	7640.0	7810	7810	25.55	7810
NW17	10875.7	11115	11133	2390.02	11115
NW18	338864.3	340160	349958	1955.31	345130
AA02	30494.0	30494	36812	3433.95	30500
AA04	25877.6	26402	33231	4086,16	28261
KL01	1084.0	1086	1096	657.59	1086
KL02	215.3	219	229	1437.17	219

Table 2. Computational results on VFP instances

probl.	LP Opt	BE-ANT	CPU sec.
WKL20S1000	152815.63	177003	655.70
WKL20S1400	98046.10	99594	250.93
WKL20S1800	96270.00	96270	675.90
WKL20S2200	96270.00	96270	4494.09
WKL30S1000	353785.66	412271	1188.79
WKL30S1400	274042.36	297906	2747.31
WKL30S1800	248205.39	286555	3741.48
WKL30S2200	241329.50	241330	2522.09
WKL40S1000	464241.24	540204	933.49
WKL40S1400	376145.98	451803	1522.61
WKL40S1800	336430.54	404128	2123.04
WKL40S2200	316794.87	349362	2546.43
VPWKL20SPC700	-	374253	327.06
VPWKL20SPC1400	96906.00	96915	2.16

The results obtained on SP instances are presented in Table 1. Columns are: *probl* problem identifier, *LP opt* optimal LP relaxation cost, *IP opt* optimal IP cost, *BE − ANT* best result obtained by BE-ANT (out of at most three runs), *CPU sec* CPU time to find the solution of the column before, *BCGA* best solution obtained by the GA of Beasley and Chu. As a comment to Table 1, on the bad side we may notice that the performance of BE-ANT is still inferior to that of BCGA, though not much so. On the good side we must notice that standard ACO (which we implemented) is not capable of finding a feasible solution for any of listed instances, whereas BE-ANT already has a performance comparable with that of state of the art solution methods. Moreover, as mentioned, the computational results are still preliminary and we expect that with a correct parameter settings and with a reasonable number of test repetitions BE-ANT results will further improve.

The results obtained on VFP instances are presented in Table 2. These are problem instances used in [3] and correspond to complex real-world settings (no pre-filtering on candidate views) which make the resulting instances much harder that those used in [5]. The first number in the instance name corresponds to the number of queries in the workload and the second one to the available memory. The bigger the first and the smallest the second number, the harder the instance. CPLEX was unable to solve on the reference machine to integer optimality the instances of the last two lines. For this problem the LP relaxation is quite close to integer optimality, usually within 2%. The results by BE-ANT are very interesting and at the moment are the only ones available for the last two instances.

4 Conclusions

This paper presented a new algorithm, named BE-ANT, designed for solving any combinatorial optimization problem in general, and very hard, tightly constrained instances in particular. The methodology includes in a standard ACO framework ideas taken from bounded enumeration search. Ants loose their identity, as at each step a partial solution is assigned to each ant, but the resulting framework can be applied also to problems for which the standard ACO framework is totally ineffective. The computational results, albeit very incomplete, testify the viability of the approach.

References

1. A. Atamtürk, G.L. Nemhauser, and M.W.P. Savelsbergh. A combined lagrangian, linear programming and implication heuristic for large-scale set partitioning problems. *Journal of Heuristics*, 1:247–259, 1995.
2. P.C. Chu and J.E. Beasley. Constraint handling in genetic algorithms: the set partitoning problem. *Journal of Heuristics*, 4:323–357, 1998.
3. M. Golfarelli, V.Maniezzo, and S. Rizzi. Materialization of Fragmented Views in Multidimensional Databases. Technical Report TR-001-02, Scienze dell'Informazione, University of Bologna, 2002.
4. V. Maniezzo. Exact and approximate nondeterministic tree-search procedures for the quadratic assignment problem. *INFORMS Journal on Computing*, 11(4):358 – 369, 1999.
5. V. Maniezzo, A. Carbonaro, M. Golfarelli, and S. Rizzi. An ANTS Algorithm for Optimizing the Materialization of Fragmented Views in Data Warehouses: Preliminary Results. In *Applications of Evolutionary Computing*, volume 2037 of *Lecture Notes in Computer Science*, pages 80–89. Springer Verlag, 2001.

Analysis of the Best-Worst Ant System and Its Variants on the QAP

Oscar Cordón, Iñaki Fernández de Viana, and Francisco Herrera

Dept. of Computer Science and Artificial Intelligence. E.T.S.I. Informática
University of Granada, Avda. Andalucía, 38. 18071 Granada, Spain
{ocordon,herrera}@decsai.ugr.es, ijfviana@teleline.es

Abstract. In this contribution, we will study the influence of the three components of the Best-Worst Ant System (BWAS) algorithm. As the importance of each of them as the fact whether all of them are necessary will be analyzed. Besides, we will introduce a new algorithm called Best-Worst Ant Colony System by combining the basis of the Ant Colony System with the special components of the BWAS. The performance of different variants of these algorithms will be tested when solving different instances of the QAP.

1 Introduction

In [3], a new algorithm called BWAS was intended being based on the AS algorithm [4]. In this contribution we develop an study applying the BWAS algorithm and its variants to the QAP. Besides, a new algorithm called Best-Worst Ant Colony System (BWACS) will be considered as well. Our aim is to demonstrate that these algorithms are robust as a whole and to analyze the relative importance among their distinguishing components.

This paper is structured as follows. In Section 2 and 3 the basis of the BWAS, its variants as well as the BWACS algorithm are studied. In section 4, the application of the ACO algorithms to the QAP is reviewed and the results we obtained are presented. We end up by discussing some concluding remarks and future works.

2 The Best-Worst Ant System

The BWAS model tries to improve the performance of ACO models using evolutionary algorithm concepts. The proposed BWAS uses the *AS transition rule*:

$$
p_k(r, s) = \begin{cases} \dfrac{[\tau_{rs}]^\alpha \cdot [\eta_{rs}]^\beta}{\sum_{u \in J_k(r)} [\tau_{ru}]^\alpha \cdot [\eta_{ru}]^\beta}, & if\, s \in J_k \\ 0, & otherwise \end{cases},
$$

with τ_{rs} being the pheromone trail of edge (r, s), η_{rs} being the heuristic value, $J_k(r)$ being the set of nodes that remain to be visited by ant k, and with α and β being real-valued weights.

M. Dorigo et al. (Eds.): ANTS 2002, LNCS 2463, pp. 228–234, 2002.
© Springer-Verlag Berlin Heidelberg 2002

Besides, the usual *AS evaporation rule* is used: $\tau_{rs} \leftarrow (1-\rho) \cdot \tau_{rs}$, $\forall r, s$, with $\rho \in [0,1]$ being the pheromone decay parameter. Additionally, the BWAS considers the three following daemon actions, that are analyzed in deep in [3]:

Best-Worst performance update rule. This rule is based on the Population-Based Incremental Learning (PBIL) [1] probability array update rule. The *offline pheromone trail updating* is done as follows:

$$\tau_{rs} \leftarrow \tau_{rs} + \Delta\tau_{rs}, \;\; where \;\; \Delta\tau_{rs} = \begin{cases} f(C(S_{global-best})), & if \; (r,s) \in S_{global-best} \\ 0, & otherwise \end{cases}$$

with $f(C(S_{global-best}))$ being the amount of pheromone to be deposited by the global best ant, which depends on the quality of the solution it generated, $C(S_{global-best})$.

Moreover, the edges present in the worst current ant are penalized: $\forall (r,s) \in S_{current-worst}$ and $(r,s) \notin S_{global-best}$, $\tau_{rs} \leftarrow (1-\rho) \cdot \tau_{rs}$.

Pheromone trail mutation. The pheromone trails suffer mutations to introduce diversity in the search, as done in PBIL with the memoristic structure. To do so, each row of the pheromone matrix is mutated —with probability P_m— as follows:

$$\tau'_{rs} = \begin{cases} \tau_{rs} + mut(it, \tau_{threshold}), & if \; a = 0 \\ \tau_{rs} - mut(it, \tau_{threshold}), & if \; a \neq 0 \end{cases}$$

with a being a random value in $\{0,1\}$, it being the current iteration, $\tau_{threshold}$ being the average of the pheromone trail in the edges composing the global best solution and with $mut(\cdot)$ being a function making a stronger mutation as the iteration counter increases.

Restart of the search process when it gets stuck. We will perform the restart by setting all the pheromone matrix components to τ_0 when the number of edges that are different between the current best and the current worst solutions is lesser than a specific percentage.

A simplified structure of a generic BWAS algorithm is shown as follows:

1. *Give an initial pheromone value, τ_0, to each edge.*
2. *For k=1 to m do (in parallel)*
   - *Place ant k in an initial node r. and include r in L_k*
   - *While (ant k not in a target node) do*
     - *Select the next node to visit, $s \notin L_k$, by the AS transition rule.*
3. *For k=1 to m do*
   - *Run the local search improvement on the solution generated by ant k, S_k.*
4. *$S_{global-best} \leftarrow$ global best ant tour. $S_{current-worst} \leftarrow$ current worst ant tour.*
5. *Pheromone evaporation and Best-Worst pheromone updating.*
6. *Pheromone matrix mutation and restart if condition is satisfied.*
7. *If (Stop Condition is not satisfied) go to step 2.*

3   Analysis of Best-Worst Ant System Components

In this section, a new ACO model based on the BWAS components is introduced, as well as the different variants considered from the basic BWAS and the new model.

3.1   The Best-Worst Ant Colony System and Its Variants

This new model is obtained by introducing the three components of the BWAS into the ACS. Hence, the differences between both algorithms are that BWACS considers the usual *ACS transition rule*

$$s = \begin{cases} arg \ \max_{u \in J_k(r)}\{[\tau_{ru}]^\alpha \cdot [\eta_{ru}]^\beta\}, \ if \ q < q_0 \\ S, \hspace{3.5cm} otherwise \end{cases},$$

and that the *online step-by-step updating rule* is applied during the ants trip: $\tau_{rs} \leftarrow (1-\varphi)\cdot\tau_{rs}+\varphi\cdot\Delta\tau_{rs}$, with $\varphi \in [0,1]$ being the pheromone decay parameter.

Hence, the only differences between BWAS and BWACS lie on the step 2 of the algorithm.

3.2   BWAS Variants Analyzed

The main objective of this paper is to study the influence of the three components of BWAS on its application to the QAP. With this study, we want to know if all of them are really important or some of them can be removed without negatively affecting the performance of the BWAS algorithm. Then, we will also try to establish a ranking of importance among components.

This analysis will be made from a double perspective: i) individualized analysis of components, and ii) cooperative analysis among pairs of components.

It seems that a certain interrelation exists among the three basic elements of BWAS. The updating of pheromone trails by the worst ant allows the algorithm to quickly discard areas of the search space while the mutation and the restart avoid the stagnation of the algorithm. It can seem that the latter two components can be redundant since they both have the same aim but we will see that a high cooperation arises between both.

In Table 1, all the algorithms used in the study are summarized. As can be seen, there are three different groups of algorithms: i) the first one includes the basic models: our two proposals, BWAS and BWACS, and the classical AS and ACS, considered for comparison purposes; ii) the second is composed of variants including a single component: restart, mutation or worst-update. The models AS_{+R} and ACS_{+R} are included in this group by adding the BWAS restart to the AS and ACS, respectively; iii) the third is comprised by the variants including a pair of the components. The different variants are notated by $BWAS_{-*}$ or $BWACS_{-*}$ standing * for the removed component (R, M or W).

Table 1. ACO models considered.

Parameter	Meaning
AS	Ant System
ACS	Ant Colony System
$BWAS$	Best-Worst Ant System
$BWACS$	Best-Worst Ant Colony System
AS_{+R}	AS with BWAS restart
ACS_{+R}	ACS with BWAS Restart
$BWAS_{-R-W}$	BWAS without restart and worst ant trail updating
$BWAS_{-M-W}$	BWAS without mutation and worst ant trail updating
$BWAS_{-R-M}$	BWAS without restart and mutation
$BWACS_{-R-W}$	BWACS without restart and worst ant trail updating
$BWACS_{-M-W}$	BWACS without mutation and worst ant trail updating
$BWACS_{-R-M}$	BWACS without restart and mutation
$BWAS_{-R}$	BWAS without restart.
$BWAS_{-W}$	BWAS without worst ant trail updating.
$BWAS_{-M}$	BWAS without mutation.
$BWACS_{-R}$	BWACS without restart.
$BWACS_{-W}$	BWACS without worst ant trail updating.
$BWACS_{-M}$	BWACS without mutation.

4   Experiments Developed and Analysis of Results

In this section, the application of ACO to the QAP is reviewed and the experiments developed and the analysis of the results obtained are reported.

4.1   Application of the ACO Algorithms Considered to the QAP

The QAP [2] is among the hardest combinatorial optimization problems. All the QAP instances used in our experimentation have been obtained from the QAPLIB [5]. We have chosen two instances of different sizes from each of the four existing classes [6] in order to perform a fair comparative study. These instances are respectively: tai50a, tai60a, nug20, sko72, bur26a, kra30a, tai50b and tai80b.

When applying the different ACO algorithms selected to solve the QAP, the next steps have to be considered: i) as in [6], we not considering the heuristic information in the transition rule. ii) the 2-opt local search algorithm considered in [6] is used.

4.2   Parameter Settings

The ACO models shown in Table 1 have been used to solve the eight QAP instances. Each model has been run 10 times in a 1400 MHz AMD Athlon processor computer. The parameter values considered are shown in Table 2, with the ones associated to AS and ACS taken from [6] and the BWAS and BWACS ones from [3]. The latter parameter values have not been obtained from

Table 2. Parameter values considered

Parameter	Value
Number of ants	$m = 5$
Maximum run time	$Ntime = 600\ seconds$
Pheromone updating rules parameter	$\rho = 0.2$
AS offline pheromone rule updating	$f(C(S_k)) = \frac{1}{C(S_k)}$
ACS offline pheromone rule updating	$f(C(S_{global-best})) = \frac{1}{C(S_{global-best})}$
Transition rule parameters	$\alpha = 1,\ \beta = 0$
ACS transition rule parameters	$q_0 = 0.98$
Initial pheromone amount	$\tau_0 = 10^{-6}$
$BWAS$ **parameters**	
Pheromone matrix mutation probability	$P_m = \{0.1, 0.15, 0.3\}$
Mutation operator parameter	$\sigma = 4$
5%	
Local search procedure parameters	
Neighbor choice rule	best improvement
Number of iterations	1000

a deep study, and only some preliminary experiments with different mutation probability values (0.1, 0.15 and 0.3) have been done.

4.3   Analysis of Results

Tables 3 and 4 collect a summary of the obtained results. Table 3 compares the algorithms two by two. Each cell a_{ij} shows the percentage of cases in which algorithm i has outperformed algorithm j. We will say that an algorithm i is better than another algorithm j for a problem instance p if the average error[1] obtained by i for p is smaller than that obtained by j. Notice that the values in Table 3 are symetric ($a_{ij} = 100 - a_{ji}$) in all cases but not in those where there have been draws between both algorithms.

A general classification of the models is shown in Table 4 which summarizes the values shown in Table 3. While the first column contains the name of the model, the other three columns collect the number of algorithms regarding which the model has obtained better, worse or similar results, respectively.

If we compare the results of the basic algorithms (AS, ACS, BWAS and BWACS), $BWAS$ and $BWACS$ present the best behavior. In every case, the best error was obtained by a BWAS model or by any of its variants.

Analyzing the results of the BWACS and its variants, we see how a great trade-off exists among its components. Notice that the elimination of any of them makes the algorithm worsen. For the eight instances, the BWACS has outperformed all its variants. However, we do not find the same in the BWAS. There are instances in which some variants of the BWAS overcome the basic

---

[1] Error stands for the percentage difference between the average cost obtained in the performed runs and the cost of the best solution known for the instance.

Table 3. Pair comparisons between ACO model

Model	AS	ACS	BWAS	BWACS	AS$_{+R}$	ACS$_{+R}$	BWAS$_{-R-W}$	BWAS$_{-M-W}$	BWAS$_{-R-M}$	BWACS$_{-R-W}$	BWACS$_{-M-W}$	BWACS$_{-R-M}$	BWAS$_{-R}$	BWAS$_{-W}$	BWAS$_{-M}$	BWACS$_{-R}$	BWACS$_{-W}$	BWACS$_{-M}$
AS	-	62	0	0	50	50	0	0	37	25	37	62	0	0	0	50	25	37
ACS	37	-	37	25	50	37	37	12	62	50	37	87	37	25	37	50	37	50
BWAS	87	62	-	37	87	50	75	37	87	75	50	100	75	37	50	87	62	62
BWACS	87	75	37	-	87	62	75	50	100	75	50	100	75	37	62	87	50	50
AS$_{+R}$	37	50	0	0	-	50	12	12	37	37	25	62	0	12	12	62	25	37
ACS$_{+R}$	50	62	37	25	50	-	37	12	75	50	37	87	37	25	37	62	25	37
BWAS$_{-R-W}$	87	62	0	0	75	50	-	25	62	62	37	100	50	12	12	62	50	50
BWAS$_{-M-W}$	87	87	37	25	75	75	50	-	62	87	62	100	37	25	37	87	50	75
BWAS$_{-R-M}$	62	37	12	0	62	25	37	37	-	37	25	100	37	0	25	50	25	37
BWACS$_{-R-W}$	50	50	12	12	50	50	25	0	62	-	25	100	25	0	25	75	25	37
BWACS$_{-M-W}$	50	62	25	25	50	50	37	12	75	62	-	100	37	12	37	87	25	50
BWACS$_{-R-M}$	37	12	0	0	37	12	0	0	0	0	0	-	0	0	0	25	12	0
BWAS$_{-R}$	87	62	0	0	87	50	25	37	50	62	37	100	-	25	12	62	50	50
BWAS$_{-W}$	87	75	25	37	75	62	62	50	100	87	62	100	50	-	62	87	75	75
BWAS$_{-M}$	87	62	25	12	75	50	50	37	75	62	37	100	62	12	-	75	50	50
BWACS$_{-R}$	25	50	0	0	25	37	25	0	50	0	0	75	25	0	12	-	0	0
BWACS$_{-W}$	62	62	12	25	62	50	25	12	75	62	50	87	25	0	25	87	-	37
BWACS$_{-M}$	50	50	12	25	50	37	25	0	62	50	25	100	25	0	25	87	25	-

Table 4. ACO models standing

Model	Best performance	Worst performance	Similar performance
BWAS	15	0	2
BWACS	15	0	2
BWAS$_{-W}$	15	1	1
BWAS$_{-M-W}$	12	2	3
BWAS$_{-M}$	12	3	2
BWAS$_{-R-W}$	11	5	1
BWAS$_{-R}$	10	5	2
BWACS$_{-M-W}$	9	5	3
BWACS$_{-W}$	10	7	0
BWACS$_{-M}$	6	9	2
BWACS$_{-R-W}$	5	10	2
ACS$_{+R}$	4	9	4
AS	4	12	1
ACS	2	11	4
BWAS$_{-R-M}$	3	13	1
AS$_{+R}$	2	13	2
BWACS$_{-R}$	1	13	3
BWACS$_{-R-M}$	0	17	0

BWAS algorithm. In spite of this, the BWAS presents better overall performance than its variants.

On the other hand, when analyzing the individual importance of each component, we see that the restart is the one giving the best results, followed by the mutation and the worst ant update. Among the combinations of two components, those not using the worst ant update achieve the best performance. Besides, those that do not consider the restart component present a very low performance. When eliminating the mutation, the results are neither as bad as if we remove the restart, nor as good as if we remove the worst ant update.

In view of these results, we can conclude that: i) in general, an appropriate trade-off exists among the three components of both BWAS and BWACS; ii) in spite of this balance, we can establish an order of importance among the components: restart, mutation, and worst update.

5   Concluding Remarks and Future Works

In this contribution, a study of all the components of $BWAS$ has been done. Besides, a new variant of BWAS called BWACS has been proposed. The performance of these algorithms and the importance of their components has been analyzed when solving eight QAP instances of different sizes and types. It has shown that the best performance have been obtained using the basic versions of BWAS and BWACS algorithms, without removing any component.

Different ideas for future developments arise: i) to study the influence of the appropriate values for the parameters, and ii) to analyze the consideration of other Evolutionary Computation aspects.

References

1. S. Baluja, R. Caruana. Removing the Genetics from the Standard Genetic Algorithm. In A. Prieditis, S. Rusell (Eds.), *Machine Learning: Proceedings of the Twelfth International Conference*, Morgan Kaufmann Publishers, pp. 38-46, 1995.
2. R.E Burkard, E. Cela. Quadratic and three-dimensional assignment problem. In M. Dell'Amico, F. Maffioli, and S. Martello (Eds.), *Annotated Bibliographies in Combinatorial Optimization*, 1996.
3. O. Cordón, F. Herrera, I. Fernández de Viana, L. Moreno. A New ACO Model Integrating Evolutionary Computation Concepts: The Best-Worst Ant System. *Proc. of ANTS'2000. From Ant Colonies to Artificial Ants: Second International Workshop on Ant Algorithms*, Brussels, Belgium, September 7-9, pp. 22-29, 2000.
4. M. Dorigo, G. Di Caro. Ant Algorithms for Discrete Optimization. *Artificial Life* 5(2), pp. 137-172, 1999.
5. E. Rainer, S. Burkard, E. Karish, F. Rendl. QAPLIB-A Quadratic Assignment Problem Library. http://www.opt.math.tu-graz.ac.at/~karish/qaplib.
6. T. Stützle, M. Dorigo. ACO Algorithms for the Quadratic Assignment Problem. In D. Corne, M. Dorigo and F. Glover, editors, New Ideas in Optimization, McGraw-Hill.

Ants and Loops

Geoffrey Canright

Telenor Research and Development
1331 Fornebu, Norway
geoffrey.canright@telenor.com

Abstract. The question of routing loop formation is examined for a class of routing schemes based on „agents" or „ants". First the notion of loop existence is discussed for the general case of routing tables with multiple, weighted entries. A simple criterion is then found which ensures loop-free routing. This criterion accounts for previous results with agent-based routing. It is however not in general met by ant-based (stochastic) routing schemes. In this case I find probabilistic arguments which conclude that, for such schemes, loops will be unlikely to form, and, if formed, unlikely to persist.

1 Introduction

Centralized network management strategies have difficulty in dealing with networks that are either too large, or too dynamic. In either case, the problem is the conveyance of enough information, quickly enough, to and from the point of central control. The alternative is decentralized schemes. For such schemes the control is „everywhere". Distributed control schemes [1],[2] can act without the need to gather a complete knowledge of the network—hence can adapt quickly to local changes. Thus one might hope that distributed control might better cope with large numbers of nodes, and/or with strongly time-varying traffic or topology changes.

Decentralized methods have their own peculiarities however. Their necessary ignorance of the global picture can have disadvantages. For instance, *routing loops* are essentially non-local things: one cannot tell, from examining a piece of a loop, that it is in fact a piece of a loop. (Of course, local examination can always detect „sufficiently small" loops—but, whatever the neighborhood size one can examine, there are always larger loops.) Hence it is of interest to study the susceptibility of decentralized routing schemes to the formation of loops.

Some of the earliest work in this area was done by Appleby and Steward (AS) [3]. They studied „mobile agents" on a simulated telephone network. Each node (considered a source, S) emitted agents which moved over the net, seeking shortest paths (more precisely, paths with the largest spare capacity) from S to all other nodes in the network. When found, these paths were entered by the agents into the routing tables (RTs) of the nodes affected. Since there was no centralized control, agents from more than one node at a time could, and generally did, write to the RTs for the same destination D.

M. Dorigo et al. (Eds.): ANTS 2002, LNCS 2463, pp. 235–242, 2002.

Schoonderwoerd et al [4],[5],[6] pointed out that this lack of coordination could give rise to routing loops. They then modified AS's agents so that agents emitted by S could only write to RTs pointing towards S, and so that only one agent at a time could work on these RTs. They found that modified agents, operating in this way, no longer gave routing loops. They also studied „ants", that is, agent-like entities, moving on the network, which could lay down „trails" (analogous to the pheromone trails of real ants), with the strength of the trail depending on the quality of the path taken. Routing with such ants is stochastic: each node's RT, for each destination, has a „weight" for each neighbor of that node. This weight reflects the accumulated trails of all ants (corrected for aging effects), and can be normalized to act as a probability for routing.

Schoonderwoerd et al found that their ants essentially never created routing loops after an initial transient period. Hence their work (i) points out the susceptibility of AS's agents to writing loops, (ii) shows a way to correct this problem, and (iii) offers the empirical observation that their particular version of ants does not form persistent loops.

It is the goal of this work to understand how loops form—and how they can be prevented from forming—during the operation of decentralized routing schemes. A good understanding should include the points (i)—(iii) made by Schoonderwoerd et al, ideally by putting them into a larger perspective. Below, I give a complete explanation for (i) and (ii) by presenting a necessary and sufficient criterion for loop-free operation. In the case of ants however—whose behavior is nondeterministic, and cannot satisfy the criterion—I will offer probabilistic arguments for (iii)—why persistent loops are unlikely. An interesting result in this direction has been given by Subramanian et al [7], who showed that a certain class of ants, operating on a *static* (and symmetric) network, gave RTs which converge, in the limit of long time, to a shortest-path tree—which is thus loop-free.

Let us now briefly establish some terminology. Following the historical precedent discussed above, the term „agents" will be used to denote messages (which may or may not include mobile code [8]) which write deterministic (ie, single-entry) RTs. „Ants" then denotes messages which use multiple, weighted entries in the RTs, and thus stochastic routing. The generic term, encompassing both, will also be „ants".

Ants traveling from source S to destination D can write to RTs pointing Backwards (towards S) or Forwards (towards D). (A variant, termed „smart" ants [2], updates RTs pointing towards all intermediate nodes in the Backward or Forward direction.) Ants can also Retrace their path, before writing to the RTs, or Not. In the following, I will consider FR ants (and agents), ie Forward-pointing with Retracing; and BN agents, Backward-pointing with No retracing. BN ants are treated in [9], which also includes a brief discussion of other pointing/retracing/routing combinations, as well as an examination of loop formation by real social insects.

Finally there is the question of routing—both for ants and for messages. Following Subramanian et al [7], „uniform" (u) routing is completely random; „regular" (r) routing follows the probabilities in the RTs; and „greedy" (g) routing will always choose the highest-weight entry in the RT. While there are reasons to consider all three strategies for ants, here I will concentrate mostly on r routing for ants. Routing of messages is discussed in the next section.

2 The Existence of Loops

For deterministic routing the question of the existence of loops is clear-cut. The entries in the RTs define a directed graph for each destination D. Let us call this graph the RT graph of D, or RTG(D). Each node in this graph has out-degree one. A loop is a closed circuit of directed arcs in RTG(D). Note that loops are a property of the RTs: this definition is independent of the nature (size, or degree of dynamicity) of the underlying network.

Now consider stochastic routing. Here—even though the weights may be very unevenly distributed—it is true that, in general, *every* neighbor of A has some nonzero weight. If we then include in the RTG a directed arc for each nonzero entry, the RTG(D) becomes the graph of the entire network. Of course, the complete network is in fact the only reasonable definition of the RTG for u routing. However there is no good motivation for the u-routing of *messages* (unlike ants). Hence u routing of messages will not be considered further.

In contrast, g routing of messages is well motivated, and has been shown to give good results in simulations [4],[5],[6]. Furthermore, g routing of messages gives an RTG with the same properties as the RTG for deterministic routing (in particular, each node has out-degree one). This makes the question of loop existence clear and unambiguous. For these reasons, I will concentrate mostly on g routing for messages.

The intermediate case, r routing of messages, is also of some interest, as (unlike u routing) it appears to have some practical motivation [10],[11],[12]. For instance, if one wishes to „piggyback" [7] ants and messages—thus economizing on the use of bandwidth—then both ants and messages must make the same routing choice; and r routing is an obvious candidate. On the other hand, as mentioned above, r routing gives an RTG which is simply the entire network. And a typical communications network—fixed or mobile—is riddled with loops, rendering the question of loop existence in the RTG meaningless. A convenient simplification is to restrict the RTG to the highest-weight entries for each D; this gives the same RTG as for g routing.

In the following, then, we will assume greedy routing of messages, so that the question of loops pertains to the RTG defined by this routing. In considering the behavior of r-routed ants on this same network, I will use the RTG as a rough indicator of the likely behavior of the ants—in those cases where the ants are sensitive to their own pheromone (i.e., F ants).

3 Agents and Loops

It is not difficult to find a necessary and sufficient criterion for loop-free routing with agents. Assume that the RTG is loop-free at some time. Assume further that agents always choose loop-free paths. Then, given these assumptions:

An RTG, initially loop-free, will remain loop-free after the action of an agent, if that agent's action results in the writing into the RTs of its complete path from S to D.

The proof is simple. Assume the contradiction, that is, an agent **a** writes its entire path from S to D, and yet leaves a loop that was not present before (Figure 1). Then **a** must have written part of the loop. We decompose **a**'s path (complete from S to D) into p (the piece prior to the loop), p' (the piece which closes the loop) and p'' (the remaining piece, from the loop to the RT destination—D for F ants, and S for B ants). There is a „loop-closing node" A which joins p' to p''. A has out-degree two: one out-arc begins the piece p''; another rejoins the loop. However this is impossible, as all nodes in the RTG have out-degree one. (Note that p'' must include at least one arc, since the destination D has no out-arcs in its own RTG, and so cannot be on the loop.)

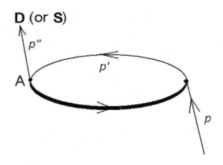

Fig. 1. Schematic for loop formation in the RTG. The heavy line is pre-existing structure; the light lines are written by a single new agent (or ant). The new path points towards D or S, for Forward- or Backward-pointing agents, respectively.

Thus the above condition is sufficient to give loop-free routing. It is also necessary. That is, while it is possible for a finite number of agents to fail to meet the above condition and still leave the RTGs loop-free, an arbitrarily large number of agents, acting over a long time, will create loops, as is easily demonstrated by example (see for example Figure 2 below).

There are two obvious ways an agent can fail to write its entire path. One is that it dies before completing the path. In this case, however, simple arguments show that, for either FR or BN agents, premature death does not give loops.

As pointed out by Schoonderwoerd et al [4],[5], there is another way for an agent to fail to write its entire path. What is needed is for *two* (or more) agents to write to the same RTG (i.e., for the same D or S) *at the same time*. For example, two FR agents can both seek and find the same D, and then, upon retracing, write to RTG(D) in a sequence of moves such that, when both agents are finally finished and have died, there is a loop in RTG(D). That is, the two writing sequences can „interfere" with one another, such that the crucial piece of p''—that is, the first arc of p'' which leaves A on the way to D—does not survive in the RTG for *either* agent's path. The same is true for BN agents writing to RTG(S).

In short, *timing* problems between simultaneously-acting agents, writing to the same RTG, can give loops. Schoonderwoerd et al. observed this problem for the FR agents of Appleby and Steward, and remedied it (without however articulating the above criterion) by turning to BR agents. Since BR agents point Backward, they point to S, which—being the source of its own agents—can keep track of how many agents are writing to its RTG (and hold that number to one). The Retracing is used to let S know that a given agent **a** is finished, so that S can safely emit another agent **a'**. Each

S then ensures that one agent at a time acts on its RTG, and the above criterion is satisfied. (Note that, for completely loop-free operation, the agents cannot be „smart": if they can point to intermediate nodes, then the simultaneous action of multiple agents cannot be ruled out.)

4 Ants and Loops

The case of ants is more difficult than that of agents (=all-or-none ants). The reason is simple: ants can never be guaranteed to write their entire paths to the RTG—such a guarantee is in contradiction to stochastic routing. Ants add weight to a path that they have found; and in general, some pieces of their path will be promoted to the highest-weight RT entry after their action, and others will not. Hence the nice remedies, with their accompanying guarantees, that we found in the previous section for agents cannot be employed for ants.

Instead we must consider probable behavior. We find some encouragement from the empirical result of Schoonderwoerd et al. [6] that „Although circular routes may appear during the very beginning of the initialization, we found that they never persist beyond this period." Here we find the strong word „never": it says that *persistent* loops are *extremely* unlikely for their ant-based routing scheme (in the present notation, BN ants, with r routing for the ants, and g routing for the messages).

It is of interest to try to understand this empirical observation, and to extend that understanding to other types of ants as well. I will present probabilistic arguments for the case of FR ants in this section. The arguments for BN ants take a different (and longer) form, since the BN ants do not smell their own pheromone. Hence these arguments are presented elsewhere [9]. I will assume for all ants, as previously assumed for agents, that individual ants will always take loop-free paths. Also let us recall that we focus on the RTG formed solely from the highest-weight entries in the RTs.

From our discussion of agents in the previous section, we can extract a criterion for an FR ant to form a loop. These ants leave source S on their way to destination D. If they reach D, they retrace their path, along the way updating the RT pointing towards D at each node visited. I assume only positive pheromone; hence the only effect such an ant can have is to enhance the weight of the taken path, relative to the other options at that node, for the destination D.

FR ants will then form a loop in the (previously loop-free) RTG if they fail to promote the crucial piece of p'' to highest weight, while subsequently promoting the entire piece p' to highest weight. We offer a schematic picture of such a development in Figure 2.

Note that the pictures in Figure 2, and the qualitative arguments advanced below, are also applicable if the destination D is replaced with an intermediate node, and the RTs pointing to that node are being updated by a „smart" ant. Differences between „plain" and „smart" ants will turn up at the quantitative level, but not at the qualitative level of argument considered here.

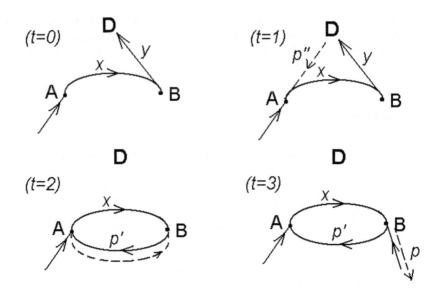

Fig. 2. Schematic for loop formation by an FR ant. Pre-existing structure, including pieces x and y, is shown at *(t=0)*. Subsequent time ordering is as indicated. Dashed arrows are paths followed by the ant on its way back from destination D. Those that are promoted to highest weight in the RTG become solid lines pointing towards D. A loop forms when p'' is not promoted, but p' is.

Now we attempt to evaluate the likelihood of this loop-forming scenario. FR ants are sensitive to pheromone levels that are reflected in the RTG—that is, thinking of D-seeking ants as a „species", FR ants are sensitive to their own species' pheromone. This means that the RTG depicts not only the *deterministic* routing for g-routed messages, but also the *likeliest* routing for r-routed, D-seeking ants. So we see from Figure 2 that, in order for the loop to form, the FR ant must, in its forward travel, choose the piece p' (upon leaving B) over the likeliest way y to D from B. It must, in addition, choose the piece p'', upon leaving A, over the most likely A-exit x. That is, the ant must, for each of the final two sections of its journey, choose legs of less-than-highest probability. The crucial point here is that the loop-forming scenario *requires* two choices of the ant which *must* be of less than the highest probability, while saying nothing whatsoever about the other choices. Since we neither require nor know anything about these other choices, we consider ensembles of behaviors, under the constraint that the two crucial loop-forming choices are taken. The ensemble of loop-forming choices then has, on average, a lower probability that the ensemble which does not form loops.

In short: for FR ants, we find information in the RTG itself (to which FR ants are sensitive) telling us that the loop-formation scenario involves two correlated unlikely events. For BN ants—which do not „feel" the RTG(S)—more complicated reasoning gives the result [9] that the loop-formation scenario requires the correlation of two events, at least one of which must be unlikely.

We now consider the stability of loops formed by ants. Again we must use probabilistic arguments. We assume an ant has formed a loop, and ask what are the likely effects of subsequent ants on this loop. First note that ants cannot affect a loop without actually reaching at least one node on it—since ants only write to the RTs of the nodes they visit. It must also reach D, since it writes to RTs upon Retracing. If it does so, it will tend to promote the arcs on its path—including that arc which it used to leave the loop. That is, it will try to promote an arc that, if successfully promoted, will break the loop. If (as is common) it uses the quality of the entire path to update, then it is most likely to promote its path—including the loop-breaking link—if it has wasted little time *in* the loop. Thus *every* ant which can affect the loop, will, at one node (at least) of the loop, attempt to break the loop by promoting a different arc; and those ants likeliest to affect the loop will not likely reinforce many links in the loop itself.

A similar conclusion may be reached for BN ants: every ant that affects the loop will try to break it, at least one point on the loop. It is interesting to note that this is true even for ants following the same path as that which formed the loop. That is, subsequent ants (FR or BN) which follow the route p''-p'-p will work to demote x (by promoting p'') and break the loop. More colloquially, we can say that if the first ant that formed the loop has actually detected „good routes", subsequent development will open the loop.

5 Summary and Conclusions

In this paper I have examined the susceptibility of distributed routing schemes—coarsely categorized as „ants" and „agents"—to the formation of routing loops. For agents—which employ deterministic, rather than stochastic, routing—a simple criterion was found, which is both necessary and sufficient for loop-free operation. This criterion gives a clear and concise explanation for previous experience with loops and all-or-none routing agents. It also gives a clear guideline for future efforts in the same direction.

For ants, employing stochastic routing, the abovementioned criterion for loop-free operation cannot hold; there is thus no reasonable way to guarantee completely loop-free behavior for ants. Instead one must hope to make loops „highly unlikely". I have offered probabilistic arguments towards two points: why the *formation* of loops is unlikely; and why a loop, once formed, is not likely to *persist*. These arguments, while far from rigorous, offer some understanding for a phenomenon which has been observed in simulations, but not explained—namely, that small, unintelligent ants, possessed only of local information, should still manage to enforce (with very high probability) a non-local property: the absence of loops.

These arguments could and should be supplemented further, both by more thorough theoretical considerations, and by the testing of their ideas and conclusions with further simulations. Certainly, if ants are ever to be turned loose on the real networks that we use and depend upon, we must be able to trust them to do a good job; and that trust must be grounded in a good understanding of *why* they will do so.

Acknowledgements. I thank Ruud Schoonderwoerd for helpful discussions on these questions.

References

1. Hayzelden, A.L.G., and Bigham, J. (eds.). *Software Agents for Future Communications Systems.* Springer, Berlin, 1999.
2. Bonabeau, E., Dorigo, M., and Theraulaz, G. *Swarm Intelligence: From Natural to Artificial Systems.* Oxford University Press, New York, NY, 1999.
3. Appleby, S., and Steward, S. Mobile software agents for control in telecommunications networks. *BT Technology Journal,* 12 (April 1994), 104-113. Reprinted as Chapter 11 of Ref. [1].
4. Schoonderwoerd, R., Holland, O., Bruten, J. Ant-like agents for load balancing in telecommunications networks. *Proceedings, 1st Intl. Conf. Autonomous Agents* (1997).
5. Schoonderwoerd, R., Holland, O., Bruten, J., and Rothkrantz, L. Ant-based load balancing in telecommunications networks. *Adapt. Beh.* 5 (1997), 169-207.
6. Schoonderwoerd, R., and Holland, O. Minimal agents for communications network routing: the social insect paradigm. Chapter 13 of Ref. [1].
7. Subramanian, D., Druschel, P., and Chen, J. Ants and reinforcement learning: a case study in routing in dynamic networks. *Proc. IJCAI-97* (1997), 832-838.
8. Milojcic, M. (ed.). Mobile agent applications. *IEEE Concurrency* (July-Sept 1999), 80-90.
9. http://p2p.nta.no/cgi-bin/Zope/Z/uploads/antsandloops.pdf.
10. Di Caro, G., and Dorigo, M. Two ant colony algorithms for best-effort routing in datagram networks. *Proceedings of 10th IASTED Intl. Conf. on Parallel and Distributed Computing and Systems (PDCS'98),* Anaheim: IASTED/ACTA Press (1998), 541–546.
11. Heusse, M., Snyers, D., Guérin, S., and Kuntz, P. Adaptive agent-driven routing and load balancing in communication networks. *Adv. Complex Systems* 1 (1998), 237-254.
12. Di Caro, G., and Dorigo, M. AntNet: distributed stigmergetic control for communications networks. *J. Artif. Intell. Res.* 9 (1998), 317-365.

Candidate Set Strategies for Ant Colony Optimisation

Marcus Randall and James Montgomery*

School of Information Technology, Bond University
Gold Coast, Queensland, 4229, Australia
{mrandall,jmontgom}@bond.edu.au

Abstract. Ant Colony Optimisation based solvers systematically scan the set of possible solution elements before choosing a particular one. Hence, the computational time required for each step of the algorithm can be large. One way to overcome this is to limit the number of element choices to a sensible subset, or candidate set. This paper describes some novel generic candidate set strategies and tests these on the travelling salesman and car sequencing problems. The results show that the use of candidate sets helps to find competitive solutions to the test problems in a relatively short amount of time.

1 Introduction

The Ant Colony Optimisation (ACO) [1] meta-heuristic is a relatively new optimisation technique based on the mechanics of natural ant colonies. It has been applied extensively to benchmark problems such as the Travelling Salesman Problem (TSP), the job sequencing problem and the Quadratic Assignment Problem (QAP). In addition, work on more complex problems that have difficult constraints in such areas as transportation and telecommunications, has been undertaken [1]. Since ACO is a constructive meta-heuristic, at each step ants consider the entire set of possible elements before choosing just one. Hence, the vast majority of an ant algorithm's runtime is devoted to evaluating the utility of reachable elements and so ACO techniques can suffer from long runtimes if attention is not paid to constructing appropriate subsets of elements from which to choose. There has been little work done in this area, despite the fact that this can potentially improve the efficiency of ACO, especially for large real world problems. The way that this is achieved is via *candidate set strategies*.

Candidate set strategies have traditionally only been used as part of a local search procedure applied to the solutions generated by ACO. However, the strategies developed for local search heuristics such as 2-opt and 3-opt are inappropriate for use in the construction phase of the ACO algorithm, and it is only in later improvements of Ant Colony System (ACS) that candidate set strategies were applied as part of the construction process [2,3]. The most common candidate set used for the TSP is *nearest neighbour*, in which a set of the k nearest

---

* This author is a PhD scholar supported by an Australian Postgraduate Award.

M. Dorigo et al. (Eds.): ANTS 2002, LNCS 2463, pp. 243–249, 2002.

cities is maintained for each city. Ants select from this set first and only if there are no feasible candidates are the remaining cities considered. This approach has been particularly useful on larger problems (more than 1500 cities) [2]. In some instances, maintaining sets of less than 10 cities can be sufficient to contain all the links in the optimal solution [4]. Bullnheimer, Hartl and Strauß [5] also use a nearest neighbour candidate set their ant system for the vehicle routing problem. Another static approach for geometric problems is to create a candidate set based on the Delaunay graph, augmented with extra edges to provide sufficient candidates [4].

Both the nearest neighbour and Delaunay graph candidate set approaches can be easily generalised for problems in which each element has a relationship with each other element, such as the TSP. However, for problems that do not exhibit strong relationships between elements, such as in the QAP where facilities relate to locations but not to other facilities, these techniques are difficult to apply.

Each of these techniques uses static candidate sets generated *a priori* (i.e. candidate sets that are derived before, and not updated or changed during, the application of the ACO meta-heuristic). In contrast, *dynamic* candidate set strategies require the sets to be regenerated throughout the search process. Although used routinely in iterative meta-heuristics like Tabu Search (TS) [6], their use in ant algorithms has only been suggested by Stützle and Hoos [7]. As much of the power of ACO comes from the use of adaptive memory (pheromone trails), it is likely that using dynamic candidate set strategies will lead to further improvements in both solution quality and computational runtime. This paper outlines generic dynamic candidate set strategies for a wide variety of common combinatorial optimisation problems. In addition, we test some appropriate generic strategies on the TSP and the car sequencing problem (CSP) [8]. An extended version of this paper is also available [9].

This paper is organised as follows. Section 2 gives a description of some generic candidate set strategies. Section 3 outlines the computational experiments and Section 4 has the concluding remarks.

2   Generic Candidate Set Strategies

Candidate sets strategies can be broadly divided into two approaches: *static*, in which candidate sets are generated *a priori* and used without change throughout the search process, and *dynamic*, in which candidates sets are regenerated during the search. Static approaches are more problem specific and suitable for simpler problems such as the TSP. Dynamic approaches are more easily generalised across different problem types and hence, their development and refinement is necessary to solve complex problems.

The TS meta-heuristic was the first to make use of dynamic candidate list strategies [6]. Greedy Randomised Adaptive Search Procedures (GRASPs) [10], another constructive approach, also use a type of dynamic candidate set that is highly similar to the first strategy described below. We describe how the TS strategy of *elite candidate set* and a new general purpose strategy, *evolving set* can be applied to ACO. Both of these approaches maintain separate candidate sets for each element in the solution, as in the static approaches used previously in ACO.

1. *Elite Candidate Set.* Initially, the candidate set is established by considering all possible elements and selecting the best k, where k is the size of the set, based on their probability values (see Dorigo and Gambardella [2]). This set is then used for the next l iterations of the algorithm. The rationale of this approach is that a good element now is likely to be a good element in the future.
2. *Evolving Set.* This is similar to the Elite candidate set strategy and follows an important aspect of the TS process. Elements that give low probability values are eliminated temporarily from the search process. These elements are given a "tabu tenure" [6] in a tabu list mechanism. This means that for a certain number of iterations, the element is not part of the candidate set. After this time has elapsed, it is reinstated to the candidate set. The threshold for establishing which elements are tabu can be varied throughout the search process depending on the quality of solutions being produced.

3   Computational Experience

Two problem classes were used to test the generic candidate set strategies, the TSP and the CSP. The CSP is a common problem in the car manufacturing industry [8]. In this problem, a number of different car models are sequenced on an assembly line. The objective is to separate cars of the same model type as much as possible in order to evenly distribute the manufacturing workload. The TSP problem instances, from TSPLIB, are gr24, hk48, eil51, st70, eil76, kroA100, d198, lin318, pcb442 and att532. The CSP problem instances are n20t1, n20t5, n40t1, n40t5, n60t1, n60t5, n80t1 and n80t5, and are available online at http://www.it.bond.edu.au/randall/carseq.tar.

Our experiments are based on the ACS meta-heuristic. The computing platform used to perform the experiments is a 550 MHz Linux machine. Each problem instance is run across 20 random seeds and consists of 3000 iterations. The ACS parameter values used are: $\beta = -2$, global pheromone decay $\alpha = 0.1$, local pheromone decay $\rho = 0.1$, number of ants $m = 10$, $q_0 = 0.9$.

For both problem types, a control strategy and an ACS with a static candidate set were run in order evaluate the performance of the generic dynamic strategies. The control ACS is simply ACS without any candidate set features. The static set strategy for the TSP and CSP is nearest neighbour. For the CSP, the separation distance between each pair of cars is calculated to produce the nearest neighbour static set. In our experiments, we set $k = 10$ for both TSP and CSP.

3.1   Code Implementation

This section describes the mechanics of implementing the candidate set strategies within the ACS framework. In particular, the heuristic for establishing and varying quality thresholds is described and problem specific details are given.

Elite Candidate Set. Elite candidate set as described in the TS literature regenerates its candidate set after a predetermined number of iterations or if the quality of elements in the set falls below a critical level. Our implementation uses the former strategy, as it stores element quality at the time it generates the candidate set. This makes it difficult to judge whether the *current* quality of elements has dropped sufficiently to necessitate the regeneration of the set. However, initial testing has shown that our method ensures that elite candidate set runs quickly, while having minimal impact on the cost of solutions generated. A candidate set may persist across iterations. Elite candidate set has two control parameters: the size of the set (expressed as a constant number of elements) and the refresh frequency (expressed as a (fractional) number of iterations). The values used in the experiments are a set size of 10 and a refresh frequency of 0.5 iterations.

Evolving Set. At each step, the Evolving set strategy examines only those elements that are reachable by an ant during that step to determine their tabu status. Elements whose tabu tenure has expired cannot be immediately reincluded on the tabu list. Elements can, and in these experiments were, placed on the tabu list for more than one iteration. Hence, Evolving set can serve to focus the search over a number of iterations, rather than just within a single iteration. In addition to the tabu threshold, which is adjusted by the algorithm, Evolving set has only one parameter, the tabu tenure. For these experiments this was set at 15 iterations. Tabu tenures smaller than one iteration were used initially but found to be less effective.

A simple heuristic has been developed for adjusting the tabu threshold periodically between preset upper and lower bounds. An initial threshold is established by calculating all the elements' probabilities and selecting a threshold value such that 50% of elements are above it. The upper bound is set such that 10% of elements are above it, while 10% of elements are below the lower bound. The mean cost of solutions produced is used as an approximate measure of the overall quality of the population of solutions. It is recorded after the first iteration and subsequently updated each time the threshold is adjusted. The threshold is adjusted every 20 iterations by examining the proportion of solutions with a cost better than the previously recorded mean. If there are proportionally more solutions with an improved cost, the algorithm is assumed to be in an improving phase and the threshold is raised. If the reverse is the case, the threshold is lowered to allow greater exploration to take place. The new threshold is related to the old by Equation 1.

$$r \leftarrow r + \frac{m_b - m_a}{m} \cdot s \tag{1}$$

$$s = \begin{cases} r_{max} - r & \text{if } m_b - m_a > 0 \\ r - r_{min} & \text{if } m_b - m_a < 0 \\ 0 & \text{otherwise} \end{cases} \tag{2}$$

m is the total number of solutions, m_b and m_a are the number of solutions with better and poorer costs than the previous mean respectively, s is a scale factor determined by Equation 2, and r_{min} and r_{max} form a lower and upper bound on the threshold.

Problem Specific Implementation Details. For the TSP, an ACS element is represented by a city. The cost of an element(city) is simply d_{ij} where i is the previous city and j is the current city.

For the CSP, an element is represented by a car. At each step, each ant adds a new car to its production line schedule. In order to calculate the element/car cost, all of the previous cars must be examined in relation to the new car. If any of these cars are of the same model type as the new car, the separation penalty (i.e. cost) for that model type is recalculated.

3.2   Results

The results are given in Table 1. Results are presented as the Relative Percentage Deviation (RPD) from the best known cost, calculated according to $\frac{c-c_{best}}{c_{best}} \times 100$, where c is the cost of the solution and c_{best} is the best-known cost for the corresponding problem. The minimum ("Min"), median ("Med") and maximum ("Max") measures are used to summarise the results as they are non-normally distributed. Given the high consistency of results for CPU time (in seconds), only the median CPU time is reported.

For the TSP, the Elite candidate set strategy appears to offer the best performance in terms of solution cost, while the static candidate set strategy offers the best increase in speed. The Evolving set strategy also produces better solutions than normal ACS in less time, but its solutions are generally more expensive than those produced by Elite candidate set and the static candidate set strategy. It is important to note that the time used by the candidate set strategies is highly dependent on the values of their control parameters.

Further experiments were carried out in which the static set and Elite candidate set strategies were run for equivalent time as the control strategy. These found that their respective performances on cost could be improved if they were run for more iterations, although the static set's performance was still worse than Elite candidate set's.

For the CSP, Evolving set appears to perform best in terms of cost, but is actually slower than normal ACS. Larger candidate set sizes may improve the performance of the static set, which was often found to contain no usable elements requiring all elements to be examined. In contrast, Elite candidate set regularly regenerates its candidate set and does not suffer from this problem. However, Elite candidate set did not perform well on the CSP in terms of solution cost.

4   Conclusions

This paper has described some generic candidate set strategies that are suitable for implementation within ACO. This has been a preliminary investigation and

Table 1. Results for each strategy applied to the TSP and CSP

Strategy	Problem Instance	TSP Cost (RPD) Min	Med	Max	CPU Time	Problem Instance	CSP Cost (RPD) Min	Med	Max	CPU Time
Control	gr24	0.0	0.5	5.0	17	n20t1	20.7	56.0	86.2	9
	hk48	0.0	0.4	3.4	68	n20t5	29.3	33.3	68.0	9
	eil51	0.5	2.6	5.2	77	n40t1	42.5	58.6	73.3	31
	st70	1.6	3.3	9.5	145	n40t5	27.0	32.4	47.2	31
	eil76	1.5	3.8	8.4	170	n60t1	113.8	141.8	162.5	68
	kroA100	0.0	1.2	3.5	293	n60t5	24.2	33.4	50.0	67
	d198	1.0	1.9	3.2	1144	n80t1	44.2	57.3	73.0	120
	lin318	8.8	12.3	15.9	2948	n80t5	24.7	34.3	47.2	119
	pcb442	20.1	23.9	27.7	5772					
	att532	20.2	26.4	31.6	8384					
Static	gr24	0.0	0.5	5.3	15	n20t1	5.2	14.7	25.9	15
	hk48	0.0	0.6	2.4	30	n20t5	49.3	50.7	52.0	15
	eil51	0.5	2.6	5.2	32	n40t1	70.5	84.2	96.6	41
	st70	0.3	2.0	7.9	45	n40t5	33.2	39.5	50.3	38
	eil76	1.1	2.1	3.5	49	n60t1	196.1	209.9	273.0	79
	kroA100	0.2	1.1	6.4	68	n60t5	30.1	37.5	50.2	79
	d198	1.6	3.0	4.7	164	n80t1	67.0	80.2	109.4	136
	lin318	2.6	7.7	13.1	332	n80t5	40.0	53.0	63.9	138
	pcb442	7.0	11.0	15.6	438					
	att532	6.3	10.3	13.3	633					
Elite Candidate Set	gr24	0.0	0.5	4.4	10	n20t1	70.7	95.7	136.2	6
	hk48	0.0	0.3	2.2	38	n20t5	68.0	97.3	174.7	5
	eil51	0.5	3.1	6.1	43	n40t1	80.8	124.0	176.0	22
	st70	0.3	2.0	9.2	83	n40t5	90.9	119.3	127.0	22
	eil76	1.3	3.4	5.9	98	n60t1	227.6	257.6	323.7	54
	kroA100	0.0	0.8	5.8	168	n60t5	107.8	129.2	133.6	55
	d198	0.8	1.5	2.1	686	n80t1	101.2	151.4	210.6	107
	lin318	1.9	3.5	5.1	1770	n80t5	92.0	123.6	154.8	109
	pcb442	3.1	5.8	8.4	3518					
	att532	3.8	4.9	6.2	5075					
Evolving Set	gr24	0.0	0.6	5.0	9	n20t1	6.9	12.9	22.4	23
	hk48	0.0	0.1	3.4	27	n20t5	10.7	10.7	12.0	17
	eil51	0.7	2.6	5.6	28	n40t1	3.4	13.4	21.9	86
	st70	0.7	3.9	7.3	47	n40t5	5.1	5.7	10.2	72
	eil76	1.1	2.8	6.7	62	n60t1	68.4	78.0	90.1	198
	kroA100	0.0	0.5	4.6	98	n60t5	2.1	2.8	4.6	174
	d198	0.7	1.8	4.0	355	n80t1	10.0	15.9	22.7	358
	lin318	3.2	5.0	6.5	954	n80t5	0.5	1.8	3.9	307
	pcb442	13.8	17.4	20.4	1861					
	att532	13.5	17.1	22.8	2883					

it is likely that other candidate set mechanisms, apart from the ones described and tested herein, are possible.

The results indicate that dynamic candidate set strategies can be applied quite succesfully in an ACO setting to combinatorial optimisation problems such as the TSP and CSP. However, each strategy's performance across the two problem types is not consistent. Further investigation into why certain candidate set strategies perform well on some types of problem and poorly on others needs to be carried out. The effects of different control parameter values also need to be more fully explored.

It is conceivable that the results could have been improved by the application of a local search procedure. This has not been done as the primary purpose of this study is to investigate how candidate set strategies can be applied in a general way to constructive meta-heuristics.

In order for ACO meta-heuristics to be used routinely for practical optimisation problems, further empirical analysis of these candidate set techniques needs to be undertaken. Future work will involve studying how the contents of these dynamic candidate sets change with time using these strategies across different problem types.

References

1. Dorigo, M., Di Caro, G.: The Ant Colony Optimization Meta-heuristic. In Corne, D., Dorigo, M., Glover, F. (eds.): New Ideas in Optimization. McGraw-Hill, London (1999) 11-32
2. Dorigo, M., Gambardella, L.M.: Ant Colonies for the Traveling Salesman Problem. BioSystems **43** (1997) 73-81
3. Stützle, T., Dorigo, M.: ACO Algorithms for the Traveling Salesman Problem. In Miettinen, K., Makela, M., Neittaanmaki, P., Periaux, J. (eds.): Evolutionary Algorithms in Engineering and Computer Science. Wiley (1999)
4. Reinelt, G.: The Traveling Salesman: Computational Solutions for TSP Applications. Springer-Verlag, Berlin (1994)
5. Bullnheimer, B., Hartl, R.F., Strauß, C.: An Improved Ant System Algorithm for the Vehicle Routing Problem. Sixth Viennese workshop on Optimal Control, Dynamic Games, Nonlinear Dynamics and Adaptive Systems, Vienna, Austria (1997)
6. Glover, F., Laguna, M.: Tabu Search. Kluwer Academic Publishers, Boston (1997)
7. Stützle, T., Hoos, H.: Improving the Ant System: A Detailed Report on the MAX-MIN Ant System. Darmstadt University of Technology, Computer Science Department, Intellectics Group., Technical Report AIDA-96-12 - Revised version (1996)
8. Smith, K., Palaniswami, M., Krishnamoorthy, M.: A Hybrid Neural Network Approach to Combinatorial Optimisation. Computers and Operations Research **73** (1996) 501-508
9. Randall, M., Montgomery, J.: Candidate Set Strategies for Ant Colony Optimisation. School of Information Technology, Bond University, Australia, Technical Report TR02-04 (2002)
10. Feo, T.A., Resende, M.G.C.: Greedy Randomized Adaptive Search Procedures. Journal of Global Optimization **6** (1995) 109-133

Dynamic Wavelength Routing in WDM Networks via Ant Colony Optimization

Ryan M. Garlick[1] and Richard S. Barr[2]

[1] CSE Dept., Southern Methodist University, Dallas, TX 75275 USA
[2] EMIS Dept., Southern Methodist University, Dallas, TX 75275 USA

Abstract. This study considers the routing and wavelength assignment problem (RWA) in optical wavelength-division-multiplexed networks. The focus is dynamic traffic, in which the number of wavelengths per fiber is fixed. We minimize connection blocking using an ant-colony-optimization (ACO) algorithm that quantifies the importance of combining path-length and congestion information in making routing decisions to minimize total network connection blocking. The ACO algorithm achieves lower blocking rates than an exhaustive search over all available wavelengths for the shortest path.

A wavelength-routing all-optical network consists of wavelength-crossconnect nodes interconnected by fiber links. A set of individual network demands are routed from origin to destination (O-D) nodes, across these fiber links. The higher transmission capacity of all-optical networks is accomplished, in part, by sending multiple signals simultaneously through the same fiber-optic cable using wavelength-division multiplexing (WDM), which transmits multiple data streams simultaneously on different frequencies, or wavelengths.

1 Problem Definition

In wavelength-routed WDM networks, lightpaths are established between nodes and can span multiple fiber links in the network. A *lightpath* is realized by allocating a wavelength on each link comprising a path between two nodes. In the absence of wavelength-conversion equipment, a lightpath must occupy the same wavelength on all of its links, a property known as the *wavelength continuity constraint*.

Once a network is designed, the routing and wavelength-assignment problem (RWA) determines a set of lightpaths to carry the designated communications traffic. In the *static RWA problem*, all lightpath requests are known in advance, and the objective is to minimize the network resources required to satisfy all demands. The focus of this study is the *dynamic RWA problem*, in which lightpath requests arrive dynamically, and the number of wavelengths is limited. The objective is to minimize connection blocking. The optimal RWA problem is NP-complete [2], and thus is suited to heuristic methods.

M. Dorigo et al. (Eds.): ANTS 2002, LNCS 2463, pp. 250–255, 2002.
© Springer-Verlag Berlin Heidelberg 2002

Although the RWA problem has been studied extensively, this paper introduces a new algorithm for evaluating potential routes based on length and congestion information. Ant-colony optimization (ACO) as a method for routing and wavelength assignment on all-optical networks is introduced, and provides valuable insight on the length-versus-congestion tradeoffs. While ACO has been applied to the static RWA problem, this is its first use for the dynamic case. ACO is used to test the hypothesis that occasionally choosing slightly longer paths with less congestion improves blocking performance. ACO provides an effective testing platform for investigating the efficacy of unconstrained dynamic routing, by using ants that prefer paths with lower levels of network traffic.

This study decreases blocked requests by quantifying the importance of using congestion information. When confronted with a shorter path carrying more traffic or a slightly longer path with less congestion, the question of how much additional path length is acceptable to avoid congestion is examined.

2   Background and Previous Research

In [4], ACO is applied to the static routing and wavelength assignment problem and provides a background for the ACO approach to the dynamic RWA presented in this paper. In each algorithmic time step, ants move from each demand origin to each destination. Varela introduced backtracking, with each ant keeping a "tabu" list [6] of previously visited nodes. Backtracking avoids dead-ends and cycles—an approach adopted in the ACO algorithms in this paper. When an ant finds itself blocked, it pops its current location from a list of visited nodes and attempts to proceed from the previous location. This ability requires each ant's memory to contain a list of nodes visited in order.

Each ant in [4] maintains its own type of pheromone, and while ants are attracted to their own pheromone, they are repulsed by the pheromone of other ants in order to obtain even loading. The best results are achieved through a global update wherein ants are increasingly repulsed by paths on which more ants have traversed. Maintaining an ant and pheromone type for each connection request is time consuming, however.

All work in this paper focuses on extensions of work done on the dynamic RWA problem. The previous studies have focused on k-shortest-path-based routing schemes [1,3]. More recent developments have incorporated congestion information into routing decisions [5,8,12].

Previous research tested the effectiveness of incorporating congestion information by testing a k-shortest-path (ksp) algorithm against a single-shortest-path strategy [7]. This work showed that a ksp algorithm achieves lower blocking than a strategy in which only a single path is available.

Chan and Yum [8] compare two routing strategies: routing connection requests on the shortest path with available capacity and on the least-loaded route from source to destination. Since the latter paths may be significantly longer, the shortest-routing strategy almost always provided lower blocking. A goal of this paper is to combine these two strategies into a routing algorithm that chooses

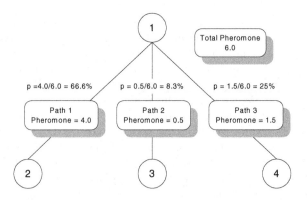

Fig. 1. Example path selection probabilities for an Ant at Node 1

short paths with low congestion, in an effort to improve performance over each strategy individually.

3   Ant-Colony Optimization

In the ACO algorithm introduced in this paper, an ant's "life" begins randomly at either the origin or destination node of the demand. It proceeds until it finds the corresponding destination or origin node, using a selected available wavelength. Each ant chooses its wavelength according to parametric rules such as most-used or random selection. At the completion of its search, the ant deposits pheromone along the path. In addition, each ant has memory of all nodes previously visited. Subsequent ants proceed similarly, choosing each vertex in their search paths based probabilistically on the level of pheromone on the connecting link, as shown in Figure 1 and described below.

We use the following notation: N is number of ants per connection request; L the set of links available from the current node; ϕ the normalized weight of length vs. weight of number of available wavelengths; l a link compromising a path P; l_c the total capacity in wavelengths of link l; l_a the set or available wavelengths on link l; and ψ_l the level of pheromone on link l. The probability γ that an ant will take a path l is the pheromone on that path normalized over the pheromone on all links available from the current node is $\gamma_l = \psi_l / \sum_{i \in L} \psi_i$.

Pheromone is deposited on a per-demand basis. The pheromone matrix is reset once the final selection of wavelength and route is made for a connection request. This requires only one type of pheromone and avoids much of the overhead found in the implementation of the static case in [4], which requires running times on the order of hours. Even loading is achieved by having more pheromone deposited on paths that fewer previously routed O-D pairs occupy.

The shortest path found with the highest pheromone is selected as the best and final route for the demand. It is important to note that this may not be the shortest available path.

A global pheromone update is performed after each ant completes a route. The shortest path found receives pheromone in inverse proportion to its length. We also favor paths that have the fewest conflicts with demands already routed. Therefore, a component of pheromone update includes more pheromone for paths with more available wavelengths. Global update is assigned based on the following equations.

The sum of available lane quantity ratios for a path P is defined as $A_P = \sum_{l \in P} \frac{l_a}{l_c}$, with the mean available lane ratio for a path P of $M_P = \frac{A_P}{|P|}$. The pheromone value on link l at time-step t, given as ψ_l^t, is updated according to Equation 1 where the scalar parameter ϕ, $0 \leq \phi \leq 1$, controls the emphasis on path length versus available-lane ratio.

$$\psi_l^{t+1} = \psi_l^t + \frac{\phi}{|P|} + M_P(1 - \phi), \forall l \in P \tag{1}$$

Although each ant initially chooses a wavelength, the final wavelength selection is not made until all ants have completed a tour from source to destination for this connection request. The best route after N ants is found. Among the contiguous wavelengths available along this path, one is selected based on the most-used, first-fit, or random wavelength selection policy.

4   Computational Experiments and Analysis

For purposes of performance comparison, network blocking is the primary focus. Hereafter, general references to the performance of a particular heuristic refer to the percentage of blocked connection requests calculated during a simulation.

In all tests, connection-request arrival and duration rates follow a Poisson distribution with a mean of λ. All tests were conducted for 5×10^5 demands at each Erlang. Load is measured in Erlang for the entire network, as in [1]. If network traffic is modeled at 50 Erlang, and the 51st connection request arrives, a random existing connection is broken and its resources freed. Testing was conducted on the well-known 21-node, 26-link ARPA-2 network [7], with each edge in the network having an assumed capacity of 16 wavelengths.

Parametric Tests. The first test conducted concerned the wavelength-selection method. Random wavelength selection provided the lowest blocking at 50 ants, but performance for random selection peaked at this N. At 200 ants, most-used was the preferential method, outperforming first-fit and random at all traffic levels. In both tests, differences in blocking were small and details are omitted for brevity. However slight the differences in performance between wavelength selection methods, most-used was the method employed in all subsequent ACO tests.

The next set of tests concerned measuring blocking at several levels of N, for a fixed ϕ of 0.50. Performance will only improve up to a certain number of ants, although reductions in blocking percentages were seen at 200 ants. Significant processing is required with $N = 200$, however this increased processing load is easily distributed, a strength of the ACO algorithm. Results of 32, 50, and 200

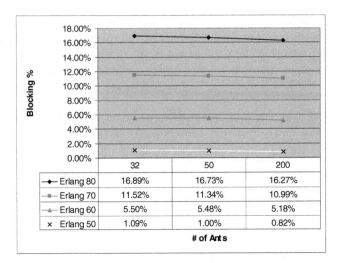

Fig. 2. Blocking % vs. N, for $\phi = 0.5, l_c = 16$, and *most-used* selection rule

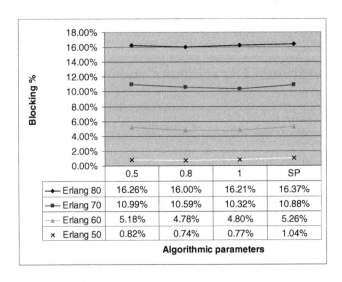

Fig. 3. ϕ and Shortest-Path Comparison, for $N = 200, l_c = 16$, and *most-used* rule

ants are presented in Fig. 2 for four traffic levels. Two hundred ants provided the best performance in all situations, and is the N value used for comparisons with other algorithms in the next section.

Comparisons with Published Algorithms. Mokhtar and Azizoglu present several heuristic RWA algorithms in [7]. They achieve the best results through an exhaustive search over all wavelengths for the shortest available path from source to destination. Their exhaustive search provided lower blocking than methods employing static routing, and provides a benchmark for ACO algorithms in this paper. The ACO algorithm at various parameters was compared with this algorithm, with the results displayed in Fig. 3. Using 200 ants, ACO provides the best results, outperforming the shortest available path at every traffic load. A ϕ value of 0.80 provided the best results in all but one case. At 70 Erlang, using no congestion information provided the best results by a slight margin. These results seem to indicate that while short paths are important, the best solution incorporates congestion information in selecting the route for each connection request. With a $\phi = 1$, congestion information is ignored, and the algorithm searches exclusively for the shortest path. However, since an initial pheromone value (ψ) of 1.0 is present on all edges in the network, ants may not find the shortest possible path.

References

1. Hui, Z., Jue, J., and Mukherjee, B. "A Review of Routing and Wavelength Assignment Approaches for Wavelength-Routed Optical WDM Networks," Optical Networks, January 2000.
2. Zhang, X. and Qiao, C. "Wavelength Assignment for Dynamic Traffic in Multi-fiber WDM Networks," ICCCN '98, pp. 479-585, 1998.
3. Stern, T.E. and Bala, K., "Multiwavelength Optical Networks." Addison-Wesley, 1999.
4. Navarro-Varela, G. and Sinclair, M. "Ant-Colony Optimisation for Virtual-Wavelength-Path Routing and Wavelength Allocation," Proc. Congress on Evolutionary Computation (CEC'99), Washington DC, USA, July 1999, pp. 1809-1816.
5. Li, L. and Somani, A., "Dynamic Wavelength Routing Using Congestion and Neighborhood Information." IEEE Trans. Networking, Vol 7, No. 5, Oct. 1999.
6. Glover, F., Laguna M., and Laguna F., Tabu Search. Kluwer Academic Publishers, 1997.
7. Mokhtar, A. and Azizoglu, M. "Adaptive Wavelength Routing in All-Optical Networks," IEEE/ACM Transactions on Networking, Vol. 6, No. 2, April 1998.
8. Chan, K. and Yum, T.P., "Analysis of least congested path routing in WDM lightwave networks." INFOCOM '94. Networking for Global Communications, 13th Proceedings IEEE, 1994. pp. 962-969.
9. Banerjee, D. and Mukherjee, B. "A Practical Approach for Routing and Wavelength Assignment in Large Wavelength-Routed Optical Networks." IEEE Journal on Selec. Areas in Comm., vol 14, No. 5, June 1996.
10. Colorni, A., Dorigo, M. & Maniezzo, V. "Distributed optimization by ant colonies," Proc. First European Conference on Artificial Life, Paris, France, pp. 134-142, 1991.
11. Dorigo, M. and Gambardella, L.M., "Ant-Colony System: A Cooperative Learning Approach to the Travelling Salesman Problem." IEEE Transactions on Evolutionary Computation, pp. 53-66.
12. Karasan, E. and Ayanoglu E., "Effects of Wavelength Routing and Selection Algorithms on Wavelength Conversion Gain in WDM Optical Networks." IEEE Trans. Networking, vol 6, pp. 186-196, April 1998.

Homogeneous Ants for Web Document Similarity Modeling and Categorization

Kok Meng Hoe, Weng Kin Lai, and Tracy S.Y. Tai

MIMOS, Technology Park Malaysia
57000 Kuala Lumpur, Malaysia.
{menghk, laiwk, sytai}@mimos.my

Abstract. The self-organizing and autonomous behavior of social insects such as ants presents an interesting and powerful metaphor for applications in the retrieval and management of large and fast growing amount of online information. The explosive growth of web documents has increasingly made more difficult and costly the manual task of organizing the documents into meaningful categories by human experts. Hence, it is desirable that some degree of automation be incorporated into the classification process to enable better scalability and prevent human classifiers from being overwhelmed by the deluge of information. This paper presents a preliminary investigation of applying a homogeneous multi-agent clustering system based on the self-organization behavior of the ants to the high-dimensional problem of web document categorization. A description of the text processing needed to obtain significant document features is included. The system will be evaluated on multi-class online English documents obtained from a popularly used search engine.

1 Introduction

Recent research into measuring the size of the Internet had estimated that it is doubling in size every year [1], [2]. In [1] that the lower bound number of publicly accessible web documents had increased from 320 million as of December 1997 to 800 million as of February 1999. Even higher growth rate was reported in [2], which estimated the number of unique online documents at 2.1 billion as of July 2001, and is expected to reach 4 billion in about seven months.

Such an explosive growth of online content has increasingly made more difficult and costly the task of organizing and categorizing documents to expedite search and retrieval. Currently, this immense task is still performed by human experts [3], but it is unlikely that such a manual and laborious categorization system will be scalable to the potential enormity of the Internet. Hence, the demand for an (semi-)automated process to assist in this arduous task will surely increase in tandem. Nevertheless, to effectively organize documents, we must first be able to identify like-content documents and subsequently cluster them into representative groups.

Document clustering has been generally defined as the operation of grouping together similar documents into multiple classes, and can be used to obtain a global or

M. Dorigo et al. (Eds.): ANTS 2002, LNCS 2463, pp. 256–261, 2002.

local (query-based) analysis of a document collection [4]. Clustering would facilitate subsequent categorization and indexing, thus improving search in large information repositories such as in Internet search engines and portals. Most traditional clustering algorithms either use a partitional or hierarchical approach to optimize predefined similarity or dissimilarity measures based on geometrical distance measures or probabilistic density estimations [5].

Before documents can be clustered, they are analyzed using text-processing techniques [4] to extract a representative set of features for each document. The feature extraction process basically involves finding the vector of words or set of descriptors that best describes a document. In the case of textual documents, words taken directly from the document are augmented with weights while disregarding the linguistic context variation at the morphological, syntactical, and semantic levels of natural language. The extracted *word-weight* vectors, known as a *bag of words* representation of the document, is usually of high dimensions i.e. over 10,000 words.

In recent studies [6], several species of ants have been observed to exhibit „intelligent" behavior locally that contributes to the emergence of global structures or patterns as a result of their collective actions. This autonomous and self-organizing behavior, or *Swarm Intelligence*, have inspired research into new forms of computational models which have been used in a variety of applications, ranging from exploratory data analysis, network routing, job-scheduling, to graph partitioning. In this paper, we investigate the use of an ant-inspired clustering algorithm to categorize English web documents of different interest domains.

2   Ant Colony Models for Data Clustering

Data clustering is an explorative task that seeks to identify groups of similar objects based on the values of their attributes [5]. Clustering works on the inherent characteristics of objects in the data and attempts to discover distinct boundaries to divide the data set into meaningful partitions. In studying the *Messor sancta* ants, Deneubourg et al. [7] discovered that worker ants exhibit „clustering" behavior by collecting their dead and piling the corpses to form „cemeteries". They derived a mathematical model that generalized the ants' behavior into two simple actions: i) picking up an isolated item, and ii) dropping the item where more similar items are present. Assuming the ants carry one item at a time and there is only one item type, they defined the actions as the following probabilistic functions, P:

$$\text{Picking up probability, } P_p = \left(\frac{k_1}{k_1 + f} \right)^2 , \tag{1}$$

$$\text{Dropping probability, } P_d = \left(\frac{f}{k_2 + f} \right)^2 . \tag{2}$$

f is the fraction of items in the neighborhood of the agent, while k_1 and k_2 are threshold constants. P_p is high when $f \rightarrow 0$, i.e. density of items in the agent's vicinity is low. Oppositely, when the density of items around the agent is high i.e. $f \rightarrow 1$, P_d is high. Hence, isolated items will be moved by ants until more dense regions are found.

Deneubourg et al.'s [7] model was later extended by Lumer & Faieta (LF) [8] to include a distance function, d between data objects, hence removing the assumption of type homogeneity and generalizing the model for exploratory data analysis. In the latter model, the inherent similarity of higher-dimensional data objects are mapped onto a lower-dimensional space. In their algorithm, agents move randomly on a 2D grid and decides to pick up (if unladen) or drop (if laden) based on a local density function, $f(o_i)$ which measures the average similarity of the discovered object i with other objects j in its neighborhood, N. Given a constant, α, the distance between object i and j, $d(o_i, o_j)$ and the cells in a neighborhood, $N(c)$, $f(o_i)$ is defined as:

$$f(o_i) = \begin{cases} \dfrac{1}{|N(c)|^2} \sum_{o_j \in N(c)} \left[1 - \dfrac{d(o_i, o_j)}{\alpha} \right] & \text{if } f > 0 \\ 0 & \text{otherwise.} \end{cases} \qquad (3)$$

In this paper, we employ a colony of homogenous ant-like agents similar to [8], but with a reformulation of the local density to use a *similarity* instead of a distance measure. We then apply our algorithm (see next section) to the high-dimensional and non-trivial problem of clustering English web documents obtained from a popular Internet search engine.

3   Homogeneous Multi-agent System for Document Clustering

Our multi-agent system based on an artificial ant colony and environment consists of three main components: (i) the colony of ant-like agents, (ii) the items for clustering (in this case, feature vector of web documents), and (iii) a square 2-D discrete space or grid, which we call the *clustering workspace*. Agents do not fall off the *toroidal* workspace and moves only one step in any direction at any time unit i.e. moving from its original cell to an *unoccupied* adjacent cell.

Only a single agent and/or a single item are allowed to occupy a cell at a time. An agent occupying any cell, c on the clustering space immediately perceives a neighborhood of 8 adjacent cells i.e. $N(c) = 8$. When an *unladen* agent encounters an item o_i at cell c, it decides to either pick up or ignore o_i with a probability P_p based on a local density function, $g(o_i)$ which determines the similarity between o_i and other items o_j, where $j \in N(c)$. If an agent *laden* with item o_i lands on empty cell c, then it calculates a probability P_d based on the same function $g(o_i)$ and decides whether to drop o_i or keep carrying it. Unlike f (see Eq. 3) which uses a distance measure and an additional parameter α, the function $g(o_i)$ directly uses a similarity measure and is defined as follows:

$$g(o_i) = \frac{1}{N(c)} \sum_{o_j} S(o_i, o_j) , \qquad (4)$$

where S is the similarity between two object o_i and o_j.

To model the inherent similarity within documents, we use the *cosine* measure,

$$S_{\cos}(doc_i, doc_j) = \frac{\sum_{k=1}^{r} f_{i,k} \times f_{j,k}}{\sqrt{\sum_{k=1}^{n}(f_{i,k})^2} \times \sqrt{\sum_{k=1}^{m}(f_{j,k})^2}} , \qquad (5)$$

where r is the number of common terms in doc_i and doc_j, n and m is the total number of terms in doc_i and doc_j, respectively, and $f_{a,b}$ is the frequency of term b in doc_a. A useful property of $S_{\cos}$ is that it is invariant to large skews in the weights of document vectors, but sensitive to common concepts within documents.

The picking up, P_p and dropping, P_d probabilities of the agent are replicated here for completion. The algorithm used follows closely that proposed in [8].

$$\text{Picking up probability, } P_p(o_i) = \left(\frac{k_1}{k_1 + g(o_i)} \right)^2 , \qquad (6)$$

$$\text{dropping probability, } P_d(o_i) = \begin{cases} 2g(o_i) , & \text{if } g(o_i) < k_2 \\ 1 , & \text{if } g(o_i) \geq k_2 \end{cases} , \qquad (7)$$

where k_1 and k_2 are threshold constants.

4   Experimental Results

As experimental data, we identified 84 web pages from 4 different categories—Business, Computer, Health and Science. The documents were randomly retrieved from the *Google* web directory within the corresponding categories. After text processing, the collection of web documents yielded 17,776 distinct words.

We used a 30×30 toroidal grid for the clustering workspace and 15 homogeneous agents. The maximum number of iterations, t_{max} for the algorithm was set to 300,000. The experiment was repeated with different initialization points, and values of k_1 (see Eq. 6) and k_2 (see Eq. 7) in [0.01, 0.2] with an increment of 0.05 for each run. Figure 1 shows the results for $k_1 = 0.01$ and $k_2 = 0.15$ at progressive time-steps of t. Upon initialization, $t = 0$, the 86 documents (see Fig. 1a) and 15 agents (not shown) were placed at random cells on the workspace. To better visualize the results, we marked the location of each document on the workspace with its actual category.

After 50,000 iterations (see Fig. 1b), clusters of mixed-categories of documents were formed. Since the workspace is continuous, clusters occupying the upper-left and bottom-right regions can be considered as a single cluster. The size of clusters varied from a minimum size of 6 documents to a maximum of about 40.

At t = 300,000, four distinct clusters of near-homogeneous type were obtained (see Fig. 1d). The majority of documents in each cluster originated from a single category, and documents from different categories were grouped into different clusters. Table 1 presents the quality of the induced clusters according to the measures of *entropy* and *purity* [9]. The number of clusters equaled the actual number of web document categories, although 8 documents remained scattered.

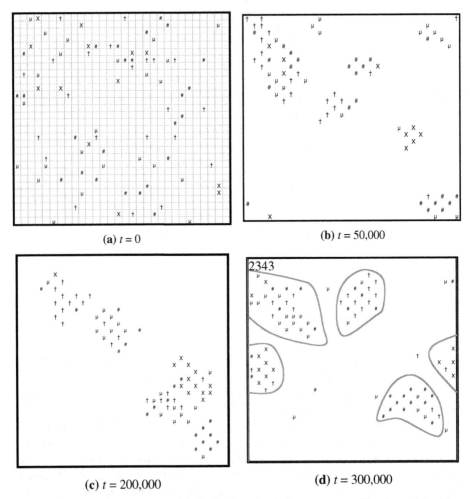

Fig. 1. (a-d) Two-dimensional spatial distribution of web documents on a 30×30 toroidal grid at different time steps, t. The different categories of web documents are marked with symbols † for Business, X for Computers, μ for Health and # for Science

Table 1. Summary of clustering quality. Cluster number are as indicated in Figure 1.

Cluster	Entropy	Purity	Majority Class
1	0.84	0.54	Health
2	0.65	0.63	Business
3	0.63	0.65	Computer
4	0.57	0.71	Science
Overall	**0.63**	**0.56**	-

5 Concluding Remarks and Future Directions

In this paper, we presented the findings of our initial study on using a multi-agent system based on the collective behavior of social insects i.e. ants, in clustering web documents retrieved from a popular search engine. Unlike earlier work, we introduced the direct application of the cosine similarity measure for determining the local density of web documents. Our experiments show that the proposed multi-agent system is able to induce near-homogeneous clusters, which number corresponds to the actual number of document classes. The results obtained, although not on par with the classification ability of human experts, do demonstrate the potential of ant-like multi-agent systems in handling complex and high-dimensional data clustering.

In conclusion, there are still areas in the current system that have to be improved before it can be benchmarked against well-established document clustering methods. Among those we identify, and which will be in our future work, are a larger perceivable time-dependent neighborhood for agents, formulation of a stopping criterion based on homogeneity and spatial distribution of clusters, and introduction of pheromone traces on objects to deter random relocation and encourage cluster formation.

References

1. Lawrence, S., Giles, C.L.: Accessibility of Information on the Web. *Nature*, 400. (1999)
2. Cyveillance: Sizing the Internet. A Cyveillance Study (2000)
3. Yahoo! Web Directory: http://www.yahoo.com
4. Baeza-Yates, R., Ribeiro-Yates, B.: *Modern Information Retrieval.* ACM, NY (1999)
5. Duda, R.O., Hart, P.E.: *Pattern Classification and Scene Analysis.* Wiley, NY (1973)
6. Bonabeau, E., Dorigo, M., Theraulaz, G.: *Swarm Intelligence: From Natural to Artificial Systems.* Oxford University Press, New York, NY (1999)
7. Deneubourg, J. L., Goss, S., Franks, N.R., Sendova-Franks, A., Detrain, C., Chretien, L.: The Dynamics of Collective Sorting: Robot-like Ants and Ant-like Robots. *Proc. Int. Conf. Sim. of Adap. Behavior: From Animals to Animats.* MIT, MA (1990)
8. Lumer, E.D., Faieta, B.: Diversity and Adaptation in Populations of Clustering Ants. *Proc. Int. Conf. Sim. of Adap. Behavior: From Animals to Animats.* MIT, MA (1994)
9. Steinbach, M., Karypis, G., Kumar, V.: A Comparison of Document Clustering Techniques. *KDD Workshop on Text Mining* (2000)

Parallel Ant System for the Set Covering Problem

Malek Rahoual[1], Riad Hadji[1], and Vincent Bachelet[2]

[1]Laboratoire LSI, Départ. Informatique, Faculté Electronique et Informatique, USTHB
BP 32 El Alia, Bab Ezzouar, Alger, Algérie
rahoual@wissal.dz, hadji@lifl.fr
[2] Fucam
Chaussée de Binche 151, 7000 Mons, Belgique.
bachelet@fucam.ac.be

Abstract. The Set Covering Problem (SCP) represents an important class of NP-hard combinatorial optimization problems. Actually, the exact algorithms, such as branch and bound, don't find optimal solutions within a reasonable amount of time, except for small problems. Recently, the Ant systems (AS) were proposed for solving several combinatorial optimization problems. This paper presents a hybrid approach based on Ant Systems combined with a local search for solving the SCP. To validate the implemented approach, many tests have been realized on known benchmarks, and, an empirical adjustment of its parameters has been realized. To improve the performance of this algorithm, two parallel implementations are proposed.

Introduction

The Set Covering Problem (SCP) is one of the most studied NP-hard combinatorial optimization problems. Many of the real world problems can be described as a SCP [8]. Several methods were proposed in the literature to solve the SCP. The exact algorithms are able to find an optimal solution at reasonable run-time for problem's instance with up to 400 rows and 4000 columns [1][6]. However, for larger instances to be solved, heuristics, metaheuristics [2][8] and hybrid approaches [2][5][8] successfully applied. Recently, the ant system has been proposed as a new metaheuristic for combinatorial optimization problems [4].

In this paper, we present a study that demonstrates the ability of the Ant System (AS) to solve the SCPs. A hybrid algorithm [7] based on the AS and a local search heuristic is proposed. The local search based heuristic is used as an intensification tool and as a means to improve the ant's solution quality. For the parameter settings, the algorithm has been tested on standard benchmark [9]. The tests have revealed that our implementation is found the optimal solution or the best-known solution on some instances within a reasonable run-time. In order to improve the solutions quality and reduce the execution time, we have implemented two parallel approaches [3][10] of our hybrid algorithm (*AntsLS*). The first approach duplicates the algorithm on several processors units, which progress independently from each other. In the second approach, each ant is considered as a process, which is assignable to one processor unit. Experimental tests confirm that these parallel approaches offer good performances comparing to sequential algorithms.

M. Dorigo et al. (Eds.): ANTS 2002, LNCS 2463, pp. 262-267, 2002.

1 The Set Covering Problem

Given a zero-one matrix (a_{ij}) of m-rows and n-columns; a n-dimensional positive integer vector ($cost_j$) corresponding to the costs of the columns, the problem consists of extracting a subset of columns with a minimal cost such as all the rows are covered. A row i is said to be covered by a column j if the entry a_{ij} is equal to 1. The variable x_j indicates whether column j belongs to the solution ($x_j = 1$).
The SCP can be stated as follows:

$$\text{Minimize} \sum_{j=1}^{n} (cost_j * x_j) \tag{1}$$

Subject to the constraints:

$$\sum_{j=1}^{n} a_{ij} x_j \geq 1 \qquad i=1...m \tag{2} \qquad\qquad x_j \in \{0,1\} \quad j=1...n \tag{3}$$

2 Ant System Optimization for the SCP

In AS, each ant constructs a solution in an iterative manner, taking into account the past experience registered in the system memory (pheromone) and a greedy algorithm. For solving the SCP, we use a greedy heuristic based on the ratio cover value divided by cost ($cov\_val_j/cost_j$) [8].

$$cov\_val_j = \sum_{i \in cov(j,s)} min\_cost(i) \tag{4}$$

cov(j,s) is the set of lines which are covered by the column j and not covered by the solution S, and min_cost(i) is the minimum cost of the columns that cover the line i.

For a particular ant, the solution representation is a n bits vector (n stands for the number of columns of the instance). Let S denote a solution. If the column j belongs to an ant solution, then the jth bit takes the value 1 otherwise 0. An ant solution cost is stated as:

$$f(S)= \sum_{j=1..n} cost_j * S[j] \tag{5}$$

The transition rule of an ant is the probability that an ant chooses one column when constructing a solution. The transition rule we use is established as follows [6]:

$$P_k(j)=(Phero[j]^\alpha * H[j]^\beta)/ \sum_{i=1..n} (Phero[i]^\alpha * H[i]^\beta) \quad \text{if } j \notin S_k \qquad 0 \text{ else} \tag{6}$$

S_k is the set of columns belonging to the partial solution of the kth ant), Phero[j] is the pheromone deposed on the column j, α and β are parameters determining the relative importance of pheromone with regard to the greedy heuristic criterion $H[j]=cov\_val_j/cost_j$.

When all ants have generated their solution, the pheromone is updated in order to influence the next iteration. Effectively, each ant k deposits a precise quantity of pheromone on each column j, belonging to the just-constructed covering S_k in the following way [6]:

$$\text{Phero}[j] = (1-\rho) * \text{Phero}[j] + \sum_{i=1..m} \Delta\text{Phero}_k[j] \tag{7}$$

with $\Delta\text{Phero}_k[j]=1/f(S_k)$ if $j \in S_k$ and 0 otherwise, and ρ is a parameter that determines the decreasing rate of pheromone ($0<\rho<1$).

In the first step of the algorithm (*Ants*) proposed, the pheromone is initialized. The initialization rate is chosen arbitrarily but is identical for each column of the treated problem. The search process is composed in a series of iterations. In each iteration, each ant constructs a feasible cover. The artificial ant starts the search from an empty solution (containing no column). Then, for each step, it adds a column to the solution with respect to the equation (6). When an artificial ant constructs a feasible cover, the redundant columns are eliminated. When all ants have finished, the pheromone updating is made according to the equation (7) and the best solution found is updated too (*Ant-best*). At the end, if the artificial ants reach the pre-established rate convergence (*rate_conv*) then the search process is stopped and the best-found solution is returned. Otherwise, new search iteration is started up.

In order to test the efficiency of the implemented algorithm *Ants*, experimentations have been made on SCP instances extracted from the OR-library [9]. For each instance, we have realized 10 tests on a Pentium II-700 Mhz station.
We have observed that the algorithm *Ants*, has rarely found the best-known solution. Statistically, the algorithm has found the best solution for only 2 tested instances among 65. On the other hand, the average rate deviation is 5.66. This lack of efficiency is the resulty of the shortcomings in the intensification search [4][6]. In order to compensate this drawback, we have coupled the *Ants* algorithm with a local search heuristic.

3   The Ant System Hybridized with a Local Search Based Heuristic

To improve the solutions found by the ants, a local search based heuristic is applied. This algorithm (*LS*) uses a neighborhood function which works as follows: in the first step, D columns are chosen to be modified, so if a chosen column belongs to the solution then it will be dropped, otherwise it will be added. In the second step, the cover is repaired by adding columns j randomly with respect to the heuristic $cost_j/cov\_val_j$ and so that $cost_j$ is less than or equal to the maximum cost in S.
One can observe that the *LS* finds the optimal solution or the best-known solution on 20% of the tested benchmarks [9]. Furthermore, these results are better than those found by the *Ants* algorithm.
The first step of the hybrid algorithm *AntsLS* proposed initializes all the used parameters. At the second step, the ants searching process is started. At each iteration, each ant constructs a solution with regards to the equation (6). When a feasible cover is realized, each ant applies a local search on its solution. When the ants reach a convergence threshold, its value is reasonably set to 98%, the ants searching process is stopped.

As a result of the experimentation, one can notice that the algorithm hybridizing AS and the local search approximates the optimal solution for 55% of little-scaled instances and finds the pseudo-optimal solution for 60% of SCP large-scaled instances [9]. These results show clearly that the combination of AS with local search lead to improvement with regard to local search or AS only. On the other hand, we notice that the run-time was multiplied by 2. This increase is due to the application of local search heuristic by each artificial ant. To reduce the run-time and improve the solution quality, we propose two parallel implementations [3][10] of the latter algorithm *AntsLS*.

4   Parallel Implementation of the Proposed Algorithm *AntsLS*

4.1   Multi-start Independent Parallel Search

In this parallel implementation, several searches based on the *AntsLS* algorithm are launched simultaneously. This parallel approach, that does not need communications between the different search processes, is well fitted to the architectures devoted to high-level flow communication such as a network of workstations. In this implementation (fig. 1), a master process creates several copies based on *AntsLS* algorithm. It gets back all the solutions found by the search processes and returns the best solution found. To measure the efficiency of this algorithm, we have done several tests on the SCP large-scaled instances (more than 500 lines and 5000 columns) [9]. The algorithm was implemented with PVM/C. It was run on a network that connects Pentium II-700 MHz machines. For each test, we have launched 40 processes on 40 machines. For each process, we have used 20 artificial ants.

We notice that the multi-start independent parallel search improves appreciably the solutions quality found compared with *AntsLS* algorithm. Actually, the solution quality was improved for the instances: E3, F5, G2, G3, G5, and H1 [9]. On the other hand, we have got the best-known solution for the instances: E3, F5, and G5. Moreover, the average deviation rate is 0.74%, the average Speedup is 37.17 and the average efficiency is 92.93%. The obtained results are explained by the fact that the probability to reach the best solution was multiplied by 40 (the number of parallel processes). On the other hand, the parallel search efficiency is explained by the fact that the communication rate is very reduced. Actually, the communication is performed when the slave processes are created and when the results are send to the master process.

4.2   The Parallel Ants

In this parallel implementation, one copy of the *AntsLS* algorithm is used. In this algorithm, each artificial ant is considered as a separate search process. At the step in which the ants constructs their solution, all the ants process are launched at the same time. Each ant process is set on an independent processor. The master process sends the necessary information (pheromone) to each ant. Then, each one, independently from each other; constructs its solution. Thus, the master process gets back the solutions as each ant process finished. When all the solutions are received, the master process updates both the pheromone and the best solution, then it restarts the ants processes after it returns them the new value of the pheromone. The search process is stopped when the ants reaches a predetermined convergence threshold. The algorithm implementation is shown in the figure 2.

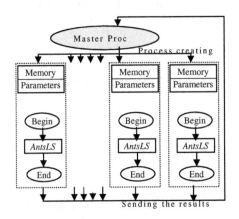

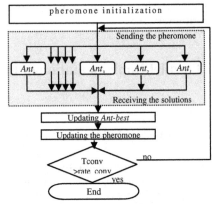

Fig. 1. Multi-start independent parallel search of the *AntsLS* algorithm.

Fig. 2. Parallel ants in the *AntsLS* algorithm.

To measure the efficiency of this algorithm, we have done several tests on one SCP's instance of each problem class [9]. For each performed test, we used 20 artificial ants launched on 20 machines. The table 1 shows the results of the performed tests.

Table 1. Results of the search: parallel ants.

I	Tp	Ts	S	E	I	Tp	Ts	S	E	I	Tp	Ts	S	E
41	20	159	7,95	39,75	C1	66	1 003	15,20	75,98	A1	48	556	11,58	57,92
51	35	264	7,54	37,71	D1	59	837	14,19	70,93	B1	47	482	10,26	51,28
61	29	122	4,21	21,03	E1	75	1 252	16,69	83,47	F1	91	1229	13,51	67,53

I: instances [9]; Tp: parallel time in second; Ts: sequential time in second; S: speedup; E: efficiency in %

In this parallel implementation, we were interested to the obtained gain of the parallel run-time. The average speedup is equal to 10.95 and is not uniformly distributed according to the treated instances. This value increases in proportion with the treated scale problem. This can be explained by the fact that the communication time is very high compared with the global run-time in the little scale instances.

5 Conclusion

In this paper, we have presented a study concerned with the developpement of AS-based hybrid metaheuristic for solving the SCP.

Several algorithms (*Ants*, *LS*, *AntsLS*) and two parallel versions of *AntsLS* have been implemented and strongly tested. We aimed at incrementally enhance our algorithms in order to reach better results. According to the experimentations related in the literature, the parameter setting is a hard and crucial key for efficiency. Nevertheless, the hybrid algorithm yields better results then *Ants* and *LS* used alone. Indeed, *AntsLS* approximates the optimal solution for 55% of little-scaled instances and finds the pseudo-optimal solution for 60% of SCP large-scaled instances [9].

In order to limit the computational time and to refine the results of sequential approaches, parallel algorithms have been developed. They neatly improve the quality of the solution (average deviation is of 0,74% on the PVM/C platform) and reduce the overall execution time of the algorithm.

However, we assume that this work should benefit from further work. At the moment, we are refining parameter setting and we are testing a new hybridization schemes involving AS.

References

[1]   E. Balas & M. C. Carrera. A dynamic subgradient-based branch and bound procédures for Set Covering. *Managment Sciences Research Report n° MSSR 568*, GSIA, Carnegie-Mellon University, October 1991. Revised May 1995, to appear in *Operations Research*.

[2]   J. E. Beasley & P. C. Chu. A genetic algorithm for the Set Covering Problem. *European Journal of Operational Research*, 95(2): 393-404, 1996.

[3]   B. Bullnheimer, G. Kotsis & C Strauss. Parallelization Strategies for the Ant System. High Performance Algorithms and Software in Non Linear Optimization, *Applied Optimization Series*, Vol. 24, Kluwer Dordrecht, 87-100, 1997.

[4]   M. Dorigo, and L. M. Gambardella. A Study of Some Properties of Ant-Q. *Proceeding of PPSN IV – Fourth International Conference on Parallel Problem Solving from Nature*, H.-M. Voigt, W. Ebeling, I. Rechenberg And H.-S. Schwefel (Eds.), Springer-Verlag, Berlin, pp. 656-665, September 22-27, 1996.

[5]   A. V. Eremeev, A. A. Kolokov & L. A. Zaozerskaya. A hybrid algorithm for Set Covering Problem. *Proc. of International Workshop "Discrete Optimization Methods in Scheduling & Computers –Aided Design- Minsk"*, 123-129, 2000.

[6]   M. L. Fisher & P. Kedia. Optimal solution of Set Covering/Partitioning Problems. *Mathematical Programming Study*, 36: 674-688, 1990.

[7]   R. Hadji, M. Rahoual, E-G. Talbi, and V. Bachelet. Ant Colonies for the Set Covering Problem. *Abstract Proceeding of ANTS'2000*, pp. 63-66, Brussels, September 7-9, 2000.

[8]   E. Marchiori & A. Steenbeek. An evolutionary algorithm for large scale Set Covering Problems with application to airline crew scheduling. *Real World Applications of Evolutionary Computing*, Springer-Verlag, LNCS 1083: 367-381, 2000.

[9]   J.E. Beasley. OR-Library: distributing test problems by email. *Journal of the Operational Research Society*, 41:1069-1072, 1990. http://mscmga.ms.ic.ac.uk/info.html.

[10]   E-G. Talbi, O. Roux, C. Fonlupt & D. Robillard. Parallel Ant Colonies Optimization Problems. *Biosp3 Workshop on Biologically Inspired Solutions to Parallel Processing Systems, in IEEE Ipps/Spdp'1999*.

Real-World Shop Floor Scheduling by Ant Colony Optimization

Andre Vogel, Marco Fischer, Hendrik Jaehn, and Tobias Teich

Chemnitz University of Technology, Dept. of Economics, Chair BWL VII*
D-09107 Chemnitz, Germany
{andre.vogel,marco.fischer,hendrik.jaehn,t.teich}
@wirtschaft.tu-chemnitz.de

Abstract. Manufacturing Control Problems are still often solved by manual scheduling, that means only out of the workers experience. Modern algorithms, such as Ant Colony Optimization, have proved their capacity to solve this kind of problems. Nevertheless, they are only used exceptionally in real world. There are two main reasons for that. Firstly, an ant-based scheduling tool has to fit into the organizational structures of today's companies, i.e. it has to be coupled with the Enterprise Resource Planning-system (ERP-system) used in the company, in order to ensure that the capacity of the colonies search is used as efficiently as possible. The second reason is the size of the real world shop floor scheduling problems. In order to be able to deal with that problem, the authors propose a continuously operating Ant Algorithm, which can easily adapt to sudden changes in the production system.

1   Introduction to the Problem

Shop floor scheduling solves the machine allocation problem. It is a much examined, $\mathcal{NP}$-hard problem.

In Operations Research, models for these kind of problems and heuristics to solve them had been developed. But it was still not possible to handle shop floor scheduling as an integral whole. The situation changed firstly due to the introduction of evolutionary methods and other meta-heuristics like Simulated Annealing or Treshold Accepting. A number of attempts have been made to solve hard combinatorial optimization problems in the field of production control using Genetic Algorithms [2], [13].

The objective of this work was to compare the developed Ant Algorithm - which will be described in the third section - to the results achieved by a Genetic Algorithm tested on the same real-world data.

2   The Ant Algorithm

The used algorithm, which is based on several works on this topic ([1], [3], [4], [6] and [11]), has already been tested successfully e.g. in order to deal with the widely

---

* phone: ++49-371-531 42 52

M. Dorigo et al. (Eds.): ANTS 2002, LNCS 2463, pp. 268–273, 2002.

known Job Shop Scheduling Benchmark problems [9]. Job Shop Scheduling is closely related to the problem considered here.

In order to use the algorithm also for Real-World Shop Floor Scheduling problems, some adaptions and changes within the algorithm are necessary.

To store the pheromone, a Position-Operation-Pheromone-matrix (P-O-P-matrix) is employed (see figure 1). Within that, the operations are stored by an accessary operation number (onr). Pheromone values $(\tau_{pos,onr})$ are allocated to the fields of the matrix during the solution process of the algorithm. The pheromone values of an operation depend on the position within the solution sequence (permutation). The pheromone values are changed by the pheromone update within each iteration according to the new solutions.

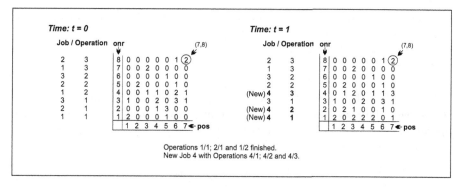

Fig. 1. Position-operation-pheromone-matrix

If a costumer inquiry is considered, a copy of the problem is used to generate an answer. A new job is set up after the costumer had accepted an offer. The new data is handed over to the Ant Colonies (by using an XML-File[12]) , which are updating the P-O-P-matrix and start the search again.

Due to the filling up of the gaps, which are result of the finished jobs, and the dynamic structure of the P-O-P-matrix connected with that, an allocation table is required (see figure 2), which establishes the relationship between the operations within a job and the operation number (onr).

It is necessary to reset the pheromone values due to the continuously incoming jobs/operations that have to be scheduled.

In case, a best solution is found in the end, all the operations of that solution receive a certain additional amount of pheromone, appropriate to their position in the solution. Because of that, during the next search, it is a lot more attractive to schedule that operation again in the position because that position of the operation already proved success. Thereby, the amount of the pheromone that has to be delivered again, depends on two factors. On the one hand, it depends on the time that remains between the current time (tc) and the attainment of the delivery date (td_j). That means, operations within a job having an earlier

Job (J$_n$)	Operation (O$_i$)	Operation-number (onr)
1	3	7
2	2	5
	3	8
3	1	3
	2	6
4	1	1
	2	2
	3	4

Allocation Table (t=1)

Fig. 2. Allocation table (time: t=1)

delivery date are considered more important in scheduling than those operations of jobs having a relatively greater temporal buffer (tf_j).

$$tf_j = te_j - tc \tag{1}$$

The end of the job time window (te_j) results out of the difference between the delivery date (td_j) and the sum of the processing durations of the remaining operations of a job (see equation 2).

$$te_j = td_j - \sum_{k \in O_{jk}} p_{jk} \tag{2}$$

$$O_{jk} \in \mathcal{PO}$$

A second factor is the necessity of the increase of jobs with a higher priority, and wit that of the operations belonging to these jobs, in comparison to other jobs: This is considered within the initial pheromone values.

In case, the P-O-P-matrix has been developed and the initial values of the pheromone are initialized, the algorithm starts. Each ant chooses the next operation (onr), according to the set of feasible operations $(\mathcal{PO})$, which can be processed in the next step, as follows:

$$P_{pos,onr} = \frac{[\tau_{pos,onr}]^\alpha \cdot [\eta_{pos,onr}]^\beta}{\sum_{h \in \mathcal{PO}} [\tau_{pos,h}]^\alpha \cdot [\eta_{pos,h}]^\beta} \tag{3}$$

$$P_{pos,onr} = \frac{(\sum_{k=1}^{pos} [\gamma^{pos-k} \tau_{k,onr}])^\alpha \cdot [\eta_{pos,onr}]^\beta}{\sum_{h \in \mathcal{PO}} (\sum_{k=1}^{pos} [\gamma^{pos-k} \cdot \tau_{k,h}])^\alpha \cdot [\eta_{pos,h}]^\beta} \tag{4}$$

$$\eta_{pos,onr} = \frac{1}{\frac{\sum_j \frac{tf_j}{(tf_j)^q}}{tf_j}} \tag{5}$$

$$tf_j = te_j - tb_j \tag{6}$$

In order to be offered a detailled description of the used equations (3) and (4), consider [8] und [5].

Within the solution search, the heuristics value ($\eta_{pos,onr}$) is calculated with regard to equation (5) in the following. The effect of the urgency on the selection can be influenced by the parameter q. The greater q the higher is the probabilitiy of the selection of an operation disposing of a quite small temporal buffer.

The next technologically possible operation of a job is part of $\mathcal{PO}$, that means, part of the quantity of the operations that can be scheduled. The start of the time window of the following operations (tb_j) of a job results from the present time (tc), because the previous operations have already been completed and the next possible operations ($onr \in \mathcal{PO}$) of a job could theoretically be scheduled immediately:

$$tb_j = tc$$

The amount of the pheromone of an operation at a position is restricted by certain upper- (τ_{max}) and lower limits (τ_{min}).

$$\tau_{min} < \tau_{pos,onr} < \tau_{max}$$

This idea is based on the works of Stuetzle [10] within which a higher solution quality for known benchmark-problems has been achieved.

After all ants of all colonies generated a solution in this way, the global pheromone update is carried out. Therefore, an elitist-strategy is applied. This means that the ant having generated the best solution (mean lateness L) is allowed to add pheromone according to:

$$\tau_{pos,onr} = \tau_{pos,onr} + \rho * \frac{1}{L}$$

In this case, pheromone is added to the positions of an operation in the current best solution within the P-O-P-matrix.

After every generation, the evaporation is carried out in every cell of the matrix, as follows:

$$\tau_{pos,onr} = (1 - \rho) \cdot \tau_{pos,onr} \tag{7}$$

Parameter ρ regulates the evaporation rate. The smaller ρ is, the longer the pheromone informations are passed on.

3   Computational Results

The shop floor scheduling problem was tested on real world data provided by a German engineering company. The plant is organized according to the workshop principle, which means that machines are grouped according to the operations that have to be carried out using this ressource rather than according to the products that are manufactured using this machines.

In order to be able to dispose of real world problem sizes and data, the following numbers were recorded within a period of two months. The release date, the finishing date and the due date of each job, as well as the date and the duration of all occured disturbances in the manufacturing process (machine breakdowns, material defects, illnesses of workers, etc.) were observed.

Because of comparing the reached solution quality, we used the mean flow time and the mean lateness of jobs. For the comparison of the results achieved by the means of the Ant Algorithm and other methods we used a priority rule scheduler (PRIORULE) and GACOPA which is a Genetic Algorithm using continuous co-evolution of the parameter settings and parallelization [7]. The priority rule scheduler was allowed to calculate a number of single, standard priority rules (e.g.: shortest processing time, earliest due dates etc.) and combinations of these rules. The best out of the achieved results was used. Manual scheduling achieved a mean lateness of 2.4 days and a mean flow time of 13.2 days. These values were set to 100% to ease the comparison.

Table 1. Relative results (manual scheduling = 100%)

scheduler	mean lateness	mean flowtime
MANUAL	100.0 %	100.0 %
PRIORULE	95.3 %	97.9 %
GACOPA	55.3 %	85.0 %
ANTS	61.8 %	87.7 %

The following parameters were employed for the tests: $\alpha = 1; \beta = 1; \rho = 0,01; c = 0,8; \gamma = 0,8; q = 2$. The results presented in table 1 show that ants perform very well, but not as good as the GACOPA, the results are compared with.

4   Conclusion

As a result it can be stated that modern scheduling tools for real world shop floor scheduling allow companies to mobilize a lot of reserves in the production system at comparably low cost. The quality of the GA solution wasn't achieved, but it was the first work completed in this field with ACO. The authors further research will be directed towards improving the presented approach by means of solution quality, reducing of computation time and improving the parameter settings.

Acknowledgements. This work has been supported by the Collaborative Research Center 457 and the CBS GmbH.

References

1. Bonabeau, E. Dorigo, M. Theraulaz, G.: Swarm Intelligence - From Natural to Artificial Systems, pp. 69-71, Oxford University Press, New York, NJ, 1999.
2. Davis, L.: Job Shop Scheduling with Genetic Algorithms. in: Grefenstette, J., Editor, Proceedings of an International Conference on Genetic Algorithms and their Applications, pp. 136-140, Hillsdale, Lawrence Erlbaum Associates, 1985.
3. Deneubourg, J.-L., et. al: Self-Organization Mechanisms in Ant Societies (II): Learning in Foraging and Division of Labour. Experientis Suppl. 54, pp. 177-196, 1987.
4. Dorigo, M.: The Ant System: Optimization by a colony of cooperating agents, in: IEEE Transactions on Systems, Man, and Cybernetics 26, (1), pp.29-41, 1996.
5. Fischer, M., Vogel, A., Teich, T., Fischer, J.: A new Ant Colony Algorithm for the Job Shop Scheduling Problem. in: Proceedings of the Genetic and Evolutionary Computation Conference, San Francisco, 2001.
6. Gambardella, L.M., Agazzi, G.: MACS-VRPTW: A Multiple Ant Colony System for Vehicle Routing Problems with Time Windows, in: Corne, D., Dorigo, M., Glover, F.(Eds.) - New Ideas in Optimization, pp.63-76, 1999.
7. Kaeschel, J., Meier, B., Fischer, M., Teich, T.: Real-World Applications: Evolutionary Real World Shop Floor Scheduling using Parallelization and Parameter Coevolution, in: Proceedings of the Genetic and Evolutionary Computation Conference, Las Vegas, 2000.
8. Merkle, D., Middendorf, M.: An Ant Algorithm with Global Pheromone Evaluation for Scheduling a Single Machine , in: Cagnoni, S., et al. (Eds.) - Real-World Applications of Evolutionary Computing, Proceedings of EvoWorkshops 2000, Edinburgh, LNCS 1803, pp.281-296.
9. OR-library: http://graph.ms.ic.ac.uk/info.html, 2002.
10. Stuetzle, T., Hoos, H.: MAX-MIN Ant System for the Traveling Salesman Problem, 1997.
11. van der Zwaan, S., Marques, C.: Ant colony optimization for job shop scheduling. in: Proceedings of the Third Workshop on Genetic Algorithms and Artificial Life (GAAL 99), 1999.
12. World Wide Web Consortium (Eds.): Extensible Markup Language (XML), http://www.w3.org/XML, 24.03.2001.
13. Yamada, T., Nakano, R.: A Genetic Algorithm with Multi-Step Crossover for Job Shop Scheduling Problems. in: Proceedings of an International Conference on GAs in Engineering Systems: Innovations and Applications (GALESIA 1995), pp. 146-151, 1995.

Simulation of Nest Assessment Behavior by Ant Scouts

E. Şahin and N.R. Franks

[1] IRIDIA, Université Libre de Bruxelles
CP 194/6, Avenue Franklin Roosevelt 50, B-1050 Bruxelles, Belgium
esahin@ulb.ac.be
[2] Centre for Behavioural Biology, School of Biological Sciences, University of Bristol,
Woodland Road, Bristol BS8 1UG U.K.
Nigel.Franks@bristol.ac.uk

Abstract. The scouts of *Leptothorax albipennis* colonies find and assess new nest sites, when their current nests become uninhabitable. Observations of these scouts have suggested that they assess, among other things, the integrity of the internal periphery and the size of the potential nest site. The hypothesis that the scouts use a 'Buffon's needle algorithm' to estimate the nest size is supported by experiments. In this paper, we present a behavioral model for the nest assessment of the scouts. This behavior is implemented on an *ant-bot*, a simulated scout model, to study the assessment process. We present the simulation results obtained from this model by systematically varying the behavior and analyzing how well the integrity of the periphery and the size of the nest was evaluated. The results indicate that the accuracy of these two evaluations requires conflicting exploration behaviors, and an optimal behavior requires a compromise in the accuracy of both.

1 Introduction

Biological systems are excellent examples of how, seemingly complex, decisions can be obtained through simple behaviors that implement rules of thumb or clever procedures, that have evolved. Modeling the decision making process from direct observations is a common methodology in biology. Although these studies are essential for the understanding of the biological systems, we believe that constructing mechanical models that replicate the results obtained from biological systems, is likely to improve our understanding of these systems. The construction process not only brings to the surface the small design details that may have been skipped during modeling, but it also allows one to vary the parameters of the constructed model to study the other varieties of the model that the evolution had not chosen.

2 Nest Assessment in *Leptothorax albipennis*

Colonies of *Leptothorax albipennis*, a small monomorphic myrmicine ant species, inhabit small flat crevices in rocks. When the current nest becomes uninhabit-

M. Dorigo et al. (Eds.): ANTS 2002, LNCS 2463, pp. 274–281, 2002.

able, the scouts explore the environment to find and assess new nest sites. These ant scouts assess potential nest sites before they attempt to initiate an emigration of the whole colony. In their assessment, the integrity of the inner wall of the potential nest site, and the floor area of the nest site seem to constitute two important criteria. Mallon and Franks[1,2] observed the visits of individual scouts to new sites. They have reported that scouts tend to make more than one visit to a new site before attempting to initiate the emigration of their entire colony. During their visits, the scouts spent a considerable part of their time exploring the internal periphery of the site, while making seemingly random explorations of the central part of the nest, Fig. 1. No significant differences were found between the duration of the first (second) visits to nests of different sizes[2]. It is also observed that in their second visits, the scouts "briefly but significantly slow down" as they cross their first visit trails. Based on these observations and many others, Mallon and Franks[1] suggested that the scouts lay an individual-specific pheromone trail during their first visit, and that they use the intersection frequency of their path with this pheromone trail during their subsequent visits to estimate the floor area of the nest. They pointed out that, this strategy is consistent with the Buffon's needle method, a technique in computational geometry to estimate π empirically, that can be adapted to measure space.

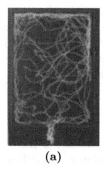 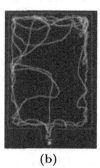

(a)   (b)

Fig. 1. Two trails of a scout visiting a new nest site as traced by an overhead camera. (a) shows the trail of the first visit, (b) shows the second visit.

They tested this hypothesis by tracing the visits of scouts to different potential nest sites in the laboratory environment. They counted the intersections of traces between the first and subsequent visits separately within the central region and the peripheral region of the new nests. The results obtained were consistent with the Buffon's needle method. Apart from the Buffon's needle method, they have also tested whether the ants use (a) the internal perimeter of the nest, (b) the 'mean, free-path-length algorithm' to assess the size of the nests. However, the experiments showed that (1) scouts were able to choose a standard-size nest over a half-size one with the same internal perimeter and, (2) a partial barrier placed inside a standard-size nest did not affect the assessment of the nest.

While exploring a nest, the component of the behavior to check the internal perimeter of a nest might be in conflict with the component of the behavior for the measurement of the nest area. First, the ant will spend less time exploring the central part of the nest decreasing the accuracy of the size assessment. Second, it may be possible that the pheromone trail at the periphery can cause problems for the implementation of 'Buffon's needle algorithm'.

This paper attempts to tackle these issues by constructing a simulation that mimics the environment, the ant and its behavior model for assessing new nests. By varying a parameter of the exploration behavior, the simulation allowed us to study the dynamics of the assessment process for achieving an optimal assessment of a new nest.

In the rest of the paper, we will first present the model for the simulation of the ant and its environment. Second, we will describe the exploration behavior of the ant model proposed for nest assessment. Third, we will describe the experiments carried out and the results obtained. Finally, the results are discussed and future directions for the work are outlined.

3   Simulation

We have modified YAKS[1], a free mobile robot simulator, to study the nest assessment process ant scouts. The simulator is designed to simulate a physical mobile robot, Khepera [3] (K-Team, Switzerland), by sampling the sensory readings from a real robot [4]. Although it is not designed to simulate ants, it is preferred since it models the interactions between the agent (robot) and the environment in a realistic way. The simulation operates in 2-D.

3.1   The Ant-Bot

The *ant-bot*, sketched in Fig. 2-(a), is created as a model of the scout ants. For this, the original Khepera robot model of the simulator is modified. The ant-bot has four infrared proximity sensors placed in the front to imitate the short-range sensing ability of the ant with its movable antennae, Fig. 2-(b). It is also equipped with a "pheromone nozzle" and a "pheromone detector", both located at the center of the body, the former for laying and the latter for detecting the pheromone in the environment.

3.2   The Nests

Three different nest designs, are shown in Fig. 3. These nests are created by walls as a closed rectangular space. Unlike the real nests, used in the experiments of Mallon and Franks [1], the entrances are omitted to remove the possibility of the ant-bot leaving the nest prematurely[2]. The small rectangle shown under the

---

[1] Available at http://www.ida.his.se/ida/~johanc/yaks/

[2] The scouts seem to spend a certain duration of time during their visits.

(a)   (b)

Fig. 2. (a) Sketch of the ant-bot. The circle represents the body. The two elongated rectangles placed on the left and right part of the body denote the wheels of the robot. The four small rectangles on the upper part of the figure shows the placement of the infrared proximity sensors. The concentric circles drawn at the center of the robot indicate the pheromone nozzle and detector. (b) *Leptothorax albipennis*.

nest, marks the position of the entrance. Within the environment the ant-bot is drawn as a circle with a line connecting its right and left and right wheels.

The nest in Fig. 3 (a), shows the standard-size nest used in our experiments. The nest design (b) shows a smaller nest which has half the size of the standard-size nest. The nest shown in (c) is a standard-size nest with a partial barrier placed at the center of the nest. The real ants can detect that nest (b) is too small, yet they are not confused by (c), the standard-size nest with a partial barrier. They respond to (c) as to the standard-size nest (a).

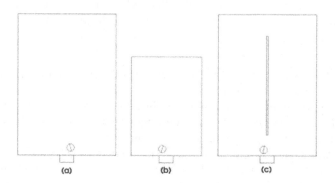

(a)   (b)   (c)

Fig. 3. The three types of nests considered for the experiments: **(a)** standard-size nest, **(b)** half-size nest, and **(c)** standard-size nest with a partial barrier.

4   Nest Assessment by the Ant-Bot

The ant-bot makes two visits to a new nest. In each visit, it starts its exploration above the entrance at a random alignment. During the first visit, it lays pheromone along its path. During its second visit, instead of laying pheromone, the ant-bot senses the pheromone layed during its first visit, and uses this information to estimate the size of the nest. In both visits, the ant-bot uses the same exploration behavior.

The exploration behavior uses the infrared proximity sensors to drive the ant-bot creating exploration trails that seem to be similar to those observed in ant scouts. The behavior is parametrized in such a way that it can generate a continuum of trails that can range from wall following to random exploration.

4.1   Exploration Behavior

The exploration behavior uses the readings obtained from the four infrared proximity sensors to drive the two motors. The ant-bot is controlled by setting the speed of its left and right wheels (m_l and m_r), which are calculated as

$$m_l = (1 - |\bar{r}|) * 0.25 - \bar{r}$$
$$m_r = (1 - |\bar{r}|) * 0.25 + \bar{r}.$$

When $\bar{r} = 0$, the ant-bot moves forward. It turns left when $\bar{r} = 1$, and right when $\bar{r} = -1$. Here, $\bar{r}$ is defined as

$$\bar{r} = \begin{cases} -1 & : & r+n < -1 \\ r+n & : & -1 < r+n < 1 \\ 1 & : & r+n > 1 \end{cases}$$

where n is a random number between -0.4 and 0.4 and r is defined as the value of the 'rotational activation'. The change in r is calculated as

$$\Delta r = -0.9r + 0.3(1-r)(w_l + 1.5I_4 + 1.2I_3) - 0.3(1+r)(w_r + 1.5I_1 + 1.2I_2).$$

where I_i denotes the infrared readings, with a value between 0 (no object) and 1 (very close object), where $1 < i < 4$ is the index. Here, w_l, w_r represent the 'perceived presence' of the wall on the right and left side respectively. The first term on the right of the equation guarantees that when no wall is perceived and the infrared readings are all zero, then any rotational activation will decay to zero in time. The second term raises the rotational activation towards 1 in proportion to the amount of wall perceived on the left side and the infrared readings from the right side. The third term tries to pull down the rotational activation to -1 in a similar way.

The variables, w_l and w_r, indicate the presence of the peripheral wall on the left and right side of the ant-bot respectively and the change in them are defined as

$$\Delta w_l = -0.1w_l + \gamma(1 - w_l)I_1 - 0.7w_l(I_2 + I_3)$$
$$\Delta w_r = -0.1w_r + \gamma(1 - w_r)I_4 - 0.7w_r(I_2 + I_3).$$

The first term on the left side causes the perceived presence of a wall to decay to zero when no objects are sensed. The second term, increases the perceived presence of the peripheral wall by the activations of infrared sensing on that side. The third term diminishes the perceived presence of any wall if the front sensors become active, to raise the priority of avoidance. The parameter γ controls the perceived presence of the wall. When the parameter $\gamma = 0$, both w_l and w_r decay to zero, and stay there. For nonzero values of γ the perceived presence of wall becomes stronger.

The exploration behavior defined above can generate exploration patterns ranging from random exploration (that is moving while avoiding the walls), to wall following, by varying γ. When $\gamma = 0$, the wall sensing part of of the behavior is removed, and the robot moves in a random way, while avoiding any obstacles on its way. As γ is increased, the wall sensing becomes active creating a attraction towards the wall. As the attraction grows larger, the robot tends to stay closer to the walls and become less likely to move into the central part of the nest. Figure 4 shows three different exploration patterns achieved by different values of γ.

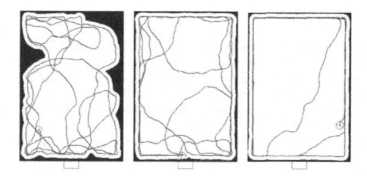

Fig. 4. Different trails can be obtained by varying γ. Three trails from the exploration of a standard-sized nest for 10000 time steps, are shown. These trails are obtained for $\gamma = 0.0, 0.3, 1.0$, from left to right. Increasing γ beyond 1 tends to make the attraction towards the wall so strong that it may overcome the obstacle avoidance component of the behavior, causing the ant-bot to crash into the walls. The uncovered periphery is marked as a dark region inside the walls of the nest.

4.2   Evaluating the Nest Assessment

The assessment of a nest by the ant-bot is evaluated, using two measures: namely, how accurate the floor area is estimated, and how well the integrity of the nest perimeter is checked.

Measuring the size of the nest. The size estimation is done by the ant-bot. The pheromone sensor, denoted as p, returns 0 or 1 reporting the absence or existence of pheromone under the ant-bot. This reading is processed by leaky-integrator:

$$\dot{p} = -0.1\bar{p} + 0.9(1 - \bar{p})p$$

that generates a smoother sensory signal. The Buffon's needle algorithm is approximated, by counting the rising edge crossings of this signal with a threshold of 0.5. In the rest of the article, we will use the term 'Buffon count' to denote the number of these crossings counted during the second visit of the ant-bot.

Measuring the periphery coverage. The success of the ant-bot at checking the periphery of the nest, is defined as the area between the inner region covered by the pheromone trail of the ant-bot and the periphery. This evaluation is done by the simulator after the first visit of the ant-bot. The dark regions between ant-bot's trail and the inner periphery, Fig. 4, shows the unchecked periphery for three different explorations.

5   Results

The three nest types, shown in Fig. 3 are used in the experiments. For each nest type, the ant-bot made two visits to the nest: the first, lasting for 10000 time steps; the second, lasting for 7500 steps. In each visit, the ant-bot began its exploration in front of the entrance, which is indicated by a small rectangular block below the nest. The initial position of the ant-bot was kept constant except that its initial orientation was varied within ∓ 15 degrees of the wall.

We have evaluated the nest size estimates of the ant-bots, and the amount of uncovered periphery while varying γ from 0 to 1. For each value of γ, twenty nest assessments are made by the ant-bot. The median of the Buffon count and the uncovered periphery is plotted with respect to γ and the interquartile range is shown as error bars.

Figure 5-(a) plots the median Buffon count for the different types of nests, with respect to γ. Two points are worth noting. First, even with the Buffon's algorithm in operation at the periphery, where trails are less random, for $\gamma < 0.3$, the ant-bot can reliably distinguish between a standard-size and half-size nest. Second, the barrier placed inside a standard-size nest, did not affect the size assessment.

It is interesting to note that, in Fig. 5-(a), at high gamma values the line for the Buffon count in half sized nests dips below the line of the other two larger nests. This occurs because the pheromone trails get "crowded" at the periphery, blending into fewer thicker trails. In the half-sized nest, the ant-bot has the time to make more "rounds" causing more blending than the standard-size nests, hence it makes fewer Buffon counts.

Figure 5-(b) plots the median percentage of covered periphery with respect to γ. It can be seen that, as expected, the amount of covered periphery increases with γ. This suggests that periphery coverage in conflict with the accuracy of the size evaluation.

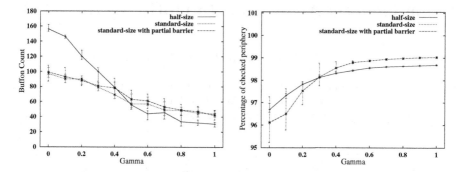

Fig. 5. Median **(a)** Buffon count and **(b)** percentage of covered periphery. The error bars indicate the interquartile range. Note that the percentage of the covered periphery will always be less than 100 since the body of the ant-bot does not touch the periphery.

6 Conclusions

We proposed a model of nest assessment in scout ants. The model shows that an exploration behavior that combines obstacle avoidance and wall following, with the addition of a high amount of noise, is sufficient both to generate similar trails to those of the real ants and to enable them assess a nest site accurately. The analysis shows that the exploration behavior has to be tuned to optimize the nest assessment, since the accuracy of nest size measurement, and the completeness of the periphery coverage require conflicting strategies.

References

1. Mallon, E.B. and N.R. Franks: Ants estimate area using Buffon's needle. Proc. Royal Society London B, Vol. 267 London, (2000), 765-770.
2. Mugford, S.T., E.B. Mallon and N.R. Franks: The accuracy of Buffon's needle: a rule of thumb used by ants to estimate area. Behavioral Ecology, Vol. 12. (2001), 655-658.
3. Mondada, F., E. Franzi and P. Ienne: Miniaturisation: A tool for investigation in control algorithms. In T. Yoshikawa and F. Miyazaki, eds., Proceedings of the Third International Symposium on Experimental Robotics. Springer-Verlag, Berlin, (1993), 501-513.
4. Miglino O., H.H. Lund, and S. Nolfi: Evolving mobile robots in simulated and real environments. Artificial Life, (2) 4. (1995), 417-434.

Using Genetic Algorithms to Optimize ACS-TSP

Marcin L. Pilat and Tony White

School of Computer Science, Carleton University,
1125 Colonel By Drive, Ottawa, ON, K1S 5B6, Canada
{mpilat,arpwhite}@scs.carleton.ca

Abstract. We propose the addition of Genetic Algorithms to Ant Colony System (ACS) applied to improve performance. Two modifications are proposed and tested. The first algorithm is a hybrid between ACS-TSP and a Genetic Algorithm that encodes experimental variables in ants. The algorithm does not yield improved results but offers concepts that can be used to improve the ACO algorithm. The second algorithm uses a Genetic Algorithm to evolve experimental variable values used in ACS-TSP. We have found that the performance of ACS-TSP can be improved by using the suggested values.

1   Introduction

Models created based on natural systems have been successfully used to solve NP-hard combinatorial optimization problems. The Ant Colony Optimization (ACO) meta-heuristic [2] is a generic framework for ant-based optimization algorithms. ACO algorithms, such as Ant System (AS) [5] and Ant Colony System (ACS) [4], were successfully used to solve instances of the Travelling Salesman Problem (TSP) [7], and other combinatorial optimization problems [3].

The ACS-TSP algorithm [4] produced better results on many TSPs compared to some genetic algorithms (GA), simulated annealing (SA), and evolutionary programming (EP) [1]. The goal of our research was to improve the performance of ACS-TSP by augmenting it with ideas from Genetic Algorithms (GAs) [6].

We look at two modifications to the ACS-TSP algorithm. The first algorithm, called ACSGA-TSP, uses a population of genetic ants modified by a GA. We present a comparison between the performance of ACSGA-TSP and ACS-TSP and provide some ideas that can be used to improve the ACSGA-TSP algorithm. The second algorithm uses a Meta GA to evolve the optimal values of experimental parameters used in ACS-TSP. We discuss improvements that can be made to the ACS-TSP algorithm based on our findings.

2   ACS-TSP Algorithm

The ACS-TSP algorithm is a modified AS-TSP algorithm where the update of the pheromone trail happens locally while an ant is building its trail and globally by the best performing ant. The tour nodes are chosen based on a more

M. Dorigo et al. (Eds.): ANTS 2002, LNCS 2463, pp. 282–287, 2002.

complicated transition function that is either deterministic (to promote use of graph knowledge) or probabilistic (to promote exploration).

The performance of ACS-TSP was experimentally found to be better than that of AS-TSP [4]. We have implemented the ACS-TSP algorithm as defined by [1] and were able to verify the results of [4].

3   ACSGA-TSP Algorithm

We propose the ACSGA-TSP algorithm as a Genetic Algorithm modification to ACS-TSP. The algorithm uses a GA to evolve a population of genetically modified ants to improve the performance of the ACO algorithm.

In the original ACS-TSP algorithm, the ants used constant global parameter values. We have augmented each ant with its own value of ACS-TSP parameters. Each GA ant was encoded by a 21 bit string composed of three parameters (β, ρ, and q_0) encoded using 7 bits each. The allowed value ranges were: integer value 0-127 for β and double value 0-1 for ρ and q_0.

Ant chromosomes were initially randomized at ant creation. The crossover operator used a simple single point crossover scheme on whole chromosomes of two parents to create two children. A bit-wise mutation operator was able to mutate each bit of a chromosome based on a given probability.

In each iteration of the algorithm, four GA ants were selected from the ant population. This selection was done using a tournament selection algorithm of size 4. Each of the four selected ants were then asked to build their TSP tours.

Each GA ant stored its values of the three encoded parameters. The β and q_0 parameters were used by the ants to choose the next city to visit. Local update of the pheromone trail was done by each ant using the value of its ρ parameter.

Once the tours were completed, the algorithm checked to see whether a new tour has been found, as in ACS-TSP. The global update of the pheromone trail was done by the ant that produced the best overall tour using the encoded value of the ρ parameter.

Fitness was calculated as the length of the generated tour. After the pheromone update, the best two selected ants were crossed over to produce two children. Mutation was then performed on the offspring and the worst two selected ants were replaced by the children. This kept the population size constant and provided pressure for the population to improve its performance.

Starting location of each ant was randomized at the start of each iteration. The behavior of the ants is influenced by the pheromone trail left during a run of the algorithm, thus, the performance of later ants can be biased. This could be solved by restarting the algorithm with the same population.

We have used a population of 20 ants in our experiments. ACSGA-TSP was run for a number of iterations such that the total number of ant tours built was equivalent to that of the ACS-TSP algorithm we have used. Probability of crossover was set to 0.9 and probability of mutation to 0.01.

Results of the experiments are given in Table 1. The algorithms shared similar results. The standard deviation of the numerical solutions found by ACSGA-TSP

Table 1. Comparison of results obtained from ACSGA-TSP and ACS-TSP algorithms. Results given are averages of best tour lengths over 10 runs (eil51, ft70) or 5 runs (kroA100) and standard deviation values for the results. Iteration values were adjusted to yield the same total number of ant tours in each algorithm.

Problem	ACS AV	STD	iter.	ACSGA AV	STD	iter.
eil51	428.7	2.45	2K	432.4	4.18	10K
kroA100	21712	316	3K	21948	92	15K
kroA100	21614	310	5K	21544	250	25K
ft70	41144	452	4K	40868	379	20K

was on average smaller than that of ACS-TSP. This was especially noticeable for larger problems.

From our results, the ACSGA-TSP algorithm does not guarantee finding the optimal solution for a problem. The ant population converges to experimental variable values that produced the best results during the algorithm run. At that point, it is difficult for the ants to improve the algorithm since the individuals can be trapped at local minima. Pheromone trails due to good solutions can be erased by worse performing ants. Thus, in its current form, ACSGA-TSP would not outperform the ACS-TSP algorithm.

The rate at which good solutions were found was observed to be quicker using ACSGA-TSP, as seen in Table 2. Quick convergence to good solutions is thus a desirable characteristic of the ACSGA-TSP algorithm. For large problems, where optimal solutions are intractable or not desired, this algorithm provides good solutions faster than the ACS-TSP algorithm, as shown in Table 2. This characteristic is mainly due to the variety in the population of ants which facilitates early exploration of the search space.

Bad performing ants can disrupt the search for an optimal solution, but can also be used to improve it. In the ACS-TSP algorithm, each ant uses the same variable values, thus a wrong choice of trails can lead other ants into computing bad tours. Ants with abnormal variable values producing bad solutions can erase the trails of other, better performing ants by saturating the graph with their own pheromone information. We conjecture that if implemented properly, this can help to improve the performance by helping out good solutions stuck at local minima of the search space.

Table 2. Comparison of results obtained from ACSGA-TSP and ACS-TSP algorithms on the large TSP instance rat783. Results given are averages of best tour lengths over 3 runs, standard deviation values for the results, and the number of iterations. In (a), the algorithms were run such that the same number of tours were explored as in ACS-TSP runs. In (b), the algorithms were run for the same amount of time as in ACS-TSP runs.

Algorithm	Average	STD	iterations
ACS-TSP	12181	135	100
ACSGA-TSP (a)	10057	201	500
ACSGA-TSP (b)	10323	80	186 (3min)

4   Meta ACS-TSP Algorithm

The proposed Meta ACS-TSP algorithm is a meta-level Genetic Algorithm running on top of ACS-TSP. The GA is used to evolve the optimal parameter values used in the ACS-TSP algorithm. We have taken the task of verifying the optimality of the specific values used by Dorigo and Gambardella [4].

The Meta ACS-TSP algorithm is a wrapper around the ACS-TSP algorithm. It uses a population of encoded values of the ACS-TSP parameters. Each individual in the population can be thought of as a separate instance of the ACS-TSP algorithm with unique parameter values.

Individuals were encoded as bit-strings of length 12 with experimental variables β, ρ, and q_0 encoded using 4 bits each. The length of 4 bits for each variable was chosen in order to decrease the search space. Value of the β parameter was an integer in range 0-15. Values of the ρ and q_0 variables were doubles in range 0-1. In our initial experimentation, we were interested in value range of approximately 10 increment units. We conjecture that a smaller digitization of the value would not improve results of the algorithm.

Pseudocode of the Meta ACS-TSP algorithm is given in Fig. 1. The selection process was done by a tournament selection algorithm with tournament of size 4.

```
 1: for each generation do
 2:    choose 4 individuals randomly from the population
 3:    for each of 4 chosen individuals do
 4:       run ACS-TSP given β, ρ, and q_o value encoded in each individual and record
          the result as the fitness of the individual
 5:    end for
 6:    choose 2 individuals with best fitness from chosen 4
 7:    produce 2 children by crossover or copy from 2 chosen best individuals
 8:    mutate the 2 children
 9:    replace 2 worst individuals from chosen 4 in the population with the 2 children
10: end for
```

Fig. 1. Pseudocode of the Meta ACS-TSP algorithm.

We have used a single point crossover operator that treated each encoded variable as an atomic unit. Thus, our three variable chromosome contained four crossover points. Using this specialized crossover operator the values of the variables were inherited by the children from one of their parents. The operator did not modify the parent values thus creating a greater probability of passing useful and well performing genetic material to the next generation. In our experiments, we have used a crossover probability of 0.9.

Our mutation operator modified only one of the encoded variables by incrementing or decrementing the variable value by 1. This resulted in small changes between generations while still allowing for slow climbing toward local optima. We have used a mutation probability of 0.2 with our mutation operator.

Table 3. Results of Meta ACS-TSP algorithm runs on TSP/ATSP instances. Average values for each problem are averages over 8 runs (eil51), 4 runs (eil76), and 3 runs (kroA100, p43, ry48p, ft70). Best values are best fitness (tour) results over all the runs of a problem. Overall average values are calculated from the table data. The simulations were run for 1K-3K iterations depending on the problem and setup.

Problem	AV β	AV ρ	AV q_0	AV Fit.	Best β	Best ρ	Best q_0	Best Fit.	Optimal
eil51	6.25	0.127	0.495	428.5	6	0.13	0.4	426	426
eil76	6.75	0.128	0.5	542.5	6	0.13	0.4	538	538
kroA100	4.67	0.145	0.503	21513	5	0.13	0.53	21308	21282
p43	2	0.363	0.927	5628	2	0.2	0.93	5620	5620
ry48p	6.67	0.173	0.3	14584	6	0.07	0.47	14495	14422
ft70	8.33	0.34	0.767	40569	8	0.73	0.87	39804	38673
Average	**5.78**	**0.187**	**0.669**	N/A	**5.5**	**0.232**	**0.6**	N/A	N/A

Table 3 summarizes our results. For most problems, the average values of the experimental variables were similar to the best values. The differences between the averages of the average results and averages of the best results were very small, thus we used the average results in our analysis. Statistical analysis was done on most of the runs and the deviation from the average values in a population was reasonable.

The most unique problem was the ATSP instance p43. It had the lowest β value, and the highest ρ and q_0 values. The variability in results between different problems would lead us to believe that the optimal values of the experimental variables are unique to the problem.

We conjecture that there is no magic value that will make ACS-TSP yield the optimal solution for every TSP/ATSP instance. From our experimentation using the Meta ACS-TSP algorithm, we propose that the suggested values given in Table 4 can be used instead of those of Dorigo and Gambardella [4] in order to yield better solutions with the ACS-TSP algorithm.

Table 4. Comparison of average values of variables found by Meta ACS-TSP and the values used by Dorigo and Gambardella in ACS-TSP [4]. Meta ACS-TSP results are from Table 3. Values we feel should yield best solutions are listed as suggested values.

	β	ρ	q_0
Meta ACS-TSP Average	5.78	0.187	0.669
Meta ACS-TSP Best	5.5	0.232	0.6
ACS-TSP	2	0.1	0.9
Suggested Values	6	0.2	0.7

5   Conclusion

We have tried to improve the performance of the ACS-TSP algorithm by proposing the ACSGA-TSP algorithm which introduced a genetic approach to ACS-TSP. Our results have shown that the algorithm does not perform as well as ACS-TSP with respect to finding the optimal solution. Quick convergence and low variability of ACSGA-TSP can be an advantage for time constrained problems.

The ACSGA-TSP algorithm can be improved by a more complex GA model. A hybrid system of ACS-TSP and ACSGA-TSP can be used to take advantage of the early convergence of ACSGA-TSP and optimal results of ACS-TSP. This hybrid system would initially execute ACSGA-TSP and eventually switch to ACS-TSP by turning off the GA. It would be interesting to apply GA models to other ACO algorithms and other combinatorial optimization problems.

We used the Meta ACS-TSP algorithm to evolve experimental variable values used in ACS-TSP. The algorithm offered results that enabled us to suggest variable values alternate to those used by Dorigo and Gambardella [4].

We have only run Meta ACS-TSP on six small TSP/ATSP instances. Running the algorithm on more and larger TSP/ATSP instances would allow for better statistical analysis of the results and improved suggested variable values. The values we suggested should also be tested to see if they statistically improve the performance of ACS-TSP over the original values.

References

1. Bonabeau E., Dorigo M., Theraulaz G. Swarm Intelligence: From Natural to Arftificial Systems. New York: Oxford University Press, 1999.
2. Dorigo M., Di Caro G.: The ant colony optimization meta-heuristic. In D. Corne, M. Dorigo, and F. Glover, editors, New Ideas in Optimization. McGraw-Hill, 1999.
3. Dorigo M., Di Caro G., Gambardella L.M.: Ant Algorithms for Discrete Optimization. Artificial Life **5** (1999) 137-172
4. Dorigo M., Gambardella L.M.: Ant Colony System: A Cooperative Learning Approach to the Travelling Salesman Problem. IEEE Trans. Evol. Comp. **1** (1997) 53-66
5. Dorigo M., Maniezzo V., Colorni A.: The Ant System: Optimization by a Colony of Cooperating Agents. IEEE Trans. Syst. Man Cybern. B **26** (1996) 29–41
6. Holland J.H.: Adaptation in Natural and Artificial Systems. The University of Michigan Press, 1975.
7. Stützle T., Dorigo M.: ACO Algorithms for the Traveling Salesman Problem. In K. Miettinen, M. Makela, P. Neittaanmaki, J. Periaux, editors, Evolutionary Algorithms in Engineering and Computer Science. Wiley, 1999.

A Method for Solving Optimization Problems in Continuous Space Using Ant Colony Algorithm

Chen Ling[1,2], Sheng Jie[1], Qin Ling[1], Chen Hongjian[1]

[1] Department of Computer Science, Yangzhou University
Yangzhou, China

[2] National Key Lab of Advanced Software Tech, Nanjing University
Nanjing, China
yzlchen@pub.yz.jsinfo.net.cn

1 Introduction

We state our algorithm using the nonlinear programming (NLP) problem, objective function G is a given non-linear function. Constraint conditions that represented by a set of inequalities form a convex domain of R^n. We can obtain the minimal n-d hypercube that can be defined as the following inequalities: $l_i \leq x_i \leq u_i$ ($I = 1, 2, \ldots, n$). Let the total number of ants be m and the m initial solution vectors are chosen at random. All the ith components of these initial solution vectors construct a group of candidate values of the ith component of solution vector. If we use n vertices to represent the n components and the edges between vertex i and vertex $i+1$ to represent the candidate values of component i, a path from the start vertex to the last vertex represents a solution vector whose n edges represent n components. We denote the jth edge between vertices i and $i+1$ as (i,j) and its intensity of trail information at time t as $\tau_{ij}(t)$.

In each iteration, to m ants, each ant select an initial value from the group of candidate values of component i ($1 \leq i \leq n$) with the probability $p_{ij}^k(t)$:

$$p_{ij}^k(t) = \tau_{ij}(t)/\sum \tau_{i\,r}(t) \ (1 \leq r \leq m). \tag{1}$$

Then perform operations of crossover and mutation on the m values to get m new values, add them to the group of candidate values for component i and using them form m solutions of a new generation, compute the fitness of these solutions. When m ants have traveled all vertices, update the trail information of the candidate values for each component according to formula (2)

$$\tau_{ij}(t+1) = (1-\rho)\,\tau_{i\,r}(t) + \sum \tau_{ij}^k \ (1 \leq k \leq m). \tag{2}$$

But if the kth ant chooses the jth candidate value of the group for component i between (t, $t+1$), then $\Delta\tau_{ij}^k = Wf_k$ otherwise $\Delta\tau_{ij}^k = 0$, where W is a constant, and f_k is the fitness of the solution found by the kth ant. At the end of an iteration, delete m values with lower intensity of trail information in each candidate group.

M. Dorigo et al. (Eds.): ANTS 2002, LNCS 2463, pp. 288-289, 2002.

2 Adaptive Crossover and Mutation

In this method, we define F as the relative fitness of a solution, if $f_{max} \neq f_{min}$ $F=(f-f_{min})/(f_{max}-f_{min})$, otherwise $F=1$. Here f_{max} and f_{min} are the maximal and the minimal fitness of the current generation.

Suppose $x_i(1)$ and $x_i(2)$ are two initial values of component i which will have crossover, the fitness of the two solutions they belong to are f_1, f_2, define the integrate fitness of $x_i(1)$ and $x_i(2)$ as $f_{ave} = (f_1 + f_2)/2$, which will also be the integrate fitness of the result of crossover operation $x_i'(1)$ and $x_i'(2)$. Suppose the crossover probability set by the system is p_{cross}, the integrate relative fitness of $x_i'(1)$ and $x_i'(2)$ is defined as $F_{ave} = (F_1 + F_2)/2$, then the actual crossover probability for $x_i'(1)$ and $x_i'(2)$ is $p_c[x_i'(1), x_i'(2)] = p_{cross}(1 - F_{ave})$. Generate a random number $p \in [0,1]$, the crossover operation could be carried out only if $p > p_c$. In the operation, we first generate two stochastic numbers c_1 and $c_2 \in [-b, b]$ so that $c_1 + c_2 = 1$ here $b = 2 \cdot (1 - f_{ave})$. The cross results $x_i'(1)$ and $x_i'(2)$ can be obtained by the affine combination of $x_i(1)$ and $x_i(2)$:

$$x_i'(1) = c_1 x_i(1) + c_2 x_i(2) \qquad x_i'(2) = c_2 x_i(1) + c_1 x_i(2). \qquad (3)$$

In the mutation operation on a component value x, the real mutate probability p_m is determined according to the relative fitness F of the solution where x belongs to. Suppose the mutation probability set by the system is p_{mutate}, we set $p_m = p_{mutate}(1 - F)$. Generate a random number $p \in [0,1]$, mutate operation is implemented only if $p > p_m$. To ensure the mutate result $x_i' \in [l_i, u_i]$, we define d_i as: $d_i = \max\{ u_i - x, x_i - l_i \}$, and x_i' can be obtained from Eq.(4):

$$x_i' = x_i + \delta(F,t)d_i \qquad \text{if } l_i - x_i \leq \delta(F,t) d_i \leq u_i - x_i,$$

$$x_i' = x_i - \delta(F,t)d_i \qquad \text{otherwise.} \qquad (4)$$

Here, $\delta(F,t)$ is a stochastic number between $[-1,1]$, it is determined by the following formula: $\delta(F,t) = r (1-F)^{1+\lambda t}$, here r is also a stochastic number between $[-1,1]$, λ is the parameter which decides the extent of diversity and adjusts the area of local probe, it can be selected between 0.0001 and 0.0003.

References

1. Dorigo M., Maniezzo V., Colorni A. Ant system: Optimization by a Colony of Coorperating Agents, *IEEE Transactions on Systems, Man, and Cybernetics*, 26(1): 28–41, 1996.

A Nested Layered Threshold Model for Dynamic Task Allocation

Tom De Wolf, Liesbeth Jaco, Tom Holvoet, and Elke Steegmans

Dept. of Computer Science, KULeuven
Celestijnenlaan 200A, B-3001 Leuven, Belgium
{Tom.DeWolf,Liesbeth.Jaco,Tom.Holvoet,Elke.Steegmans}@cs.kuleuven.ac.be

Dynamic task allocation is an essential aspect in modeling ant behaviour [1,2]. In this paper, we propose a new (abstract) model for dynamic task allocation of agents that combines a nested layered architecture with a threshold mechanism. We apply this new model to achieve flexible ant behaviour.

To describe the actions of an agent, we describe a task allocation model based on behaviours and sub-behaviours that are associated with stimuli. At any time, an agent is performing one particular sub-behaviour within one behaviour. Changing stimuli may lead to reallocation of sub-behaviour or even the behaviour of the agent. A behaviour is a (course-grain) task an agent can perform. Behaviours can be seen as a collection of smaller, more primitive parts, called sub-behaviours.

This hierarchical setting allows an interesting allocation mechanism that allows easy reallocation of tasks without trashing between different behaviours, as well as task specialisation. Figure 1 illustrates the task allocation procedure. First the ant tries to choose a new sub-behaviour within the current behaviour. Only if this is impossible, it changes its behaviour. This two-step changing prevents the ant from switching too drastically when it perceives a sudden, only temporary change in the environment.

Adaptive behaviour necessarily depends on taking into account feedback, either from the environment, either based on the own experience of the agent. Ants constantly perceive stimuli. An ant can be seen as being controlled by a task manager, which handles the actual decision-making about the (re-)allocation

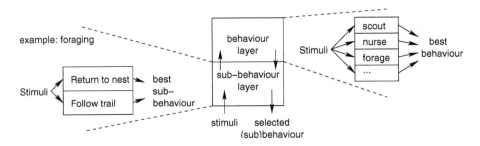

Fig. 1. The nested layered task allocation model.

M. Dorigo et al. (Eds.): ANTS 2002, LNCS 2463, pp. 290–291, 2002.

of (sub-)behaviours. As input the task manager gets the current stimuli the ant perceives, as output it gives the sub-behaviour that is most appropriate at this moment. A threshold value is associated with every behaviour and every sub-behaviour. To compute the threshold we use a statistical method based on positive and negative feedback. It is inspired by the work presented in [3]. We adapted the calculation of the relevance of an active behaviour, which can be seen in the thresholds. We want a percentile value of the threshold to make it easily comparable to the stimulus value.

$$relevance = ((corr(pos.feedback, active) - corr(neg.feedback, active) + 2) * 25$$
$$threshold = |(relevance - 100)|$$

A stimulus value is computed for each (sub-)behaviour, based on the currently perceived stimuli, which is a percentile value for the amount of stimuli that are relevant for a behaviour and are perceived at a certain moment in time. Now the threshold and the stimulus value are compared with each other to calculate the probability for a behaviour to become active. The behaviour with the highest probability is chosen as the behaviour to be executed next:

$$probability = stimulus^2/(stimulus^2 + threshold^2)$$

We have two levels of decision-making. At the first level, an appropriate behaviour must be chosen, and then the right sub-behaviour within that behaviour must be decided upon at the second level. Only when it is impossible to choose a sub-behaviour within the current behaviour, or when there are "strong indications" that it is better to change the behaviour, the ant chooses another one. This shows that our model combines two types of task allocation models [4], i.e. a horizontally layered model (e.g. to choose between sub-behaviours) and a vertically layered model (with two layers). Our architecture is thus a "nested layered model": it consists of two horizontally layered models, each of which is nested in a vertical layer. First experiments illustrate the usefulness of the model for modelling ant behaviour as well as for modelling other agent behaviour.

References

1. Samuel N. Beshers and Jennifer H. Fewell. Models of division of labor in social insects. *Annual Review Entomology*, 46:413–440, 2001.
2. Eric Bonabeau, Marco Dorigo, and Guy Theraulaz. *Swarm Intelligence: From Natural to Artificial Systems*. SFI Studies in the Sciences of Complexity. Oxford University Press, 1999.
3. Pattie Maes and Rodney A. Brooks. Learning to coordinate behaviours. In *National Conference on Artificial Intelligence*, pages 796–802, 1990.
4. Gerhard Weiss, editor. *Multiagent Systems: A Modern Approach to Distributed Artificial Intelligence*. The MIT Press, 1999.

ACO Algorithm with Additional Reinforcement

Stefka Fidanova

IRIDIA, Université Libre de Bruxelles
CP 194/6, Av. Franklin D. Roosevelt 50, 1050 Brussels, Belgium
`fidanova@ulb.ac.be`

The aim of the paper is to develop the functionality of the ant colony optimization (ACO) algorithms by adding some diversification such as additional reinforcement of the pheromone. This diversification guides the search to areas in the search space which have not been yet explored and forces the ants to search for better solutions. In the ACO algorithms [1,2] after the initialization, a main loop is repeated until a termination condition is met. In the beginning ants construct feasible solutions, then the pheromone trails are updated. Partial solutions are seen as states: each ant moves from a state i to another state j corresponding to a more complete partial solution. For ant k, the probability p_{ij}^k of movement from state i to a state j as a next state is given as:

$$p_{ij}^k = \begin{cases} \dfrac{\tau_{ij}\eta_{ij}^{\beta}}{\sum_{l\in allowed_k} \tau_{il}\eta_{il}^{\beta}} & \text{if } j \in allowed_k \\ 0 & otherwise \end{cases}, \qquad (1)$$

where τ_{ij} is a pheromone level corresponding to this movement, η_{ij} is the heuristic information and $allowed_k$ is the set of remaining feasible states.

In the beginning initial pheromone level is set to be τ_0. After all ants have completed their tours, the pheromone level is updated as follows:

$$\tau_{ij} \leftarrow \rho\tau_{ij} + \Delta\tau_{ij} \qquad (2)$$

where $0 < \rho < 1$ is a pheromone decay parameter and $\Delta\tau_{ij}$ is different for different ACO algorithms.

When we perform the ACO algorithm search stagnation may occur. This can be happen, if the pheromone trail is significantly higher for one choice than for all others. The main purpose of this research is to avoid the stagnation situations by using additional reinforcement for unused movements and to search for better solutions. If some movement is not used in current tour we use additional pheromone reinforcement as follows:

$$\tau_{ij} \leftarrow \tau_{ij} + q\tau_{max}, \qquad (3)$$

where $q \geq 0$ is a parameter and τ_{max} is a maximal amount of the pheromone. After additional reinforcement unused movements have great amount of pheromone than used one which, belong to the poor solutions and less to used movements that belong to best solution. Thus, we force ants to choose new direction of search space without repeating the bad experience. Preliminary results for the traveling salesman problem (TSP) suggest the usefulness of above diversification. Table 1 shows a comparison between our ACO algorithm with additional reinforcement and MMAS. We use a particular implementation of ACO algorithm,

M. Dorigo et al. (Eds.): ANTS 2002, LNCS 2463, pp. 292–293, 2002.
© Springer-Verlag Berlin Heidelberg 2002

known as ant system with an elitist ant. Set of TSP from TSPLIB benchmark library ($http://www.iwr.uniheidelberg.de/iw/compt/software/TSLIB95$) is used as a test problems. The selected parameters for the test are , m=25 ants, $\rho = 0.8$, $\beta = 2$ and $q = 0.2$. The maximum number of cycles was set to 20000 for all experiments and the average are taken over 25 trials. For comparison, the same program is used for both algorithms, with same parameters and same number of iterations and the difference is only in the pheromone updating. Bold numbers indicate the better results. From the achieved result of table 1, it is clear that our ACO algorithm with additional reinforcement performs better than MMAS. For future research is to develop this new ACO algorithm and test on different optimization problems.

Table 1. The results of ACO algorithm with additional reinforcement and MMAS algorithm

Instance	Add. Reinforcement		MMAS	
	Best	Average	Best	Average
d198	15780	**15780.3**	15780	15780.9
lin318	42029	42051.8	42029	42051.8
pcb442	**50785**	**507891**	50795	509421
att532	**27686**	**27706.8**	27693	27707.4
rat783	8808	**8820**	8808	8841
dsj1000	**18688548**	**18704977.1**	18708266	18736562.2
pr1002	**259167**	**259692.5**	259290	260025
vm1084	239321	**239434**	239321	239490.17
pcb1173	**56897**	56970.5	56909	**56969.4**
d1291	50801	**50820**	50801	50847.7
rl1304	252948	**253159**	252948	253528.17
rl1323	**270226**	**270580.17**	270281	270725.5
fl1400	**20161**	**20229.67**	20254	20299.17
fl1577	**22337**	**22412.83**	22349	22423.67
vm1748	**336873**	**337107.17**	337506	337772.83
u1817	**57294**	**57346.5**	57314	57458.67
rl1889	**317430**	**317743.33**	318284	319295

Acknowledgments. Stefka Fidanova was supported by a Marie Curie Fellowship of the European Community program "Improving Human Research Potential and the Socio-Economic Knowledge Base" under contract number No HPMFCT-2000-00496. This work was supported by the "Metaheuristics Network", a Research Training Network funded by the Improving Human Potential program of the CEC, grant HPRN-CT-1999-00106. The information provided in this paper is the sole responsibility of the authors and does not reflect the Community's opinion. The Community is not responsible for any use that might be made of data appearing in this publication.

References

1. Dorigo, M., Di Caro, G.,Gambardella, L.M.: Ant algorithms for distributed discrete optimization. Artificial Life **5** (1999) 137–172
2. Dorigo, M., Di Caro, G.: The ant colony optimization metaheuristic. In: Corne, D., Dorigo, M., Glover, F. (eds.): New Idea in Optimization, McGrow-Hill, (1999) 11–32

Ant Colony System for Image Segmentation Using Markov Random Field

Salima Ouadfel[1,2], Mohamed Batouche[2], and Catherine Garbay[3]

[1]Département Informatique, Université de Batna
Batna, Algérie
souadfel@yahoo.fr
[2]Equipe Vision et Infographie, Laboratoire LIRE, Université de Constantine
Constantine, Algérie
batouche@wissal.dz
[3]Laboratoire TIMC-IMAG, Institut Bonniot
38706 La Tronche Cedex, Grenoble, France
catherine.garbay@imag.fr

1 Introduction

Clustering is the process of partitioning a given set of pixels into a number of homogenous clusters based on a similarity criterion. The clustering problem is a difficult optimization problem for two main reasons: first the search space of the optimization is too large, second the clustering objective function is typically non convex and thus may exhibit a large number of local minima. In this paper we propose the use of the Ant Colony System (ACS) [1] to solve the clustering problem.

2 ACS for Image Segmentation

In our ACS based clustering algorithm, the task of each ant is to build a feasible partition by mapping pixels to clusters. The mains steps of the algorithm are:
1. Initialize pheromone trails to τ_0;
2. For each ant k
   - Built a partition by assigning pixels to clusters according to the transition rule;
   - Modify the amount of pheromone by a local updating rule;
3. Modify the amount of pheromone trails by a global updating rule;
4. If the maximum number of iterations is not reached go to step 2.

The ant chooses to assign a pixel to a cluster using the pseudo random rule. The amount of pheromone trail τ measures the desirability of the assignment of a pixel to a cluster, the heuristic function η maintains spatial regularity in the presence of the noise, and is formulated according to the MRF theory [2] as:

$$\eta(s,e) = \left(y_s - \mu_e\right)^2 * \left(1 - \frac{1}{Ns} \sum_{(s,s')neighbors} \delta(e_s, e_{s'})\right) \tag{1}$$

M. Dorigo et al. (Eds.): ANTS 2002, LNCS 2463, pp. 294–295, 2002.

where: y_s is the gray level of the pixel s, μ_e is the mean gray level of the cluster e. e_s and $e_{s'}$ are respectively the label of the pixel s and its neighbor s'. N_s is the number of neighbors of pixel s and δ is the kronecker symbol.

The pheromone trail is updated locally by each ant using the following rule:

$$\tau(s,e)=(1-\rho)\tau(s,e)+\rho\tau_0 \tag{2}$$

and globally by the best ant that finds the optimal partition using this updating rule:

$$\tau(s,e)=(1-\rho)\tau(s,e)+\rho\frac{1}{M} \quad if\ (s,e)\ is\ in\ the\ best\ partition \tag{3}$$

M is the clustering metric which measures the quality of a partition:

$$M=\sum_{s\in S}\left((y_s,\mu_s)^2+\sum_{(s,s')neighors}\delta(e_s,e_{s'})\right) \tag{4}$$

where S is the set of pixels, μ_s is the mean gray level of the cluster of the pixel s.

3   Experimental Results

Experiments were carried out for our algorithm and for the conventional k-means algorithm using different images. An example is given in Fig 1 with a muscle cell image. We observe that our method performs well and is less sensitive to noise.

Fig. 1. (A) muscle cell image. (B) our clustering result. (C) K-means result.

4 Conclusion

In this paper we have presented a novel approach to image segmentation. An adapted ACS algorithm is used in conjunction with the MRF theory to obtain an optimal partition of the image.

References

[1] M. Dorigo, and L. Gambardella. Ant Colony System: A Cooperative Learning Approach to the Traveling Salesman Problem, *IEEE Transactions on Evolutionary Computation* 1(1): 53–66, 1997.
[2] S. Geman. and D. Geman. Stochastic relaxation, Gibbs distributions and the Bayesian restoration of images, *IEEE Transactions on Pattern Analysis and Machine Intelligence* 6(6): 721–741, 1984.

Bidimensional Shapes Polygonalization by ACO

Ugo Vallone

DIS, Università degli Studi di Pavia
Via Ferrata 1, 27100 Pavia, Italy
vallone@vision.unipv.it

Polygonalization is a way to approximate curves by straight-line segments. All the polygonalization methods use thresholds to produce the extreme points of the straight lines used in the approximation [1,2,3]. The method here introduced does not use thresholds to select the points. Ant Colony Optimization (ACO) is an optimization paradigm that mimics the exploration strategy of a colony of ants [4]. Images containing bi-dimensional shapes have been considered in this work. The goal is to find a good polygonal approximation of the contours of these images. The ants travel on segments connecting the points of the contours. Two points define a path in the ant travel. The points number of the polygonal approximation (*num_steps*) is user-defined. Last point connects with the first one. The two catheti AB and BC in Fig. 1 are a better curve approximation than the hypotenuse AC. AB + BC is longer than AC.

Fig. 1. AB+BC >AC and is a better approximation

Fig. 2. Clockwise exploration: black point are unvisited

Fig. 3. B^1 is better than B

Fig. 4. Polygonalized shape

So the idea is, given a fixed number of segments, to maximize the polygon perimeter to have better approximations. Ants starting point is the topmost left point of the shape (Fig. 2). Points are numbered in a clockwise order along the contour starting from this point. The length of a path connecting two of these points is so defined:

$$pathLength_{i,j} = \max Length - length_{i,j} \qquad (1)$$

$length_{i,j}$ is the Euclidian length between the points i and j. *maxLength* is the higher value of the segments Euclidian lengths connecting two points on the contour. The probability $p_{i,j}$ to go from point i (current) to an unvisited point j ($j > i$), is so defined:

$$p_{i,j}(t) = \frac{\tau_{i,j}(t)}{\sum_{j>i} \tau_{i,j}(t)} \qquad (2)$$

M. Dorigo et al. (Eds.): ANTS 2002, LNCS 2463, pp. 296-297, 2002.

$\tau_{i,j}$ is the trail on the segment connecting i and j. t is the current cycle. In order to avoid accumulations on the final points of the contour, the selection of point j at the step st is repeated (num_steps - st) times. It is selected the point j closer to the previous point i. At the end of any cycle the ants update the trail on the segments used. Longer the travel made by the ant, lower the level of the trail left on the segments explored. The contribution of an ant to the trail level on the segments visited is so calculated:

$$\Delta\tau_{ant} = \frac{Q}{travelLength_{ant}} \tag{3}$$

where Q is a constant that controls the trail quantity left by the ant on every segment in her path and $travelLength$ is the length of the ant travel. The trail is so updated:

$$\tau_{i,j}(t+1) = \rho\tau_{i,j}(t) + (1-\rho)\Delta\tau_{i,j} \tag{4}$$

where ρ defines the relative importance between the trail at cycle t and $t+1$. $\rho \; \varepsilon$ [0-1]. $\Delta\tau_{i,j}$ is the sum of the contributions ($\Delta\tau_{ant}$) of the ants that use the connection between points i and j in the current cycle. $\Delta\tau_{i,j}(0)$ is equal to 1. After all the ants have made a travel the trail is updated according to the formula 4 and a new cycle starts. Trail, in the experiments, is limited in the interval [0.1-10] (min-max approach). The result is the more short travel made by the ants in all the cycles. This travel is then locally improved. Analyzing clockwise the polygon points 3 by 3, the middle points like B in Fig. 3 are replaced by points like B^1 that minimize the paths sum $AB+BC$ (increasing the Euclidian length). Nearly collinear points are then removed from the resulting polygon. If A, B and C are consecutive polygon points and $(AB+BC) - AC < MinInc$, where $MinInc$ is a user-defined minimum increment, point B is removed. In our experiments, ρ is 0.5, the ants number is 10, Q is 1000, num_steps is 30. The algorithm uses 30 cycles. Fig. 4 shows a bidimensional shape in gray and its polygonal approximation in white. The image contour is 353 points long. The real number of points used in the approximation is 24 ($MinInc$ is 0.8). The elaboration time is only some second. The algorithm introduced has two very good characteristic: it doesn't use thresholds to select the polygon points and it lets to decide the maximum number of points in the approximation (num_steps).

References

1. Williams, C., Bounded straigth-line approximation of digitized planar curves and lines. *Computer Graphics Image Processing*, **16:** 370–381 (1981)
2. Pavlidis, T., Horowitz, S., Segmentation of planar curves. *IEEE Trans. Comput.*, **C-23**: 860–870 (1974)
3. Wall, K., Danielsson, P.E.: A fast sequential method for polygonal approximation of digitized curves. *Comput. Vision, Graphics and Image Process.*, **28**: 220-227 (1984)
4. Stützle, T., Dorigo, M.: ACO Algorithms for the Quadratic Assignment Problem. In *New Ideas in Optimization*, McGraw-Hill (1999)

Coevolutionary Ant Algorithms Playing Games

Jürgen Branke, Michael Decker, Daniel Merkle, and Hartmut Schmeck

Institute AIFB, University of Karlsruhe
76128 Karlsruhe, Germany
{branke,merkle,schmeck}@aifb.uni-karlsruhe.de

1   Introduction and Main Ideas

In this paper, we use the idea of coevolution in the context of ant algorithms to develop game-playing strategies for the simple games of Nim and Tic-Tac-Toe.

Since these games have a rather small set of possible non-final distinct states (20 for Nim and 627 for Tic-Tac-Toe), we represent a strategy as a lookup-table, specifying for each situation the suggested move. To generate such a strategy, an ant algorithm operates on a $|S| \times |M|$ pheromone matrix, with $|S|$ being the number of possible states, and $|M|$ being the (maximal) number of possible moves per state. We require that the sum of pheromone values in each row is equal to 1. An ant produces a strategy by moving through all the rows and probabilistically (according to the pheromone values in each row) selecting a move for each state.

As default, we use two ant colonies. In every iteration, each ant colony creates k strategies based on its pheromone matrix. The strategies are evaluated in what we call a "tournament", with each ant from colony P_1 playing a game against each ant from colony P_2, with the starting player being determined at random. Of course, we would like to favor strategies that won a lot of games. However, following the ideas of Rosin an Belew [1], we also want to enforce diversity and favor strategies which won against opponents no-one else was able to beat. Therefore, a strategy's fitness is proportional to the number of games it won with the reward for each defeated opponent being inversely proportional to the number of games the opponent lost.

More specifically, let D_i be the set of opponents defeated by ant i, and let L_i be the set of opponents against which ant i lost. Then, the fitness of ant i is calculated as $f_i = \sum_{j \in D_i} \frac{1}{|L_j|}$. If the game allows for draws/ties (as is the case with Tic-Tac-Toe), these are integrated into the equation with half the weight.

Since the fitness evaluation in a coevolutionary setting is stochastic, it is not possible to tell whether the ant with the best fitness really created the best strategy. Therefore, we use fitness-proportional update and no elitism. It is important to note that during a single game, only very few out of all possible states are encountered. A strategy's fitness is therefore only representative for the decisions made in those states, and only those decisions should be updated. Let S_{ij} be the set of states encountered by ant i while playing against opponent j. Furthermore, let γ be a parameter determining the update strength and thus also the speed of convergence. Then, the pheromone values for colony k are updated proportional to fitness as follows:

M. Dorigo et al. (Eds.): ANTS 2002, LNCS 2463, pp. 298–299, 2002.
© Springer-Verlag Berlin Heidelberg 2002

for all $i \in P_k$ **do**
  for all $j \in D_i$ **do**
    for all $s \in S_{ij}$ **do**
      $\tau^k_{s\sigma(s)} \leftarrow \tau^k_{s\sigma(s)} + \gamma \frac{f_i}{|S_{ij}|}$
    end for
  end for
end for

Since the pheromone is added in only some of the rows, and with different strength, the usual pheromone evaporation rule can not be applied. Instead, at the end of each iteration, we re-normalize the pheromone values in each row to sum up to 1.

2 Empirical Evaluation

Unless stated otherwise, we used 10 ants per colony, $\gamma = 0.01$ for Nim and $\gamma = 0.02$ for Tic-Tac-Toe. To evaluate the success of a run, we create a final solution deterministically by selecting for each State $s \in S$ the move $m \in M$ with the highest pheromone value, i.e. $\sigma^*(s) = \mathrm{argmax}_m\{\tau^1_{sm}\}$.

Since for Nim there exists only one optimal strategy, it is possible to compare the coevolved pheromone matrix to a pheromone matrix representing the optimal strategy. The proposed coevolutionary ant algorithm converged to a matrix corresponding to the optimal solution in all of the 10 runs within 400 iterations.

For Nim, the optimal strategy is independent of whether a player plays first or second. Thus the question arose, whether it may be sufficient to just use a single colony, with all individuals playing against each other. To have the same number of 100 games per tournament, in the case of a single colony, we play two games for every pair of ants (including self-play), and make each pair play two games, where each player is allowed to start once. Since now all individuals update on the same pheromone matrix, in order to have the same amount of pheromone update as in the case of two colonies, we also reduced the learning rate γ to half its usual value. As it turned out, the runs with only one colony playing against itself converged slightly faster, and also to the optimum.

To confirm that the convergence to an optimal strategy is in fact due to coevolution and can not be reproduced by simple optimization, we run another set of experiments where the second colony is always playing at random. As expected, in this case the pheromone matrix of colony P_1 does not converge to the optimal solution, as the random players are not a sufficient challenge.

For Tic-Tac-Toe, during first experiments, the simple coevolutionary scheme did not always converge to the optimal strategy, mainly because during a single game, only a small fraction of all possible states are encountered. If an ant colony was not exposed to some of the states by opponents from the other colony, it simply didn't learn how to react.

However, when we introduced more diversity by requiring a minimum pheromone level or by starting from a randomly chosen state, again the coevolutionary scheme was able to produce optimal strategies.

References

1. C. D. Rosin and R. K. Belew. New methods for competitive coevolution. *Evolutionary Computation*, 1996.

GAACO: A GA + ACO Hybrid for Faster and Better Search Capability

Adnan Acan

Computer Engineering Dept., Eastern Mediterranean University
Gazimağusa, T.R.N.C. Mersin 10, Turkey
adnan.acan@emu.edu.tr

1 Introduction

Considering the similarities and characteristics differences between ant colony optimization (ACO) and evolutionary genetic algorithms (GAs), a novel hybrid algorithm combining the search capabilities of the two metaheuristics, for faster and better search capabilities, is introduced. In the GAACO approach, ACO and GAs use identical problem representations and they run in parallel. Migration occurs between the two algorithms whenever any of the them finds an improved potential solution after an iteration. Migration provides further intensification capabilities to both of the algorithms other than their own search mechanisms. In this respect, GAs support ACO by strengthening potential search alternatives for artificial ants and ACO supports GAs by exporting promising potential solutions into its population. The developed algorithm is tested on the solution of two NP-hard combinatorial optimization problems, the obtained results outperform those obtained by both of the individual algorithms when applied alone.

2 The GAACO Approach

The main objectives of the proposed hybrid algorithm are; to improve the convergence speed of both algorithms through cooperation in which both algorithms receive supervisory feedback from a strategically different search algorithm working on the same problem in exactly the same environment, and to improve the quality of the found solutions by providing new intensification strategies systematically imported into each algorithm's search procedure as they run. The results obtained with this implemented strategies demonstrate that the desired objectives are all satisfied.

The direction of migration is from the side where the improvement is made to the other. If an improved solution is found on both sides, then migration is bidirectional. Otherwise, no information exchange is made and the algorithms run in parallel independent of each other. The implemented migration strategy is as follows; K% of the best solutions from ACO site are moved into the GA population, where they do not replace any existing individuals, instead they directly enter into the mating pool and duplicated proportional to their fitness, L% of the best individuals in GAs site are used to add fitness-proportional pheromone while

M. Dorigo et al. (Eds.): ANTS 2002, LNCS 2463, pp. 300–301, 2002.

Table 1. Performance comparison of GAs, MMAS, and GAACO approaches.

Problem	Algorithm	Alg. Parameters	Scores		
			Best	Average	Worst
TSP (fl1577)	GAs	Elitism= 10%, Tour. Selection, P_c=0.8, P_m=0.05, Pop. Size=500, OX Correction.	22305	22385	23350
	MMAS	Num. ants=25, ρ=0.2, α=1, β=2.	22286	22311	22358
	GAACO	GA + MMAS Parameters with K=50, L=5, M=5.	22286	22302	22367
QAP (tai50a)	GAs	Same as above	5.06e+6	5.065e+6	5.074e+6
	MMAS	Num. ants=5. Other parameters are the same as above.	5.04e+6	5.045e+6	5.050e+6
	GAACO	GA + MMAS Parameters.	4.99e+6	5.020e+6	5.065e+6

another M% of the worst-fitness individuals are used to evaporate a constant amount of pheromone from the ACO problem domain environment.

GAACO is compared with the traditional implementations of ACO algorithms and the GAs in the solution of two widely known and difficult combinatorial optimization problems, namely the traveling salesman problem (TSP) and the quadratic assignment problem (QAP) [1,2,3].

Real-valued problem representations are used in all implementations and both ACO and GAs are coded using widely employed parameter settings. Each experiment is performed 10 times over 1000 iterations and the population size in GAs is taken as 500. The following table illustrates the improvements achieved by GAACO hybrid over ACO and GAs, the convergence speed of GAACO is also better than that of both algorithms.

References

1. Stützle, T., Dorigo, M.: ACO Algorithms for the Traveling Salesman Problem. In: Miettinen, K., Neittaanmaki, P., Periaux, J. (eds.): Evolutionary Algorithms in Engineering and Computer Science, John Wiley & Sons (1999).
2. Stützle, T., Dorigo, M.: ACO Algorithms for the Quadratic Assignment Problem. In: Corne, D., Dorigo, M., Glover, F. (eds.): New Ideas in Optimization, McGraw-Hill, (1999).
3. Maniezzo, V., Colorni, A.: The Ant System Applied to the Quadratic Assignment Problem. IEEE Transactions on Knowledge and Data Engineering, (1999).

GPS Positioning Networks Design: An Application of the Ant Colony System

Hussain Aziz Saleh

IRIDIA, Université Libre de Bruxelles
CP 194/6, Avenue Franklin Roosevelt 50, 1050 Brussels, Belgium
hsaleh@ulb.ac.be

The Global Positioning System (GPS) positioning network problem can briefly defined as follows. A number of receivers are placed at stations to be coordinated by determining sessions between these stations. A session can be defined as a period of time during which two or more receivers simultaneously record satellite signals for a fixed duration. After a certain time of observation, the receivers are then moved to other stations for further measurements. This process of sessions observation continues till the whole network is completely observed. The problem is to search for the best order in which these sessions can be organized to give the cheapest schedule i.e.,*Minimize: C(V)*, where $C(V)$ is the total cost of a feasible schedule V. In this paper the concept of the Ant Colony System (ACS) algorithm coupled with local search procedure, which is a particular instance of the ant colony optimization [3], has been applied to solving the GPS positioning network problem as follows:

1. INITIALIZATION
   - Define the sessions to be observed N, set the number of ants M ($M = N$) with a tabu list for each ant and set initial pheromone τ_0;
   - Set the control parameters; the trail parameter α, the visibility parameter β, the evaporation parameter ρ and the iteration counter K;
2. CONSTRUCTION AND SELECTION STRATEGY
   - Build up the cheapest schedule V using the probability equation (*Eq.1*).

$$P_m(i,j) = \begin{cases} \dfrac{[\tau_{(i,j)}] \cdot [\eta_{(i,j)}]^{\beta}}{\sum_{k \in S_m(i)} [\tau_{(i,k)}] \cdot [\eta_{(i,k)}]^{\beta}} & \text{if } j \in S_m(i) \\ 0 & \text{otherwise} \end{cases} \qquad (1)$$

   - Add the observed sessions by each ant to its tabu list as it proceeds, then empty the tabu list when the schedule is completed.
   - Apply the local search method on all the obtained schedules.
   - Apply the local update rule on all the obtained schedules (*Eq.2*);

   $$\tau_{(i,j)} \leftarrow (1 - \rho) \cdot \tau_{(i,j)} + \rho \cdot \tau_0 \qquad (2)$$

   - Apply the global update rule on the best found schedule (*Eq.3*);

   $$\tau_{(i,j)} \leftarrow (1 - \alpha) \cdot \tau_{(i,j)} + \alpha \cdot \Delta\tau_{(i,j)} \qquad (3)$$

   where

   $$\Delta\tau_{(i,j)} = \begin{cases} (C_{gbs})^{-1} & \text{if } (i,j) \in \text{ global-best-schedule} \\ 0 & \text{otherwise} \end{cases} \qquad (4)$$

   where C_{gbs} the cost of the best found schedule from the beginning.
   - Update the iteration counter $K = K + 1$.

M. Dorigo et al. (Eds.): ANTS 2002, LNCS 2463, pp. 302–303, 2002.

3. TERMINATION
   - The maximum number of iterations is greater than the pre-specified maximum number of iterations.

   OTHERWISE Go to (2).
   - Stop the program and declare the cheapest schedule and its cost.

It is preferable in metaheuristics to evaluate the performance of a proposed technique by comparison with an existing optimal solution. By applying the proposed ACS technique on a GPS network (with known solution) observed in New Brunswick in Canada and consists of 10 sessions [2],the same cost as the optimal schedule was obtained. The selected control parameters for this network are: $\alpha = 0.6$, $\beta = 1$, $\rho = 0.3$ and $K = 10$. To generalize the developed ACS algorithm and work with larger networks, the Malta GPS network with 38 sessions was implemented. Several benchmarks were available for this network which allowed comparisons to be made with respect to the effectiveness and computational efficiency of the proposed technique. The first benchmark, which was manually generated using the intuition and experience of the surveyors, represented the actual operating schedule with cost of 1405 minutes [1]. The other two benchmarks were the metaheuristic schedules obtained by applying simulated annealing and tabu search techniques and had a cost of 1355 minutes and 1075 minutes respectively [4]. By implementing the proposed ACS technique using the same data set, the overall cost was reduced to 895 minutes when the control parameters are $\alpha = 0.7$, $\beta = 2$, $\rho = 0.4$ and $K = 80$. Directions for future research should be the analysis of parameter settings using extra GPS networks with different types and sizes. Another important direction for current research in which the proposed ACS technique can be implemented is the optimization of ambiguity resolution in GPS data and then improving the accuracy of GPS positioning.

Acknowledgments. This research was supported by both the Syrian Ministry of Higher Education and by a Marie Curie Fellowship awarded to Hussain Saleh (CEC-IHP Contract N. HPMF-CT-2000-00494). Also, this work was supported by the "Metaheuristics Network", a research Training Network,funded by the Improving Human Potential programme of the CEC, grant HPRN-CT-1999-00106. The European Commission is not responsible for any use that might be made of data appearing in this publication.

References

1. P. Dare. Project malta' 93: The establishment of a new primary network for the republic of malta by use of the global positioning system. Technical report, Report for Mapping Unit, Planning Directorate, Floriana, Malta, 1994.
2. P. Dare and H. A. Saleh. GPS network design: logistic solutions using optimal and near optimal methods. *Journal of Geodesy*, 74:467–478, 2001.
3. M. Dorigo. *Optimization, learning and natural algorithms.* PhD thesis, Dipartimento di Elettronica e Informazione, Politecnico di Milano, Italy, 1992.
4. H. A. Saleh and P. Dare. Effective Heuristics for the GPS Survey Network of Malta: Simulated Annealing and Tabu Search Techniques. *Journal of Heuristics*, 7:553–549, 2001.

Author Index

Lecture Notes in Computer Science

For information about Vols. 1–2380
please contact your bookseller or Springer-Verlag

Vol. 2417: M. Ishizuka, A. Sattar (Eds.), PRICAI 2002: Trends in Artificial Intelligence. Proceedings, 2002. XX, 623 pages. 2002. (Subseries LNAI).

Vol. 2418: D. Wells, L. Williams (Eds.), Extreme Programming and Agile Methods – XP/Agile Universe 2002. Proceedings, 2002. XII, 292 pages. 2002.

Vol. 2419: X. Meng, J. Su, Y. Wang (Eds.), Advances in Web-Age Information Management. Proceedings, 2002. XV, 446 pages. 2002.

Vol. 2420: K. Diks, W. Rytter (Eds.), Mathematical Foundations of Computer Science 2002. Proceedings, 2002. XII, 652 pages. 2002.

Vol. 2421: L. Brim, P. Jančar, M. Křetínský, A. Kučera (Eds.), CONCUR 2002 – Concurrency Theory. Proceedings, 2002. XII, 611 pages. 2002.

Vol. 2422: H. Kirchner, Ch. Ringeissen (Eds.), Algebraic Methodology and Software Technology. Proceedings, 2002. XI, 503 pages. 2002.

Vol. 2423: D. Lopresti, J. Hu, R. Kashi (Eds.), Document Analysis Systems V. Proceedings, 2002. XIII, 570 pages. 2002.

Vol. 2425: Z. Bellahsène, D. Patel, C. Rolland (Eds.), Object-Oriented Information Systems. Proceedings, 2002. XIII, 550 pages. 2002.

Vol. 2426: J.-M. Bruel, Z. Bellahsène (Eds.), Advances in Object-Oriented Information Systems.Procedings, 2002. IX, 314 pages. 2002.

Vol. 2430: T. Elomaa, H. Mannila, H. Toivonen (Eds.), Machine Learning: ECML 2002. Proceedings, 2002. XIII, 532 pages. 2002. (Subseries LNAI).

Vol. 2431: T. Elomaa, H. Mannila, H. Toivonen (Eds.), Principles of Data Mining and Knowledge Discovery. Proceedings, 2002. XIV, 514 pages. 2002. (Subseries LNAI).

Vol. 2432: R. Bergmann, Experience Management. XXI, 393 pages. 2002. (Subseries LNAI).

Vol. 2434: S. Anderson, S. Bologna, M. Felici (Eds.), Computer Safety, Reliability and Security. Proceedings, 2002. XX, 347 pages. 2002.

Vol. 2435: Y. Manolopoulos, P. Návrat (Eds.), Advances in Databases and Information Systems. Proceedings, 2002. XIII, 415 pages. 2002.

Vol. 2436: J. Fong, C.T. Cheung, H.V. Leong, Q. Li (Eds.), Advances in Web-Based Learning. Proceedings, 2002. XIII, 434 pages. 2002.

Vol. 2438: M. Glesner, P. Zipf, M. Renovell (Eds.), Field-Programmable Logic and Applications. Proceedings, 2002. XXII, 1187 pages. 2002.

Vol. 2439: J.J. Merelo Guervós, P. Adamidis, H.-G. Beyer, J.-L. Fernández-Villacañas, H.-P. Schwefel (Eds.), Parallel Problem Solving from Nature – PPSN VII. Proceedings, 2002. XXII, 947 pages. 2002.

Vol. 2440: J.M. Haake, J.A. Pino (Eds.), Groupware: Design, Implementation and Use. Proceedings, 2002. XII, 285 pages. 2002.

Vol. 2442: M. Yung (Ed.), Advances in Cryptology – CRYPTO 2002. Proceedings, 2002. XIV, 627 pages. 2002.

Vol. 2443: D. Scott (Ed.), Artificial Intelligence: Methodology, Systems, and Applications. Proceedings, 2002. X, 279 pages. 2002. (Subseries LNAI).

Vol. 2444: A. Buchmann, F. Casati, L. Fiege, M.-C. Hsu, M.-C. Shan (Eds.), Technologies for E-Services. Proceedings, 2002. X, 171 pages. 2002.

Vol. 2445: C. Anagnostopoulou, M. Ferrand, A. Smaill (Eds.), Music and Artificial Intelligence. Proceedings, 2002. VIII, 207 pages. 2002. (Subseries LNAI).

Vol. 2446: M. Klusch, S. Ossowski, O. Shehory (Eds.), Cooperative Information Agents VI. Proceedings, 2002. XI, 321 pages. 2002. (Subseries LNAI).

Vol. 2447: D.J. Hand, N.M. Adams, R.J. Bolton (Eds.), Pattern Detection and Discovery. Proceedings, 2002. XII, 227 pages. 2002. (Subseries LNAI).

Vol. 2448: P. Sojka, I. Kopeček, K. Pala (Eds.), Text, Speech and Dialogue. Proceedings, 2002. XII, 481 pages. 2002. (Subseries LNAI).

Vol. 2449: L. Van Gool (Ed.), Pattern Recognotion. Proceedings, 2002. XVI, 628 pages. 2002.

Vol. 2451: B. Hochet, A.J. Acosta, M.J. Bellido (Eds.), Integrated Circuit Design. Proceedings, 2002. XVI, 496 pages. 2002.

Vol. 2452: R. Guigó, D. Gusfield (Eds.), Algorithms in Bioinformatics. Proceedings, 2002. X, 554 pages. 2002.

Vol. 2453: A. Hameurlain, R. Cicchetti, R. Traunmüller (Eds.), Database and Expert Systems Applications. Proceedings, 2002. XVIII, 951 pages. 2002.

Vol. 2454: Y. Kambayashi, W. Winiwarter, M. Arikawa (Eds.), Data Warehousing and Knowledge Discovery. Proceedings, 2002. XIII, 339 pages. 2002.

Vol. 2455: K. Bauknecht, A M. Tjoa, G. Quirchmayr (Eds.), E-Commerce and Web Technologies. Proceedings, 2002. XIV, 414 pages. 2002.

Vol. 2456: R. Traunmüller, K. Lenk (Eds.), Electronic Government. Proceedings, 2002. XIII, 486 pages. 2002.

Vol. 2458: M. Agosti, C. Thanos (Eds.), Research and Advanced Technology for Digital Libraries. Proceedings, 2002. XVI, 664 pages. 2002.

Vol. 2462: K. Jansen, S. Leonardi, V. Vazirani (Eds.), Approximation Algorithms for Combinatorial Optimization. Proceedings, 2002. VIII, 271 pages. 2002.

Vol. 2463: M. Dorigo, G. Di Caro, M. Sampels (Eds.), Ant Algorithms. Proceedings, 2002. XIII, 305 pages. 2002.

Vol. 2464: M. O'Neill, R.F.E. Sutcliffe, C. Ryan, M. Eaton, N. Griffith (Eds.), Artificial Intelligence and Cognitive Science. Proceedings, 2002. XI, 247 pages. 2002. (Subseries LNAI).

Vol. 2469: W. Damm, E.-R. Olderog (Eds.), Formal Techniques in Real-Time and Fault-Tolerant Systems. Proceedings, 2002. X, 455 pages. 2002.

Vol. 2470: P. Van Hentenryck (Ed.), Principles and Practice of Constraint Programming – CP 2002. Proceedings, 2002. XVI, 794 pages. 2002.

Vol. 2479: M. Jarke, J. Koehler, G. Lakemeyer (Eds.), KI 2002: Advances in Artificial Intelligence. Proceedings, 2002. XIII, 327 pages. (Subseries LNAI).

Vol. 2483: J.D.P. Rolim, S. Vadhan (Eds.), Randomization and Approximation Techniques in Computer Science. Proceedings, 2002. VIII, 275 pages. 2002.